Canadian Professional Engineering and Geoscience: Practice and Ethics

Canadian Professional Engineering and Geoscience: Practice and Ethics

Gordon C. Andrews
University of Waterloo

THOMSON

NELSON

Australia　Canada　Mexico　Singapore　Spain　United Kingdom　United States

THOMSON

NELSON

Canadian Professional Engineering and
Geoscience: Practice and Ethics

by Gordon C. Andrews

Editorial Director and Publisher:
Evelyn Veitch

Acquisitions Editor:
Anthony Rezek

Marketing Manager:
Janet Piper

Senior Developmental Editor:
Joanne Sutherland

Permissions Coordinator:
Terri Rothman

Executive Production Editor:
Susan Calvert

Copy Editor/Proofreader:
Matthew Kudelka

Indexer:
Andrew Little

Production Manager:
Renate McCloy

Creative Director:
Angela Cluer

Interior Design:
Katherine Strain

Cover Design:
Katherine Strain

Cover Image:
© Daryl Benson/Masterfile

Frontispiece Image:
NASA

Compositor:
Courtney Hellam

Printer:
Transcontinental Printing

**National Library of Canada
Cataloguing in Publication**

Andrews, G. C. (Gordon Clifford),
1937–
    Canadian professional
engineering and geoscience:
practice and ethics/Gordon
C. Andrews.

Includes bibliographical references
and index.
ISBN 0-17-641594-7

    1. Engineering—Vocational
guidance—Canada. 2. Engineering
ethics. 3. Earth sciences—
Vocational guidance—Canada.
4. Earth scientists—Professional
ethics—Canada. I. Title.

TA157.A6825 2004   620'.0023'71
C2004-902331-4

ABOUT THE COVER AND FRONTISPIECE

The cover and frontispiece illustrate recent Canadian engineering and geo-science achievements, which range from the ocean seabed to outer space. The cover photograph shows the Confederation Bridge, which opened to traffic in May 1997 and links New Brunswick and Prince Edward Island across the Northumberland Strait. The bridge is 12.9 km long, which makes it the world's longest saltwater bridge subject to ice hazards.

The Confederation Bridge was designed by Strait Crossing Inc. of Calgary, Alberta, and will survive the harsh climate of the Northumberland Strait for the next century. It has two highway traffic lanes (plus emergency shoulders). The distinctive arches provide a clearance of 60 m for seagoing vessels. The bridge approach on Prince Edward Island has 7 spans (580 m); the central bridge portion has 44 spans (11 km); the bridge approach in New Brunswick has 14 spans (1300 m). The water is as deep as 35 m at the support piers, which are protected by conical ice shields to resist the severe scouring of floating ice.

The bridge's piers, spans, and other components were prefabricated on land, moved to a jetty by a special slider system, and positioned on the site by a huge floating crane. The bridge is equipped with electronic devices that monitor stress, strain, and motion and transmit this data to universities and research facilities. In this way, the bridge's responses to ice, traffic, and earth movements can be constantly observed.

The frontispiece is a photograph of the Canadian robot arm on the International Space Station, which is in orbit about 400 km above the earth. The technical name for the robot arm is the Space Station Remote Manipulator System (SSRMS); its working name is Canadarm2, and it is a larger version of the Canadarm—the robot arm installed on the American space shuttles. The Canadarm2 is part of Canada's contribution to the International Space Station and was designed for the Canadian Space Agency (CSA) by MD Robotics of Brampton, Ontario.

The robot arm is essential for assembling the space station and identifies Canada as a key partner in the project. Canada contributed the Mobile Servicing System (MSS), which is made up of three key parts: a movable base, called the Mobile Remote Servicer Base System (MBS); the Canadarm2; and a Special Purpose Dextrous Manipulator (SPDM). Our country also contributed the Canadian Space Vision System (CSVS), which permits objects outside the space station to be located accurately by the Canadarm2 operator. The Canadarm2 was installed on the space station by Canadian Space Agency astronaut Chris Hadfield in April 2001.

The photograph shows the Canadarm2 connected to the MBS, which is mounted on one of the truss segments that form the spine of the space station. The MBS can move along the truss segments, thus extending the reach of Canadarm2. Canada will also provide additional grapple fixtures to permit Canadarm2 to move "hand over hand."

The space station is a truly international project, constructed by the United States, Russia, fifteen European countries, Japan, Brazil, and Canada. In return for Canada's contribution, Canadian researchers will have proportional use of the station. Once all of the components and solar panels have been installed, the space station will have a mass of 450 tonnes and will cover an area equal to a Canadian football field. The International Space Station has been occupied by astronauts since November 2000 and orbits the earth about every 92 minutes. It is visible with the naked eye as a moving star in the night sky. The CSA website provides links to orbit data and siting times.

PREFACE

The goal of this text is to acquaint engineers and geoscientists with the structure, practice, and ethics of their profession and to encourage them to apply ethical concepts in their professional lives. It is intended for senior undergraduate students, recent university graduates, practising professionals, and immigrants to Canada who wish to practise engineering or geoscience. It is directed to engineers and geoscientists in every branch of the profession, practising in every province or territory of Canada, and it should be of particular value to people preparing to write the Professional Practice Examination (PPE).

Engineering—including geoscience, which is a recognized branch of the profession in most of Canada—is a creative, enjoyable, and rewarding profession with a long history and a bright future. The profession played a key role in establishing Canada as a nation, and it will be even more important in the new millennium. Professional engineers and geoscientists are essential to our future prosperity and to meeting the challenges of international competition.

The profession will also be essential for addressing pollution, waste management, and other environmental problems. These problems, which are side effects of industrialization, are already threatening the Canadian way of life. Global problems such as overpopulation, inadequate housing, and food and energy shortages will also be felt in Canada. Many of these social problems may have engineering solutions, and all of them have ethical implications.

ORGANIZATION

This text is organized into five parts. The first four describe professional practice and ethics; the fifth is a single chapter for readers preparing to write the PPE.

Part I: Introduction to Engineering

The first two chapters introduce the reader to engineering and geoscience and give an overview of the profession, its history, the provincial and territorial Acts that regulate engineering in Canada, and the legal bodies that enforce those Acts. These chapters provide essential basic information for someone entering the engineering profession, such as the standards for admission, the significance of the Code of Ethics and the Engineer's Seal, and the definition of professional misconduct. This part includes two case histories describing tragic incidents that provided the impetus for regulation of the profession.

Part II: Professional Practice

These three chapters describe professional practice in industry, management, and consulting. They discuss the role of the professional as well as typical problems in professional practice. A chapter on hazards, liability, standards, and safety provides extremely important advice for managing these critical issues; it includes two case histories involving hazards and safety.

Part III: Professional Ethics

These nine chapters deal with professional ethics. They introduce ethical theory and Codes of Ethics and show how those theories and codes should be applied to typical ethical problems in industry, management, and private practice. This part also reviews many ethical problems associated with the environment, as well as the duty to report unethical behaviour, also called "whistleblowing." Chapters are devoted to computer ethics, to fairness and equity in the profession, and to the role played by provincial Associations in investigating professional misconduct. The Associations have the power to discipline unethical, negligent, or incompetent practitioners. This part provides more than twenty case studies of ethical dilemmas, with solutions suggested by the author, as well as eight case histories. Some of these case histories are tragic, others are cautionary, and others are inspiring; each teaches an important ethical lesson that was learned at great cost.

Part IV: Maintaining Professionalism

These two chapters describe how engineers can maintain their competence in a rapidly changing world. The continuing professional development (CPD) programs, which are mandatory in some provinces, are discussed in detail. Specific advice is given to engineers and geoscientists considering graduate studies. The author explains the vital importance of engineering societies in developing and communicating new ideas, and how to find and enrol in a suitable engineering society. Charitable, honorary, and student engineering societies are also discussed. This part closes with descriptions of the ceremonies of the Iron Ring and the more recent Earth Sciences Ring, and the rituals that accompany them.

Part V: Exam Preparation

As an aid to readers preparing to write the Professional Practice Examination (PPE), the final chapter describes the examination and suggests a new approach—the EGAD strategy—for writing essay-type exams. This chapter includes twenty-five ethics exam questions, with answers suggested by the author.

FEATURES

This comprehensive and readable text follows a logical sequence for studying professional practice and ethics. It is appropriate for individual study, for classroom use, or as a reference for practising engineers and geoscientists. The following features help the text achieve its goal:

- A logical, readable style.
- Comprehensive coverage of the topic, from basic to advanced concepts, suitable for every province and territory in Canada.
- More than twenty realistic case studies that pose ethical problems and ask the reader to suggest the appropriate course of action. The author then recommends a solution.
- Twelve detailed case histories in which engineering practices that deviated from ethical standards led to disaster or personal tragedy.
- Twenty-five typical examination questions, taken from PPEs in several provinces, to assist readers who are preparing for the examination.
- A discussion of professional practice from several perspectives, including the engineer or geoscientist as employee, as manager, and in private practice.
- Guidance for young professionals in planning their careers.
- Topics for further study and discussion, in a standard format, at the end of each chapter.

In addition, the CD-ROM included with the textbook contains the following:

- Appropriate excerpts from all of the provincial and territorial Acts that regulate engineering and geoscience, including the Codes of Ethics and definitions of professional misconduct for all provinces and territories.
- A selection of Codes of Ethics for several societies, including ACM, ASCE, ASME, IEEE, and NSPE and a code proposed for NAFTA engineers.
- The NSPE guidelines for professional employees, which provide practical advice on workplace policies and procedures that are not discussed in Codes of Ethics.
- Many additional assignments and suggested readings, and a few additional case histories.
- Twenty-five additional case studies, which describe ethical dilemmas and pose questions, followed by the author's comments on the likely outcome of each case.
- Additional information on reducing design hazards and advice to entrepreneurs who want to get started in consulting or business.

ACKNOWLEDGMENTS

I would like to thank everyone who provided assistance, guidance, or advice in preparing this manuscript, as well as the many people who gave permission to publish copyright material. Your help is very much appreciated.

I would also like to say a special word of thanks to many people with whom I have enjoyed a professional relationship over the years, and who have freely shared ideas and given inspiration. The list is too long to reproduce in this short space, but many University of Waterloo friends and colleagues come immediately to mind. In addition, many friends and colleagues from provincial Associations, coast to coast; from the Canadian Council of Professional Engineers (CCPE); and from the Canadian Council of Professional Geoscientists (CCPG) have assisted me by providing information about their professional activities, and I am deeply grateful to them all.

The contribution of those who reviewed the text is gratefully acknowledged. The reviewers were Judith Dimitriu of Ryerson University, David Frost of McGill University, Gilles Y. Delisle of the University of Ottawa, Richard Thibault of the University of Sherbrooke, and Andrew Latus of Memorial University of Newfoundland.

A particular thanks is owed to CCPE staff, who provided statistics and copyright permissions. In particular, I would like to thank Deborah Wolfe, Director, Education, Outreach and Research, Karen Martinson, Manager, Research and Evaluation, Marie Carter, Director, Professional and International Affairs, and Marc Bourgeois, Manager, Communications. Much-appreciated information on the history and development of CCPG was received from Dr. Gordon D. Williams, President, Canadian Council of Professional Geoscientists. Mr. Harry McBride, Deputy Director, Professional Ethics, APEGBC, provided useful advice on the text, and on the 25 case studies in Appendix CD-F, and his assistance is gratefully acknowledged.

Dr. Monique Frize, P.Eng. O.C., contributed Chapter 13, and her valuable efforts are very much appreciated. Dr. Frize is a Professor in the Faculty of Engineering, University of Ottawa and Carleton University, and was formerly the NSERC/Nortel Chair for Women in Science and Engineering (Ontario).

This new text owes its existence to the professional staff at Thomson Nelson. I am indebted to Evelyn Veitch, Editorial Director, Higher Education, and Anthony Rezek, Acquisitions Editor, for their initiative in encouraging and sponsoring this book. I would like to say a very special thanks to Joanne Sutherland, Senior Developmental Editor, who communicated with me regularly during the writing and revision of each chapter, and who provided the right balance of personal motivation and editorial coercion to meet the publication deadlines. Susan Calvert, Executive Production Editor, provided essential guidance for the textbook production, and the precise review by copy editor Matthew Kudelka was greatly appreciated. I am also indebted to Terri Rothman, Freelance Permissions Coordinator, who was very resourceful in obtaining photographs and copyright permissions.

Finally, I would like to acknowledge my thanks and appreciation to my wife, Isobelle, whose support and companionship sustained me through the writing of this manuscript.

Gordon C. Andrews
Waterloo, Ontario
April 2, 2004

CONTENTS

CD-ROM CONTENTS

Appendix CD-A
List of Provincial and Territorial Engineering/Geoscience Associations

Appendix CD-B
Excerpts from the Provincial and Territorial Engineering/Geoscience Acts and Regulations

Appendix CD-C
Codes of Ethics for Some Engineering Societies

Appendix CD-D
NSPE Guidelines to Professional Employment for Engineers and Scientists

Appendix CD-E
Additional Assignments and Further Reading

Appendix CD-F
Additional Case Studies

Appendix CD-G
Miscellaneous Articles of Interest

Part One
Introduction

Chapter 1
Introduction to the Profession

Welcome to engineering—a challenging, creative, and rewarding profession! Engineering is one of Canada's largest and most diverse professions. The traditional engineering disciplines (such as Civil, Chemical, Electrical, Industrial, and Mechanical Engineering) have welcomed the newer disciplines (such as Aerospace, Communications, Computer, Software, Electromechanical, Mechatronics, and Systems Design Engineering) to make engineering a very broad, authoritative, and progressive profession. Moreover, most provinces and territories include geoscience, geology, and geophysics in the laws regulating engineering (or they regulate geoscience separately, as an equivalent profession). In this text the terms *engineers* and *engineering* are used for brevity, but generally refer to practitioners in all of the above branches and disciplines.

The engineering profession plays a central role in generating wealth and in improving the quality of our daily lives. Engineering achievements are found almost everywhere, hidden in the many basic devices that assist, guide, inform, and protect us. For example, the graceful structure of a bridge, the digital accuracy of electronic equipment, and the sleek lines of a new automobile illustrate excellence in engineering design. Good engineering is especially evident in the goods and services that are essential to our high standard of living, such as pure and abundant tap water, dependable electricity, and safe and efficient elevators, machinery, and aircraft. These devices and systems, all of which were designed by engineers, illustrate the ingenuity, competence, and diverse interests of the people who enter the profession.

This text describes the key elements of engineering practice and ethics, and the importance of the profession to society. In this introductory chapter, we begin by examining the history and structure of the profession.

A BRIEF HISTORICAL NOTE

Engineers played a vital role in the development of modern civilization. The roots of the engineering profession are seen in ancient achievements such as the building of the pyramids and the discovery of iron. The first people to be identified clearly as engineers were military engineers. Roman armies marched with a complement of engineers, who devised and operated weapons and built the roads and fortifications that were needed to wage war. In fact, the terms *engineer* and *ingenuity* come from the same Latin root, *ingenium*, which means talent, genius, cleverness, or native ability. Over the centuries, nonmilitary or *civil* engineers emerged to design and supervise the construction of roads, bridges, canals, and irrigation systems, all of which are necessary for a productive society.

The Renaissance gave us Leonardo da Vinci (1452–1519), who was a brilliant artist, scientist, and engineer. His notebooks contain designs for helicopters, submarines, gear trains, and other devices that were not invented by others until centuries later. The advent of printing from movable type, in the fifteenth century, resulted in broader dissemination of knowledge. The scientific and mathematical discoveries that followed were guided by familiar, eminent names such as René Descartes (1596–1650) and Sir Isaac Newton (1642–1727).

Descartes established the "scientific method," whereby truth is discovered through logic and objective reasoning, without any deference to authority. The scientific method is as important today as it was in Descartes' time. This well-known method tests a hypothesis by following several steps: A hypothesis is used to predict the outcome of an experiment. An experiment is then conducted, and the experimental results are compared with the prediction to test the truth of the hypothesis. If the results confirm the prediction, then the hypothesis is proved; if the results disagree with the hypothesis, then the difference must be explained or the hypothesis must be revised. Engineering—indeed, all of modern science—is based on this simple and logical method, but it was a revolutionary idea in the seventeenth century.

Although Newton has been described as the most important scientist of all time, he was also a mathematician and an early engineer. It was he who invented calculus (an honour he shares with Leibnitz), and his treatises on optics, dynamics, and mechanics are the foundations of many engineering courses studied today. Newton discovered the law of gravity and used it to explain the motion of celestial bodies; he was therefore the first person to postulate, convincingly, the feasibility of space travel.

The scientific and mathematical revolutions of the sixteenth and seventeenth centuries led to the Industrial Revolution of the eighteenth and nineteenth centuries. In the twentieth century, the wealth generated by standardized, mass-produced products encouraged exponential growth in engineering. The twentieth century unleashed on the world much conflict, including two devastating world wars and many regional ones, but it also brought many benefits to the world's more-favoured nations: the widespread distribution of electricity; the invention of methods of mass communication such as radio, television, the telephone, and the facsimile machine (fax); and the computer revolution that presently engulfs us.

The computer has brought about profound changes in society. Calculating devices such as the abacus, the slide rule, and the adding machine have existed for centuries, but the first truly electronic computer was not built until

1945, at the end of the Second World War. The first computers were slow and expensive behemoths by today's standards; they filled rooms, yet they were capable of only primitive calculations. The invention of the transistor and large-scale circuit integration (LSI) permitted the miniaturization of electronic devices. As a result of these developments, the first commercial personal computer (the Apple II) was developed in 1977. Over the following two decades the desktop workstation evolved, yielding immense, convenient computing power. Entirely new fields of engineering are now emerging and evolving: hardware and software engineering; computer-aided design, analysis, and manufacturing; nanotechnology; computational fluid dynamics; finite-element analysis; and system and process simulation, to name only a few.

Recent university graduates may not realize how profoundly computers have changed our society. Unfortunately, many older engineers have not been able to cope with this rapid transformation and have had to make way for the new generation of computer-literate engineers, some of whom are practising in branches of engineering that did not exist a decade ago. However, although some aspects of professional practice may change, the principles of professional ethics do not. Ethics is a major focus of the later chapters of this book.

ENGINEERING AND GEOSCIENCE IN CANADA

Canada is a modern industrialized society, with all of the benefits and challenges associated with such a society. For the past decade, the United Nations Human Development Index has ranked Canada among the best countries in the world in which to live (often the very best, and always in the top eight). Many different factors have led to this happy result, but engineering and geoscience contribute directly to increasing this country's gross domestic product (GDP), which is one of the index's key measures. Our profession is important to Canada's prosperity.

The history of engineering and geoscience in Canada is an inspiring story of achievement in a harsh physical and economic climate. This text describes the present state of the profession. Also, photographs of important Canadian engineering and geoscience achievements are provided throughout the text, and Appendix CD-E lists related publications for further reading. A brief overview of the profession follows.

Regulation of Engineering

Engineering evolved into a profession over several centuries. The need for formal training of engineers became evident during the Industrial Revolution. In Britain, teenage engineers-to-be typically paid for the privilege of a five-year apprenticeship in the office of a practising engineer. Formal education eventually superseded the apprenticeship system. Over the past century, engineering has become recognized as a profession and government regulations have been imposed. Academic and experience standards today are very high: to obtain an engineering licence in Canada, an applicant typically must complete both a four-year university degree and an engineering internship. The latter typically involves an additional four years of engineering experience under the guidance of a professional engineer.

The title *professional engineer* is now restricted by law to those persons who have demonstrated their competence and who have been licensed in a provincial or territorial Association of professional engineers (in Quebec, the *Ordre des ingénieurs du Québec*).[1] This legal restriction on the title of Professional Engineer is enforced by the Associations, which also restrict the use of any abbreviation or adaptation of the term engineer (or *ingénieur*) that might imply that the person using the title is licensed.[2]

Regulation of Geoscience

Canadian geoscientists are regulated in much the same way as engineers. Alberta was the first province to regulate geoscience (specifically, geology and geophysics). Eight provinces and territories now regulate engineers and geoscientists together, as a single profession. However, some provinces regulate geoscience differently. The situation is summarized as follows:

- **Identical legislation.** Eight jurisdictions regulate engineers and geoscientists together in the same legislation: Alberta, British Columbia, Manitoba, New Brunswick, Newfoundland, the Northwest Territories, Nunavut, and Saskatchewan. The various Acts set standards for admission, practice, ethics, and discipline, which are virtually identical for engineers and geoscientists.
- **Separate legislation.** Three provinces—Ontario, Nova Scotia, and Quebec—have enacted legislation for geoscientists as an independent, self-governing profession that is equivalent to the established engineering profession. In these provinces, standards for admission, practice, ethics, and discipline are similar to those for engineers and geoscientists but are not necessarily identical.
- **No legislation.** Prince Edward Island and Yukon regulate engineering but do not regulate geoscience as a profession. These jurisdictions have few geoscientists and, fortunately, many practising geoscientists are licensed in adjacent areas.

In provinces that regulate geoscientists, the titles Professional Geoscientist, Professional Geologist, and Professional Geophysicist are restricted by law to those persons who have demonstrated their competence and who have been licensed in a provincial or territorial Association for geoscientists (in Quebec, the *Ordre des géologues du Québec*). The Associations license geoscientists and enforce the legal restrictions on the title. The Associations also restrict the use of any abbreviation or adaptation of the term Geoscientist, Geologist, or Geophysicist (or géologue) that might imply that the person using the title is licensed.

Most practice and ethics issues in engineering are independent of the practitioner's branch, subdiscipline, or specialty. This text therefore addresses the many branches of engineering, geoscience, geology, and geophysics as a single, unified profession. Although for brevity's sake this book uses the general term "engineer," discussions always encompass geoscientists, geologists, and geophysicists. (Chapter 2 discusses licensing and the regulation of the profession in more detail.)

Distribution of Engineers and Geoscientists in Canada

In 2002 there were more than 160,000 licensed professional engineers in Canada (including about 7,600 geologists, geophysicists, and geoscientists, in provinces where they are regulated).[3, 4] The distribution of professional engineers across Canada is shown in Figure 1.1. The distribution is not uniform: most engineers are clustered in the industrially developed regions of Ontario

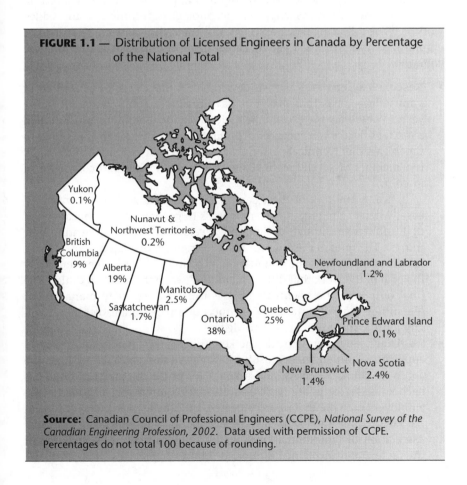

FIGURE 1.1 — Distribution of Licensed Engineers in Canada by Percentage of the National Total

Yukon
0.1%

Nunavut &
Northwest Territories
0.2%

British
Columbia
9%

Alberta
19%

Manitoba
2.5%

Saskatchewan
1.7%

Ontario
38%

Quebec
25%

Newfoundland and Labrador
1.2%

Prince Edward Island
0.1%

Nova Scotia
2.4%

New Brunswick
1.4%

Source: Canadian Council of Professional Engineers (CCPE), *National Survey of the Canadian Engineering Profession, 2002.* Data used with permission of CCPE. Percentages do not total 100 because of rounding.

and Quebec, with the next-largest number in the resource-rich province of Alberta. When we compare the coasts, we find that British Columbia has almost twice as many engineers as the four Atlantic provinces combined. Saskatchewan and Manitoba together have slightly fewer engineers than the four Atlantic provinces. The Yukon and Northwest Territories have very few of the nation's engineers (just over 0.3 percent). Nunavut has a small number of engineers, but they are included in the Northwest Territories' total.[5]

THE TECHNICAL TEAM

Engineers are primarily concerned with design and development, which is the creative process of converting theoretical concepts into useful applications. Engineers occasionally work alone, especially on small projects, but complex engineering projects usually require teams of people with different skills and education. Engineers are only one group in the technical spectrum, but they constitute the vital link between theory and application. The full spectrum includes research scientists, engineers, technologists, technicians, and skilled workers.[6] The following paragraphs describe the technical team. (It should be emphasized that this is a rough categorization and that exceptions are common.)

- **Research scientist.** Scientists develop ideas that expand the frontiers of knowledge—ideas that may not have practical applications for many years. A doctorate is typically the basic educational requirement, although a master's degree is sometimes acceptable. A scientist is rarely required to supervise other technical personnel except research assistants, and usually is a member of several learned societies in his or her particular field of interest.

  Basically, the task of the research scientist is to generate new knowledge, whereas the task of the engineer is to generate new applications of that knowledge through design and development (that is, creating new things). The roles of the scientist and the engineer overlap to some degree, and in some projects the boundary may be invisible. It is sometimes only the goal of the work that differentiates the two, not the actual duties. The final output of the scientist's work is usually a report to management (describing a new business opportunity) or a paper published in a scholarly journal (describing new knowledge).

- **Engineer.** Engineers provide the key link between theory and practical applications. An engineer must have extensive theoretical knowledge, the ability to think creatively, and a knack for obtaining practical results. A bachelor's degree is the basic educational requirement, although some employers may prefer a master's degree. In Canada, provincial or territorial laws require the engineer to be licensed by a provincial or territorial Association of Professional Engineers in order to practise. (Regulation of the profession is explained in more detail in Chapter 2).

Engineers are usually concerned with design—that is, with creating devices, systems, and structures for human use. However, many engineers supervise construction, testing, manufacturing, and similar activities related to design. Some engineers (and geoscientists in particular) are engaged in the discovery and utilization of natural resources. A small fraction are employed in a consulting capacity, advising others on the above activities, and many engineers are diverted into the management of companies or government agencies. These engineers are not directly engaged in design, but some aspect of creativity or design is usually still at the core of their activities.

- **Technologist.** Technologists usually work under the direction of engineers in applying engineering principles and methods to fairly complex engineering problems. The basic educational requirement is usually a diploma from a technology program at a community college, CEGEP (*Collège d'enseignement général et professionel*), or CAAT (college of applied arts and technology), although many technologists have a bachelor's degree (usually in science, mathematics, or technology). Technologists often supervise the work of others and are encouraged to have qualifications that are recognized by a technical society.

Some provincial engineering Associations are considering whether to include technologists and technicians under the Acts in their jurisdictions. Technologists and technicians are essential members of technical teams and deserve recognition for their training as well as respect for their emerging professional status. As this status grows, the boundary between engineers and technologists is likely to become less rigid. More effective use of technologists may well lead to increased productivity for everyone.

Associations of engineering technicians and technologists have been established in all ten provinces (although not in the territories) to certify the qualifications of technologists. The provincial technician/technologist associations are, in turn, members of an umbrella organization, the Canadian Council of Technicians and Technologists (CCTT), which is a federation of the ten provincial associations. CCTT was established in 1972 to coordinate activities and facilitate exchanges of information among the provincial associations of technicians and technologists. In 2003 the ten provincial associations within CCTT represented more than 40,000 registered technicians and technologists across Canada. However, since certification is voluntary, there were probably many more people actually practising as technicians or technologists. The total number is estimated to be roughly equal to the number of professional engineers practising in Canada (more than 160,000).

Certification as a technologist requires an assessment of the candidate's educational background. The candidate also must document at least two years of relevant experience. Certification is voluntary and is not required in order to work as a technician or technologist in Canada; however, only certified technologists are entitled to use the following designations: Certified Engineering Technologist (CET), Applied Science Technologist

(AScT), Registered Engineering Technologist (RET), or *Technologue Professionnel* (TP). Which one is used depends on the province in which certification is granted. In recent years the certification of technologists and technicians has been assisted by provincial Associations of Professional Engineers. In fact, the organization of the provincial technology associations and CCTT closely parallels that of the provincial engineering Associations and CCPE. (This structure is described in the next chapter.)

- **Technician.** Technicians usually work under the supervision of engineers or technologists in the practical aspects of engineering tests or equipment maintenance. The basic educational requirement is usually a diploma from a program at a community college, CEGEP, or CAAT. This program is usually shorter than for technologists. In most provinces the title Certified Technician (C.Tech.) may be awarded by the provincial association of technicians and technologists after the applicant completes the appropriate education and acquires two years of appropriate experience. Certification is not essential to work as a technician.

- **Skilled worker.** Typically, skilled workers apply highly developed manual skills to carry out the designs and plans of others. Skilled workers must be trained by master artisans, and usually it is the quality of this apprenticeship, rather than the worker's formal education, that is important. Each type of trade worker (electrician, plumber, carpenter, welder, pattern maker, bricklayer, machinist, etc.) comes under a different certification procedure, and these procedures vary from province to province.

In practice, the above categories are not quite as clearly defined as this book makes them sound; the boundaries are not rigid, nor should they be. Transition from one group to another is always possible, although it may not always be easy. Each group in this technical spectrum has a different task, and there are great differences in the skills, knowledge, and training required. In a large project all five types of technical expertise will be required, at different times. The engineer is usually the key link between theory and practice, but co-operation and mutual respect for the particular skills and knowledge of each group are essential to create a positive, productive working environment.

BRANCHES OF ENGINEERING AND GEOSCIENCE

Over the years many different branches or disciplines have been established in engineering. The most general branches are civil, electrical, mechanical, industrial, and chemical engineering. However, since university engineering programs are accredited in Canada on an individual basis, many more specialized programs, such as computer engineering and systems design engineering, have been accredited at specific institutions. In fact, as shown in Table 1.1, more than eighty different engineering and geoscience programs are presently accredited at Canadian universities. (The program titles in French are offered by universities in Quebec.)

The choice of branch is important and is usually made on entry to university. The applicant must, of course, specify the branch when applying to the provincial Association for registration as a professional engineer and for award of a licence to practise. Changes from branch to branch are possible. Since the readers of this text are likely to be senior university undergraduates or recent engineering or geoscience graduates, most will already have chosen a branch, so a detailed discussion of branches is unnecessary.

TABLE 1.1 — List of Accredited Engineering Programs at Canadian Universities

Aerospace Engineering	Agricultural Engineering
Agricultural & Bioresource Engineering	Biological Engineering
Bio-resource Engineering	Biosystems Engineering
Building Engineering	Ceramic Engineering & Management
Ceramic Engineering & Society	Chemical & Biochemical Engineering
Chemical Engineering	Chemical Engineering & Management
Chemical Engineering & Society	Civil Engineering
Civil Engineering & Management	Civil Engineering & Society
Communications Engineering	Computer Engineering
Computer Engineering & Management	Computer Systems Engineering
Electrical Engineering	Electrical Engineering & Management
Electrical Engineering & Society	Electronic Systems Engineering
Engineering Chemistry	Engineering Physics
Engineering Physics & Management	Engineering Physics & Society
Engineering Science	Engineering Systems & Computing
Environmental Engineering	Environmental Systems Engineering
Extractive Metallurgy Engineering	Forest Engineering
Génie Agroenvirronemental	*Génie Alimentaire*
Génie du Bois	*Génie Chimique*
Génie Civil	*Génie de la Construction*
Génie Électrique	*Génie Électromécanique*
Génie Géologique	*Génie Industriel*
Génie Informatique	*Génie des Matériaux*
Génie des Matériaux & de la Métallurgie	*Génie Mécanique*
Génie des Mines	*Génie des Mines et De La Minéralurgie*
Génie Physique	*Génie de la Production Automatisée*
Génie des Systèmes Électromécaniques	*Génie Unifié*
Geological Engineering	Geomatics Engineering
Industrial Engineering	Industrial Systems Engineering
Integrated Engineering	Manufacturing Engineering
Manufacturing Engineering & Management	Manufacturing Engineering & Society

Continued

TABLE 1.1 — List of Accredited Engineering Programs at Canadian Universities

Materials Engineering	Materials Engineering & Management
Materials Engineering & Society	Mathematics & Engineering
Mechanical Engineering	Mechanical Engineering & Management
Mechanical Engineering & Society	
Metallurgical Engineering & Management	Metallurgical Engineering
Mining & Mineral Process Engineering	Metals & Materials Engineering
Ocean & Naval Architectural Engineering	Mining Engineering
Petroleum Engineering	Oil & Gas Engineering
Software Engineering & Management	Software Engineering
Water Resources Engineering	Systems Design Engineering

Note: Some newer disciplines, such as Mechatronics, are awaiting CEAB accreditation.

SOURCE: Canadian Council of Professional Engineers (CCPE), *CEAB Accreditation Criteria and Procedures* (Ottawa: CCPE, 2002). Reprinted with permission of CCPE.

JOB FUNCTIONS OF PRACTISING ENGINEERS

Although the engineer's primary task is typically design and development, other tasks, jobs, and activities occupy engineers and are included in the broad spectrum of engineering practice. Table 1.2 lists the job functions reported by practising professional engineers and geoscientists working in various disciplines in Canada in 2002.[7]

TABLE 1.2 — Job Functions Reported by Practising Engineers

	Job function	Percentage of engineers
1.	Management/Administration	42%
2.	Engineering Design	33
3.	Technical Support Services	32
4.	Consulting	26
5.	Specification or Technical Writing	17
6.	Research and Development	14
7.	Operations and Production	14
8.	Program Management	14
9.	Marketing and Sales	12
10.	Contract Administration	11

SOURCE: Canadian Council of Professional Engineers (CCPE), *National Survey of the Canadian Engineering Profession, 2002*. Reprinted with permission of CCPE.

These statistics require interpretation, but lead to interesting conclusions. The data were obtained through a survey of 26,271 engineers and geoscientists practising across Canada. The respondents were asked to specify the job functions to which they devoted at least 25 percent of their time. Only functions reported by at least 10 percent of the members are indicated in this table. The table thus represents the top ten job functions performed.

Most of the tasks on the list are engineering or technical tasks (Engineering Design, Technical Support Services, Consulting, Specification or Technical Writing, Research and Development, and Operations and Production). Three of the tasks are engineering-related management tasks (Program Management, Marketing and Sales, and Contract Administration). However, the task cited most often is Management or Administration, which involves managing the activities of other people on the engineering team. This surprising observation implies that engineers should develop management skills. This point is discussed in more detail in Chapter 3.

EMPLOYMENT PROSPECTS

The number of professional engineers in Canada has remained relatively stable over the past ten years. However, a decrease is likely in the next decade as the "baby boom" generation (born in the years following the end of the Second World War) reaches retirement age. With the Canadian and American economies rebounding from the foreign wars and economic scandals of 2001 to 2003, employment opportunities for engineers are expanding. The combination of these factors may well result in a shortage of engineers in the next decade. Employment prospects for engineers therefore look very good.

This optimistic outlook is supported by the historical employment record for engineers. The unemployment rate for engineers was typically around 1 percent in the decade prior to 1982, indicating very secure employment for professional engineers. During the recession of 1982, unemployment reached a peak of 7,000 engineers, or about 6 percent of registered professional engineers.[8] It declined gradually until it was typically below 2 percent by 1997.[9] The terrorist attacks of 2001 and their aftermath, and the recessionary effects of the scandalous collapse of several giant American companies, including Enron and Worldcom, had an impact on the entire global economy, and as a result, the unemployment rate for engineers rose to about 3 percent in 2002.[10] Even so, the unemployment rate for engineers is much lower than the overall rate for Canadian workers, and although global events such as recessions and wars may change the picture, the prospects for recent graduates seem very positive, as the economy rebounds in 2003.

A BRIEF DISCUSSION OF PROFESSIONAL STATUS

Engineers have high professional status. In both Canada and the United States, opinion surveys consistently show that engineers rank near the top for honesty and integrity and that engineering is among the most attractive of all fields to enter. Clearly, the engineering profession, and engineers themselves, are held in high regard by the general public. But we need to define the term "profession" more precisely. What is a profession? How does it differ from a job? The following dictionary definition of a profession helps to answer these questions:

> Profession: A calling requiring specialized knowledge and often long and intensive preparation including instruction in skills and methods as well as in the scientific, historical, or scholarly principles underlying such skills and methods, maintaining by force of organization or concerned opinion high standards of achievement and conduct, and committing its members to continued study and to a kind of work which has for its prime purpose the rendering of a public service. (From *Webster's Third New International*® *Dictionary*.)[11]

Engineering (including geoscience) certainly requires "specialized knowledge," "intensive preparation," and "instruction in skills and methods as well as in the scientific, historical, or scholarly principles underlying such skills and methods." In fact, provincial regulations now require engineers to complete seven or eight years of formal education and relevant work experience before they can practise. This is as much as is required in medicine and law (two professions that serve as a useful basis for comparison).

Furthermore, engineering has a "force of organization," in the form of laws and regulations. These have been enacted in every province and territory in Canada (although they are still evolving in Prince Edward Island and Yukon, where geoscience is not yet a regulated profession). The Acts, laws, and regulations governing engineering include Codes of Ethics that commit practitioners to "high standards of achievement and conduct" and to "continued study and to a kind of work which has for its prime purpose the rendering of a public service" (as discussed in more detail later in this text).

In summary, engineering and geoscience easily satisfy the above definition of a profession. In fact, these two are recognized almost everywhere in Canada as a single, unified profession, or as equivalent, similar professions.

A basic fact of life for engineers is that they outnumber every other professional group except teachers and nurses. For example, there are roughly three engineers for every physician in Canada. Moreover, most engineers are employees of large companies, where they work in teams on projects. Most other professionals are self-employed and work with clients on a one-to-one basis. These differences do not affect the professional nature of engineering—opportunities exist for individual creativity, and the need for personal professional responsibility has never been greater. Although engineering teams may now be larger and more specialized than they were in simpler, bygone days, the projects are even larger and individual challenges still exist.

INTRODUCTION TO CASE HISTORIES

As Canadians, we have many spectacular engineering achievements of which we can be proud, from the construction of the transcontinental railway in 1885, to the design of the Avro Arrow in 1958, to the opening of the Confederation Bridge in 1997. We tend to take our engineering success for granted when well-designed structures and devices work properly. In contrast, when engineering structures fail, we focus our attention on the failure. We ask why it happened and how similar failures can be avoided in the future. If the failure is especially costly, be it in lives or in money, an investigation panel or Royal Commission may be formed to study the failure impartially and publicly. As a result, we often learn more from failures than from successes, although the lessons are learned at a great cost.

When reading the many case histories in this text, remember that failure itself is not necessarily evidence of unethical or incompetent practice. Many engineering projects push the limits of knowledge. Novel or experimental projects always involve some risk, and even the most determined and ethical practitioners cannot guarantee success every time. Some failures must simply be accepted.

Most of the case histories in this text are historical summaries of avoidable incidents involving engineers and/or geoscientists. Most have been chosen because they involve an ethical aspect such as negligence, incompetence, conflict of interest, corrupt practices, or the need to report such practices. All of the cases are fairly well known. Most of them relate to events that took place in Canada and that had some kind of impact on the engineering profession. Most of them involved a tragedy, but some good may come out of them, if we learn how to avoid similar tragedies in future.

CASE HISTORY 1.1

THE QUEBEC BRIDGE COLLAPSE

The first case history tells the story of the collapse of the Quebec Bridge in 1907 and the negligence that led to that collapse. This case is important because as a result of it, Canadians were made aware of the need for professional licensing of engineers.

Introduction

The Quebec Bridge, which had its official opening in 1919, is the longest cantilever span in the world, with a centre-distance between supports of 549 m (1800 ft). Although there are suspension bridges with longer spans, the massive size of the Quebec Bridge and the length of its span make it a very impressive structure. In fact, you must see it in person to fully comprehend its grandeur and the achievement it represents.

A view of the pier and main truss pin of the completed Quebec Bridge. The sentry on duty during the First World War shows the magnitude of the bridge components.

Source: The Quebec Bridge over the St. Lawrence River near the City of Quebec: *Report of the Government Board of Engineers,* Department of Railways and Canals Canada, 1919.

However, the Quebec Bridge is infamous for the many lives lost in the harrowing accidents that occurred during its construction. The *Canadian Encyclopedia Plus* summarizes these tragic losses succinctly:

Québec Bridge Disasters: Construction on the Québec Bridge, 11 km above Québec City, officially began in 1900. On 29 Aug 1907, when the bridge was nearly finished, the southern cantilever span twisted and fell 46 m into the St Lawrence R. Seventy-five workmen, many of them Caughnawaga Indians, were killed in Canada's worst bridge disaster. An inquiry established that the accident had been caused by faulty design and inadequate engineering supervision. Work was resumed, but on 11 Sept 1916 a new centre span being hoisted into position fell into the river, killing 13 men. The bridge was completed in 1917 and the Prince of Wales (later Edward VIII) officially opened it 22 Aug 1919.[12]

History

The residents of Quebec City advocated building a bridge over the St. Lawrence River as early as 1852, and a site had been chosen where the river narrowed, just upstream of the city. Designs were prepared, but serious work did not begin until 1900. The success of the cantilevered Forth Bridge, built in 1890 in Scotland, was a factor in the choice of a cantilever design for the Quebec Bridge. The Forth Bridge, the first bridge built entirely of steel, has two spans of 521 m (1710 ft) each. At the time, these were the world's longest unsupported (cantilevered) bridge spans, and they would remain so for twenty-seven years, until the Quebec Bridge was successfully completed.

At the time of the 1907 accident, four parties were directly involved in constructing the Quebec Bridge superstructure:

- the government of Canada, which had provided subsidies and a guarantee of bonds to
- the Quebec Bridge & Railway Company (known simply as the "Quebec Bridge Company"), which had responsibility for the complete structure, and which had contracted with
- the Phoenix Bridge Company in Phoenixville, Pennsylvania, to design and construct the superstructure, and which had subcontracted with
- the Phoenix Iron Company, to fabricate the steel components.

The Quebec Bridge Company employed a chief engineer, Edward Hoare, who was on site, and a consulting engineer, Theodore Cooper of New York, as well as many hundreds of erection and inspection staff. In technical terms, Cooper was highly competent: "In the extent of his experience and in reputation for integrity, professional judgement and acumen, Mr. Cooper had few equals on this continent." Early in the design work, it was decided that Cooper's decisions on technical matters would be final. Cooper had insisted on this, and such authority was extended to him in writing, in the form of a government order-in-council.[13] Yet even though he enjoyed ultimate design authority, he visited the Quebec site only while the supporting piers were being built and was never on site thereafter. Furthermore, over the many years that the bridge components were being fabricated, he visited the Phoenix Iron Company shops only three times.[14]

Norman McLure was the Quebec Bridge Company's inspector of erection. He had been appointed by Cooper with Hoare's agreement, and he received instructions from both of them. He reported to Hoare mainly on "matters regarding monthly estimates, and to Cooper on matters of construction."[15] The Phoenix Bridge Company's chief engineer was Mr. Deans, who was an experienced bridge builder but was more accurately described, after the accident, as its "chief business manager."[16] The design engineer was Mr. Szlapka, a German-educated engineer with twenty-seven years of experience in designing many similar projects. Szlapka was responsible for generating the design details and had the full confidence of Cooper.[17]

The Quebec Bridge (Phoenix design) immediately before the collapse on August 29, 1907.

Source: The Quebec Bridge over the St. Lawrence River near the City of Quebec: *Report of the Government Board of Engineers,* Department of Railways and Canals Canada, 1919.

Initial Construction

A competition for the design was held in 1898. Cooper reviewed the submitted plans and recommended the Phoenix Bridge Company's design, which showed a span of about 488 m (1600 ft) between the supporting piers. The contracts for detailed design and construction were signed, and work began in 1899. Cooper requested further investigation of the riverbed to ascertain the best locations for the supporting piers. After much study, he recommended that the piers be located closer to shore, thus lengthening the unsupported span to 549 m (1800 ft). The work was slow at first, since there was some uncertainty about the financial status of the Quebec Bridge Company and its ability to pay its contractors. Government support was assured in 1903, and it was urged that the bridge be completed before the Quebec Tercentennial in 1908.[18] The fabrication and erection of the superstructure proceeded fairly rapidly thereafter.

The First Disaster

In 1907, with the first span of the cantilever now reaching out over the water, it became obvious that some parts of the structure were deforming in unexpected ways. This was communicated to Cooper in New York. H. Petroski summarizes these fatal days concisely in his highly readable book *Engineers of Dreams:*

The south arm of the Québec Bridge had been cantilevered out about six hundred feet over the St. Lawrence River by early August 1907, when it was discovered that the ends of pieces of steel which had been joined together were bent. Cooper was notified, by letter, by Norman R. McLure, a 1904 Princeton graduate who was "a technical man" in charge of inspecting the bridge work as it proceeded, who suggested some corrective measures. Cooper sent back a telegram rejecting the proposed procedure and asking how the bends had occurred. Over the next three weeks, in a series of letters back and forth among Cooper, chief engineer Deans, and McLure, Cooper repeatedly sought to understand how the steel had gotten bent, and rejected explanation after explanation put forth by his colleagues. Cooper alone seems to have been seriously concerned about the matter until the morning of August 27 when McLure reported that he had become aware of additional bending of other chords in the trusswork and, since "it looked like a serious matter," had the bends measured; he explained that erection of additional steel had been suspended until Cooper and the bridge company could evaluate the situation.

Yet, even as McLure went to New York to discuss the matter with Cooper, Hoare, as chief engineer of the Quebec Bridge Company, had authorized resumption of work on the great cantilever. As soon as McLure and Cooper had discussed the bent chords, Cooper wired Phoenixville: "Add no more load to bridge till after due consideration of facts." McLure had reported that work had already been suspended, and so contacting Québec more directly was not believed to be urgent, but when McLure went on to Phoenixville, he found that the construction had in fact been resumed. Some conflicting reports followed, thanks in part to a telegraph strike then in progress, as to whether Cooper's telegram was delivered and read in time for Phoenixville to alert Québec.

In any event, the crucial telegram lay either undelivered or unread as the whistle blew to end the day's work at 5:30 P.M. on August 29, 1907. According to one report, ninety-two men were on the cantilever arm at that time, and when "a grinding sound" was heard, they turned to see what was happening. "The bridge is falling," came the cry, and the workmen rushed shoreward amid the sound of "snapping girders and cables booming like a crash of artillery." Only a few men reached safety; about seventy-five were crushed, trapped, or drowned in the water, surrounded by twisted steel. The death toll might also have included those on the steamer *Glenmont*, had it not just cleared the bridge when the first steel fell. Boats were lowered at once from the *Glenmont* to look for survivors, but there were none to be found in the water. Because of the depth of the river at the site, which allowed ocean liners to pass, and which had demanded so ambitious a bridge in the first place, the debris sank out of sight, and "a few floating timbers and the broken strands of the bridge toward the … shore were only signs that anything unusual had happened." The crash of the uncompleted bridge "was plainly heard in Québec," and the event literally "shook the whole countryside so that the inhabitants rushed out of their

The wreckage of the Quebec Bridge (Phoenix design) after the collapse on 29 August 1907.

Source: The Quebec Bridge over the St. Lawrence River near the City of Quebec: *Report of the Government Board of Engineers,* Department of Railways and Canals Canada, 1919.

houses, thinking that an earthquake had occurred." In the dark that evening, the groans of a few men trapped under the shoreward steel could be heard, but little could be done to help them until daylight.[19]

The Report of the Royal Commission

Within hours of the accident a Royal Commission was established to determine the cause. The commission prepared a thorough report containing lessons— learned at great cost—that have benefited structural engineers and bridge designers in Canada and around the world.[20] As G.H. Duggan later wrote:

> The report of the Royal Commission appointed to investigate the failure of the Phoenix[-designed] Bridge in 1907 is very comprehensive, and goes beyond the mere taking of evidence and the investigation of the faults of the bridge, as the Commission assembled most of the available data on other long span bridges, illustrated their important features, recorded the tests on large size compression members that had any bearing upon the work, and made a number of tests to supply some lacking experimental data of the behaviour of large compression members under stress.[21]

The report concluded that Hoare's appointment as chief engineer of the Quebec Bridge Company was a mistake. Although he had a "reputation for integrity, good judgement and devotion to duty," he was not technically competent to control the work. Regarding Deans, chief engineer of the Phoenix Bridge Company, the report concluded that his "actions in the month of August, 1907, and his judgement . . . were lacking in caution, and show a failure to appreciate emergencies that arose." However, the commission assigned most of the blame for the bridge's collapse to errors in judgment made by Cooper and Szlapka.[22]

Design and Communication Deficiencies

The commission's report identified several serious deficiencies in the bridge's design and in the construction methods followed. Specifically, the design loads were underestimated, and the engineers failed to investigate, even when very high unit stresses were known to exist in the compression members. The commission also criticized the curvature and the "splicing and lacing" (joining and cross-bracing) of the bottom chords. In any cantilever beam that is loaded downward at the end, the bottom chords are in compression. In the case of the Quebec Bridge, the curvature and the many connections reduced the chords' ability to resist buckling. The compressive chords had been designed by Szlapka and examined and approved by Cooper.

The stresses had been calculated by Szlapka based on an estimate of the total dead weight of the bridge—an estimate made by Cooper at the start of the design process. However, as the detail design progressed and as the precise shapes of the members were determined, the dead weight changed. The stresses should have been recalculated using more accurate estimates of the dead weight. This was not done. It is especially noteworthy that the bridge span had originally been specified as 488 m (1600 ft), but Cooper later recommended moving the supporting piers, which increased the span to 549 m (1800 ft). When the bridge's span was increased, the dead weight increased significantly, yet this increase was not included in the calculations. This point is explained more clearly by Petroski:

> In short, what Szlapka had done was to let stand an educated guess as to the weight of steel that the finished bridge would contain. Such guesses, guided by experience and judgment, are the only way to begin to design a new structure, for without information on the weight of the structure, the load that the members themselves must support cannot be fully known. When the loadings are assumed, the sizes of the various parts of the bridge can be calculated, and then their weight can be added up to check the original assumption. For an experienced engineer designing a conventional structure, a final calculation of weights only serves to confirm the educated guess, and so such a calculation may not even be made in any great detail. In the case of a bridge of new and unrealized proportions, however, there is little experience to provide guidance in guessing the weight accurately in the first place; a recalculation, or a series of iterated recalculations, is necessary to gain confidence in the design. . . . According to the findings of the commission, "the failure to make the necessary recomputations can be attributed in part to the pressure of work in the

designing offices and to the confidence of Mr. Szlapka in the correctness of his assumed dead load concentrations. Mr. Cooper shared this confidence." Since Cooper was well known to have a "faculty of direct and unsparing criticism," his confidence in Szlapka's design work went unquestioned.

Just as Cooper had confidence in Szlapka's work, so did the resident engineer at the construction site have confidence in the work of them both. When a construction foreman expressed serious concern over the condition of the fatal member, the resident engineer thought the matter of little importance, telling the foreman, "Why, if you condemn that member, you condemn the whole bridge." After the collapse, it was reported that the resident engineer "had confidence in that failing chord because it was to him unbelievable that any mistake could have been made in the design and fabrication of the huge structure over which able engineers had toiled for so many years." . . . The underestimation of the true weight of the bridge had actually come to Cooper's attention earlier in the design process, but only after considerable material had been fabricated and construction had begun. At this time, a recalculation of the stresses in the bridge led Cooper to consider that the error had meant that some stresses had been underestimated by 7-10 percent. All structures are designed with a certain margin of safety; he felt the error had reduced that margin to a small but acceptable limit, and so the work was allowed to proceed. In fact, some of the effects of the underestimated weights were, in the final analysis, of the order of 20 percent, and this was beyond the margin of error that the structure could tolerate.[23]

Other human failures also contributed to the collapse, and addressing them might have prevented the tragedy or lessened its consequences. Because of advancing age and declining health, Cooper had been unable to visit the construction site during the two most recent years of construction. Also, Szlapka criticized Cooper for making the bottom chords curved "for artistic reasons" and for failing to visit the Phoenixville plant where the bridge parts were being fabricated. The Royal Commission's report commented on Cooper's role and on the design deficiencies and communication problems:

Mr. Cooper states that he greatly desired to build this bridge as his final work, and he gave it careful attention. His professional standing was so high that his appointment left no further anxiety about the outcome in the minds of all most closely concerned. As the event proved, his connection with the work produced in general a false feeling of security. His approval of any plan was considered by everyone to be final, and he has accepted absolute responsibility for the two great engineering changes that were made during the progress of the work—the lengthening of the main span and the changes in the specification and the adopted unit stresses. In considering Mr. Cooper's part in this undertaking, it should be remembered that he was an elderly man, rapidly approaching seventy, and of such infirm health that he was only rarely permitted to leave New York.[24]

Cooper's distance from the construction site and his inability to travel created a communication problem that played a critical role in the days leading up to the disaster. Nevertheless, even today, when cellular telephones, fax machines, e-mail, and overnight courier service permit design work to be conducted off-site, it is unimaginable that the key consulting engineer would neglect to ever visit the construction site—especially when that engineer has ultimate technical authority of the sort that Cooper wielded.

Organizational Deficiencies

The Royal Commission also criticized the way in which the project had been organized, both by the Quebec Bridge Company and by Cooper:

> Mr. Cooper assumed a position of great responsibility, and agreed to accept an inadequate salary for his services. No provision was made by the Quebec Bridge Company for a staff to assist him, nor is there any evidence to show that he asked for the appointment of such a staff. He endeavoured to maintain the necessary assistants out of his own salary, which was itself too small for his personal services, and he did a great deal of detail work which could have been satisfactorily done by a junior. The result of this was that he had no time to investigate the soundness of the data and theories which were being used in the designing, and consequently allowed fundamental errors to pass by him unchallenged. The detection and correction of these fundamental errors is a distinctive duty of the consulting engineer, and we are compelled to recognize that in undertaking to do his work without sufficient staff or sufficient remuneration both he and his employers are to blame, but it lay with himself to demand that these matters be remedied.[25]

Problems of this nature persist even today. Moreover, in the case of the Quebec Bridge, it seems that this lesson was not fully learned by the government's Board of Engineers. When the bridge reconstruction began, that board spent more than two years and half a million dollars preparing specifications for the bridge. But then, having expended so much time and money, it expected engineering companies to prepare detailed competitive bids within four months with no remuneration.[26]

Redesign and Reconstruction

In 1908 the government of Canada, recognizing that the bridge would be a key link in the Transcontinental Railway, decided that the demolished bridge should be redesigned and reconstructed. The government established a three-person Board of Engineers to prepare plans and specifications and to supervise the reconstruction. The board's duties and powers were clearly defined.

Having reviewed the earlier plans and the report of the Royal Commission, the board adopted a modified cantilever structure with a wider base between the side-trusses and with straight lower chords. Removing the twisted steel and

TABLE 1.3 — Comparison of Design Specifications

Cantilever Bridge	Longest span	Dimensions of compressive chords	Steel cross-sectional area of chords
Firth of Forth Bridge, Scotland (1890)	521 m (1710 ft)	3.66 m (12 ft) diameter circle	0.516 m² (800 square inches)
Original design by the Phoenix Bridge Co. (1907)	549 m (1800 ft)	1.38 m (4' 6.5") high	0.543 m² (842 square inches)
Final design by the St. Lawrence Bridge Co. (1917)	549 m (1800 ft)	2.21 m (7' 3") high	1.252 m² (1941 square inches)

debris from the 1907 disaster took two years. After that, new supporting piers were built that went down to bedrock. Under the board's direction, the superstructure was designed, manufactured, and erected by the St. Lawrence Bridge Company, Ltd., of Montreal. In the new design, the compressive chords were significantly larger than in the original design. As shown in Table 1.3, the steel cross-sectional areas of the rectangular chords of the original Phoenix design were slightly greater than in the Forth Bridge, which had circular cross-sections. The Forth's circular cross-sections gave a larger resistance (moment of inertia) against buckling, but more importantly, circular sections did not require lattice-work, cross-braces, or other heavy links to stiffen rectangular plates.[27] These secondary members added a great deal of weight.

It is clear even from the scanty data in Table 1.3 that the Phoenix bridge, with less efficient compressive chords, must have been a very slender design compared to the Firth of Forth bridge. Circular chords were considered for the Quebec bridge, but rejected as uneconomical. The circular chords of the Forth Bridge could be built easily in Scotland, where the shipyards were accustomed to raising large structures; as well, Scottish shipyard workers had the knowledge and machinery needed to fabricate the curved surfaces. In North America at that time, facilities for a similar project of this magnitude were scarce. The final (St. Lawrence) bridge was designed to instil confidence in the structure; as shown in Table 1.3, the massive compressive chords were almost 2.5 times as heavy (per unit length) as those on the Forth bridge.

The Second Accident

During the reconstruction, a second disaster occurred. The original (Phoenix) erection plan had been to construct the bridge entirely in place by building each cantilever out from the river bank until the two met at mid-span. For the second (St. Lawrence) design, the erection plan was to build the cantilevers only partway out from the shore. Meanwhile, the central part of the span would be assembled onshore. At the appropriate time, it would be floated out and raised into position. On September 11, 1916, the weather and tides were suitable for floating the middle span to the bridge and raising it into place. All

The instant of collapse of the centre span of the Quebec Bridge (St. Lawrence design) on September 11, 1916.

Source: The Quebec Bridge over the St. Lawrence River near the City of Quebec: *Report of the Government Board of Engineers*, Department of Railways and Canals Canada, 1919.

went smoothly, and by midmorning the span had been lifted about 7 m above the water. At about 11 a.m., a sharp crack was heard and the centre span slid off its four corner supports into the river. Thirteen men were killed and fourteen more were injured.

An investigation conducted by the St. Lawrence Bridge Company and the Board of Engineers found that the accident was unrelated to the design and was caused by a material failure in one of the four bearing castings that supported the central span temporarily while it was being transported and hoisted. The St. Lawrence Bridge Company assumed the responsibility for the failure, a second span was constructed, and the design of the support bearings was changed from a casting to a lead "cushion." The new middle span was successfully lifted into place, over three days, in August 1917. The bridge was opened to traffic in 1918, and a formal ceremony attended by the Prince of Wales was held on August 22, 1919.

Conclusion

In the decade following the Quebec Bridge disasters, the first Acts to license professional engineers were put into law. The Ritual of the Calling of an Engineer (described in Chapter 16) was instituted, and even today the chain and Iron Rings used in that ritual are rumoured to be made from the actual steel that claimed the lives of so many men in the cold waters of the St. Lawrence. There are many lessons to be learned from Canada's worst bridge disaster, such as the importance of the following:

- Providing adequate capitalization for large-scale projects.
- Hiring capable and competent professionals.
- Defining clearly the duties, authority, and responsibility of professional personnel.
- Discussing design decisions and related technical problems openly, and listening receptively.

A view of the completed bridge from the north shore.

Source: The Quebec Bridge over the St. Lawrence River near the City of Quebec: *Report of the Government Board of Engineers,* Department of Railways and Canals Canada, 1919.

- Reviewing details, especially in the iterative task of engineering design.
- Monitoring work on the site adequately.
- Ensuring that communication is rapid and accurate.
- Providing adequate support staff and remuneration for highly skilled professional people.

Provincial regulation of engineering helps achieve these goals. The Professional Engineer's stamp on engineering plans and specifications identifies unambiguously who is responsible for the accuracy of the documents and for the computations on which they are based. These lessons were learned at great cost.

WHERE TO LEARN MORE   The two-volume book cited below describes the Quebec Bridge design and construction in impressive detail. The book is a classic of project documentation, and its purpose was undoubtedly to restore public confidence in the safety of the final structure. The book is available in most university libraries and is still well worth reading, almost a century later, by anyone interested in structural design. More suggestions for further reading about Canadian engineering history are listed in Appendix CD-E, on the CD-ROM disk included with this text.

The Québec Bridge over the St. Lawrence River near the City of Québec: Report of the Government Board of Engineers, Department of Railways and Canals Canada, printed by order of the Governor General in Council, 31 May 1919.

DISCUSSION TOPICS AND ASSIGNMENTS

(Additional assignments are found in Appendix CD-E on the CD-ROM included with this text.)

1. Discuss the boundary between engineering and science, and/or the boundary between geoscience and science. Give examples where the applied nature of engineering and geoscience helps to identify these boundaries. What differences in these disciplines would motivate a person to select a career in one over the other? What factors motivated you?

2. Compare the roles in the technical spectrum discussed in this chapter (scientist, engineer, technologist, technician, skilled worker). Using the Internet, try to discover what rewards (salary, work environment, prestige, personal satisfaction, etc.) are generally associated with each occupation. What characteristics or traits would generally be helpful for people entering each of these occupations? Considering your own characteristics and expectations, which of these roles is best for you?

3. This chapter defines the term "engineer" in a very general way. However, the term "professional engineer" has a legal definition in each province or territory (for example, see Appendix CD-B and Chapter 2). Obtain and compare at least five different general definitions of engineer. Use the Internet, dictionaries, or encyclopaedias. How well do these general definitions agree with the legal definition of professional engineer for your province or territory? Likely none of the definitions will agree precisely with the legal definition. Explain or comment briefly on the differences in the definitions. Repeat this exercise for the term "geoscientist."

4. Journalists sometimes use the term "engineering" incorrectly. A few examples follow: Radio broadcasts often refer to their technicians as "engineers", and "genetically engineered" foods are almost always developed by scientists. Similarly, journalists may refer to a "scientific achievement" when things go well, but use the term "engineering failure" when a disaster occurs. Using the Internet, search for at least three examples of the incorrect use of the terms *engineer* or *engineering* by the news media. What rules should the media follow when using these terms? If you find obvious errors or apparent bias in your examples, send an e-mail to the journalist to explain your point of view.

5. For final-year students: Examine your university course catalogue and analyze your engineering program as follows: Create a matrix and categorize each course in your program from first year to graduation. Each row of the matrix should correspond to one of your courses, and each column should correspond to an important engineering job function. Start with the job functions discussed in Table 1.2 of this chapter, but do not be restricted to that list. Add any and all skills that you consider important to the education of an engineer. Then rate each course on how it contributes to preparing you for future employment as an engineer practising in the twenty-first century. Involve your entire class to get a broad range of opinions and a consensus on the ratings. Does your program of courses prepare students properly? Summarize your conclusions and comments in a report to your faculty dean. (This project has occasionally been used as part of an exit survey by engineering students in their final year.)

NOTES

1. Throughout this text, the term "Association of Professional Engineers"—or simply "Association," when capitalized—refers to the legal entity that has been established by statute in the reader's province or territory to regulate the practice of professional engineering and/or professional geoscience. Addresses of these Associations are listed in Appendix A.

2. The term "Professional Engineering Act," or simply "Act," refers to the statute itself, which may have a slightly different name depending on the province or territory. The provincial and territorial statutes are listed in Chapter 2, and excerpts from the statutes are included in Appendix CD-B. Similarly, the term "engineer," when used in this text, also refers to geologists, geophysicists, and geoscientists in the eleven provinces and territories that regulate these specialties.

3. Canadian Council of Professional Engineers (CCPE), *Annual Report*, CCPE, Ottawa, 2002.

4. Canadian Council of Professional Geoscientists (CCPG), <www.ccpg.ca> (August 25, 2003).

5. Canadian Council of Professional Engineers (CCPE), *National Survey of the Canadian Engineering Profession*, 2002, CCPE, Ottawa, 2003.

6. G.C. Andrews, J.D. Aplevich, R.A. Fraser, and H.C. Ratz, *Introduction to Professional Engineering in Canada*, Prentice-Hall, Toronto, 2003, p. 5.

7. CCPE, *National Survey*.

8. Canadian Council of Professional Engineers (CCPE) Task Force on the Future of Engineering, *The Future of Engineering*, CCPE, Ottawa, 1988, p. 33.

9. CCPE, *The Future of Engineering*, ibid.

10. CCPE, *National Survey*.

11. *Webster's Third New International® Dictionary*, Unabridged © 1993 by Merriam-Webster, Incorporated.

12. H.A. Halliday, *The Canadian Encyclopedia Plus*, McClelland & Stewart, Toronto, 1995.

13. Canada, Royal Commission, *Québec Bridge Inquiry Report*, Sessional Paper No. 154, 7–8 Edward VII, Ottawa, 1908, p. 49.

14. Ibid., p. 50.

15. Ibid., p. 50.

16. Ibid., p. 51.

17. Ibid., p. 52.

18. H. Petroski, *Engineers of Dreams: Great Bridge Builders and the Spanning of America*, New York: Vintage Books, 1995, p. 102.

19. Petroski, *Engineers of Dreams*, pp.104–5. Copyright © 1995 by Henry Petroski. Used by permission of Alfred A. Knopf, a division of Random House, Inc.

20. Canada, Royal Commission, *Québec Bridge Inquiry Report*.

21. G.H. Duggan, *The Québec Bridge*, bound monograph prepared originally as an illustrated lecture for the Canadian Society of Civil Engineers, January 10, 1918.

22. Ibid.

23. Petroski, *Engineers of Dreams*, pp. 108–109. Copyright © 1995 by Henry Petroski. Used by permission of Alfred A. Knopf, a division of Random House, Inc.

24. Canada, Royal Commission, *Québec Bridge Inquiry Report*, p. 49.

25. Ibid., p. 50.

26. Petroski, *Engineers of Dreams*, p. 115.

27. Duggan, *The Québec Bridge*.

Chapter 2
Regulation of Engineering and Geoscience

Most countries license, regulate, or control the professions. Licensing is intended to serve three purposes: to protect the safety of the public, to restrict unqualified people from practising, and to discipline negligent or unscrupulous practitioners. In Canada the legal right to regulate the professions falls under provincial jurisdiction; similarly, in the United States the states have this regulatory power.

THE EVOLUTION OF LICENSING LAWS IN CANADA

Professional Engineering

Efforts to place engineering on the same professional footing as law and medicine began formally as early as 1887, when the Canadian Society of Civil Engineers (CSCE) held its first general meeting. The campaign to regulate the engineering profession was led by the CSCE (which in 1918 became the Engineering Institute of Canada, or EIC). In the years after Confederation, most of Canada followed the British model: engineers entered the profession after a period of apprenticeship, and few engineers were university graduates. However, from its very start the CSCE took it upon itself to establish and maintain high standards for admission to the Society, with the goal of improving professional engineering practices. Applicants were required to be at least thirty years of age and to have at least ten years of experience, which could include an apprenticeship in an engineer's office or a term of instruction in a school of engineering acceptable to the CSCE Council. Each applicant also had to show "responsible charge of work" for at least five years as an engineer, designing and directing engineering works.[1]

In spite of this early Canadian initiative, the United States was, in fact, the first country to regulate the practice of engineering. The State of Wyoming enacted a law in 1907 as a result of many instances of gross incompetence observed in a major irrigation project.[2] In Canada, the deadly collapses that occurred during the construction of the Quebec Bridge emphatically reinforced the drive to regulate the profession. (The Quebec Bridge disasters are described in Chapter 1.)

However, it would be many years before Canadian engineers overcame professional rivalries, business competition, class barriers, and other impediments and agreed on proposals to improve professional standards—and, indirectly, the status of engineers. In August 1918, at a general meeting of the CSCE held in Saskatoon, an Alberta engineer named F.H. Peters called on the Society to seek licensing legislation. In his view, engineers had developed the nation's resources but had yet to receive the remuneration and the respect they deserved.[3] At that time, the First World War was drawing to an end, and the flood of returning soldiers—some of whom had been involved in various aspects of military engineering—was dramatically increasing the number of engineers. This was depressing salaries, increasing competition, and placing quality standards at risk.

The CSCE (which had just changed its name to the EIC) drafted a model act, which was published in the *EIC Journal*. In September 1919 the *Journal* announced that 77 percent of EIC members had approved the model act by mail ballot. By the spring of 1920, all provinces except Ontario, Saskatchewan, and Prince Edward Island had passed licensing laws. In Ontario, a joint advisory committee redrafted the bill, which was passed in 1922. The laws enacted in British Columbia, Manitoba, Quebec, New Brunswick, and Nova Scotia were "closed," which meant that engineers would require a licence either to practise engineering or to use the title of Professional Engineer (P.Eng.). In Alberta and Ontario the laws were "open," which meant that the P.Eng. title was now protected; however, since licensing was voluntary, unlicensed people could still practise engineering. Alberta amended its Act to close it in 1930; Ontario closed its Act in 1937.[4]

In the years that followed, all of Canada's provinces and territories and all of the American states amended or passed licensing laws to regulate the engineering profession and the title of Professional Engineer. Prince Edward Island, in 1955, was the last province to enact closed legislation. There is a key difference between the Canadian and American engineering laws. In Canada the engineering profession is self-regulating: each province or territory has passed an Act to create an Association of professional engineers, which in turn regulates the profession. Each Association's regulations and bylaws must be approved by its governing council; the majority of that council's members are elected by, and from, the members of the Association. In contrast, in the United States the state governments license engineers directly; they also establish the regulations that engineers must follow in their work.

In some countries, there is no licensing of engineers and the term "engineer" is not regulated by law. Instead, the possession of a degree or membership in a technical society is used to gauge the person's competence. In Britain, for example, licensing is not compulsory and the title is not protected: *engineer* often means *mechanic*—the sign "Engineer on Duty" is found outside many garages. British engineering societies award the title of Chartered Engineer to members who join voluntarily and meet the Society's admission requirements.

Professional Geoscience

The licensing of professional geoscientists has followed roughly the same path as for professional engineers over the past eighty years. A brief historical outline follows:

> The engineering professions were regulated in Canada in the early decades of the twentieth century. From the outset, it was recognized that the work of many geoscientists also affected the public welfare through their involvement in oil, gas and ore reserves estimation, exploration and mining activities, and construction of major engineering works such as dams and bridges. More recently, geoscientists have become major players in the broad area of environmental practice.
>
> Initially, geoscientists whose work impacted the welfare of the public were licensed as engineers, usually as mining engineers. In Alberta, John A. Allan, a prominent geoscientist and founder of the Geology Department at the University of Alberta, took an active role in establishing the Association of Professional Engineers of Alberta (APEA) in the 1920s and became its president in the 1930s. In the 1950s, the discovery of oil and gas in Alberta focussed attention on the geoscience professions, with the result that geologists, and the practice of geology and geophysics were explicitly identified in the Engineering Act in Alberta in 1955. Separate designations for geologists and geophysicists (*P.Geol.* and *P.Geoph.*) were introduced in 1960 and, in 1966, APEA changed its name to become the Association of Professional Engineers, Geologists and Geophysicists of Alberta (APEGGA).
>
> Following the pattern in Alberta, geoscientists are now licensed in most Canadian provinces and territories by Associations of engineers and geoscientists, established by legislative acts covering the professions of engineering and geoscience.[5]

Two recent tragic events have spurred the regulation of geoscientists in much the same way that the collapse of the Quebec Bridge spurred the regulation of engineers.

BRE-X

Bre-X Minerals Limited, a mining company in Calgary, was developing a gold mine in Indonesia in the early 1990s when a monumental fraud was perpetrated; someone tampered with the core samples from the test holes. Assay results showed extremely high gold content, and Bre-X shares rose dramatically on the news. When the fraud was revealed in early 1997, the Bre-X stock value dropped sharply until it became worthless. A panic sale of other mining shares and precious-metal mutual funds resulted, causing financial disaster for many investors. Estimates of the total financial losses as a result of the Bre-X fraud run as high as $6 billion. The fraud eroded public confidence in the Canadian stock market and had a seriously damaging effect on the Canadian mining industry. (The Bre-X fraud is discussed in detail later in this chapter.)

WALKERTON

In May 2000, an easily avoidable tragedy fell on Walkerton, a small farming town in southwestern Ontario. A well supplying Walkerton's drinking water system became contaminated with *E. coli*, a deadly bacteria. Before the cause was determined and remedied, seven people died, and more than 2,300 became ill, some of them very seriously ill. A subsequent inquiry by a judge, the Honourable Dennis O'Connor, reported: "The community was devastated. The losses were enormous. There were widespread feelings of frustration, anger, and insecurity." Testimony offered at the inquiry revealed shocking incompetence, negligence, and fraud on the part of the water system's manager and his brother, the foreman. The inquiry also revealed "omissions or failures to take appropriate action" on the part of Ontario's Ministry of the Environment (MOE). The MOE is responsible for enforcing legislation, regulations, and policies concerning the construction and operation of municipal water systems. Closer monitoring and action by the MOE might have prevented the *E. coli* outbreak or at least reduced its extent.[6]

These events taught us a hard lesson—that some tasks should be performed or monitored only by competent professionals. In the wake of Bre-X and Walkerton, several provincial governments that were not already regulating geoscientists recognized the need to do so. Shortly afterwards, Ontario passed the Professional Geoscientists Act (2000), Quebec passed the Geologists Act (*Loi des géologues*—2001), and Nova Scotia passed the Geoscience Profession Act (2002). However, geoscience is still not regulated in Prince Edward Island or Yukon.

Qualified Persons—An Important Role for Engineers and Geoscientists

Before the Bre-X scandal and the Walkerton tragedy, very few government regulations specifically mentioned either engineers or geoscientists. However, politicians now recognize the value of experts in preventing such tragic events. The term "qualified person" (QP), now beginning to appear in legislation and regulations, generally refers to professional engineers or geoscientists.

For example, the Canadian Securities Administrators (CSA), an umbrella body for provincial securities regulators, has issued a document titled *National Instrument 43-101*, which came into effect on February 1, 2001. This document must be followed in Canada when information on mineral projects is being disclosed. It specifies the format for providing oral statements or written disclosures of scientific or technical information to the public concerning mineral projects. The document is extremely specific—it even lists the headings for technical reports. The CSA also states that only a QP can disclose scientific or technical information to the public regarding a mineral project. The CSA defines a QP as an individual who:

    a)   is an engineer or geoscientist with at least five years of experience in mineral exploration, mine development or operation or mineral project assessment, or any combination of these;

b) has experience relevant to the subject matter of the mineral project and the technical report; and

c) is a member in good standing of a professional association.[7]

Other people may work on the project, but if a QP is assisted by people who are not qualified (under this definition) in preparing a technical report, the QP assumes the responsibility and must take action to ensure that the information is correct. A QP must always visit the site on which the report is based.

Several provincial Associations have issued very readable guidelines defining the responsibilities of professional engineers undertaking the role of a qualified person.[8] QPs are also essential to direct the restoration of "brownfields" (contaminated properties). Several provinces are developing legislation to monitor the decontamination of brownfields.

PROVINCIAL AND TERRITORIAL ACTS

Engineering and geoscience are regulated by Acts of provincial legislatures (in the provinces) or legislative councils (in the territories). The Acts are grouped below according to how they regulate geoscience:

Combined Engineering and Geoscience Acts

The following provinces and territories regulate engineering and geoscience in the same Act:

Alberta	Engineering, Geological and Geophysical Professions Act
British Columbia	Engineers and Geoscientists Act
Manitoba	Engineering and Geoscientific Professions Act
New Brunswick	Engineering and Geoscience Professions Act
Newfoundland	Engineers and Geoscientists Act
Northwest Territories	Engineering, Geological and Geophysical Professions Act
Nunavut	Engineers, Geologists and Geophysicists Act (Nunavut) S.N.W.T.
Saskatchewan	Engineering and Geoscience Professions Act

Separate Engineering and Geoscience Acts

The following provinces regulate engineering and geoscience in separate Acts:

Nova Scotia	Engineering Profession Act and Geoscience Profession Act.
Ontario	Professional Engineers Act and The Professional Geoscientists Act.
Quebec	Engineers Act (*Loi des ingénieurs*) and Geologists Act (*Loi des géologues*).

Engineering Acts Only

In Prince Edward Island and Yukon, the engineering profession is regulated; however, the geoscience profession is not yet regulated by Act:

Prince Edward Island	Engineering Profession Act
Yukon	Engineering Profession Act

Contents of the Acts

Each of the above Acts contains the following:

- The purpose of the Act.
- The legal definition of engineering and/or geoscience.
- The procedure for establishing a provincial or territorial Association, and the purpose (or objects) of the Association.
- Standards for admission to the Association (or for the granting of a licence).
- Procedures for establishing regulations to govern professional practice.
- Procedures for establishing bylaws to govern the Association's administration and to elect a governing council.
- A Code of Ethics to guide the personal actions of the members.
- Disciplinary procedures.

Miscellaneous Notes Concerning the Acts and Associations

- Relevant excerpts from all of the Acts are included in Appendix CD-B of this text. A complete copy of each Act can be found easily by an Internet search, or may be obtained directly from the provincial or territorial Association.
- Throughout this text the terms Professional Act, provincial Act, or simply Act refer to the relevant Act or ordinance (as listed above) for the reader's province or territory.
- Similarly, the term provincial Association or simply Association refers to the Association of Professional Engineers and/or Geoscientists (or *Ordre des ingénieurs* or *Ordre des géologues*) for the reader's province or territory.
- The term "engineer" includes geologists, geophysicists, and geoscientists, unless indicated otherwise.
- In past decades, engineers were registered as members of the Associations; this reflected the idea that the member had undergone a voluntary certification process and was permitted to use the professional title. However, as explained earlier in this chapter, the right to practise is now restricted by all of the Acts to those practitioners who are licensed to practise the profession. The terms "licensed," "registered," "member," and "membership" are now equivalent and are used interchangeably in this text.

LEGAL DEFINITIONS OF ENGINEERING AND GEOSCIENCE

Definition of Engineering

To regulate the profession effectively, it is important to define the terms "professional engineer" and "the practice of professional engineering." These terms delineate the boundary between engineering and other professions (such as architecture and town planning) and between engineers and other personnel in the design and development spectrum (such as scientists and technologists). The Canadian Council of Professional Engineers (CCPE) has proposed simple, national definitions of these terms. (CCPE is described in more detail later in this chapter.) CCPE defines the practice of professional engineering as follows:

> any act of planning, designing, composing, evaluating, advising, reporting, directing or supervising, or managing any of the foregoing, that requires the application of engineering principles and that concerns the safeguarding of life, health, property, economic interests, the public welfare or the environment.[9]

The CCPE definition includes an exemption for natural scientists, as follows:

> anyone who either holds a recognized honours or higher degree in one or more of the physical, chemical, life, computer or mathematical sciences, or who possesses an equivalent combination of education, training and experience . . . from practising natural science which . . . means any act (including management) requiring the application of scientific principles, competently performed.[10]

The goals of this national definition are as follows: to achieve coordination and uniformity within the engineering profession; to enable the mobility of engineers wishing to practise engineering in different provinces and territories throughout Canada; and to enable the administration of engineering practice to be more consistent across the country. It is anticipated that each province and territory, assisted by CCPE, will eventually converge on common national definitions of these basic terms; at present, however, every province and territory defines these terms differently. All of these legal definitions of engineering and/or the practice of professional engineering are included in Appendix CD-B.

No two definitions are identical. Some provinces list the types of machinery or structures (such as railways, bridges, highways, and canals) that are within the engineer's area of practice. This makes the definition very clear and specific, but also very long and difficult to read or understand. As time passes, such lists inevitably fall out of date as some components (such as steam engines) disappear and new areas (such as engineering software) are added.

The shorter definitions are easier to understand and remember. However, they may contain terms (such as "engineering principles") that are very general and that must be interpreted and defined. In Ontario, for example, the definition of the practice of professional engineering is almost identical to the CCPE definition:

The Practice of Professional Engineering: Any act of designing, composing, evaluating, advising, reporting, directing or supervising wherein the safeguarding of life, health, property or the public welfare is concerned and that requires the application of engineering principles, but does not include practising as a natural scientist.[11]

For comparison, we could also examine the same terms as they are defined in the United States. The following definition is from the Model Law (1996 revision) prepared by the U.S. National Council of Examiners for Engineering and Surveying. This law, like the CCPE definition, serves merely as a guide for lawmaking bodies; it has no legal effect unless written into law by state legislatures:

Engineer: The term "Engineer," within the intent of this Act, shall mean a person who is qualified to practice engineering by reason of special knowledge and use of the mathematical, physical, and engineering sciences and the principles and methods of engineering analysis and design, acquired by engineering education and experience.[12]

The above definitions are far from identical but do show some similarities. All of them, however, use terms such as "engineering principles" and "engineering sciences"; the result is circular definitions unless further explanations are provided. The difference between *engineering* principles and *scientific* or *technological* principles hinges on the purpose and depth of the study, and is explained by the Canadian Engineering Accreditation Board. (CEAB is the CCPE committee that accredits Canadian university engineering programs.)

The CEAB criteria for accreditation state: "Engineering science subjects normally have their roots in mathematics and basic sciences, but carry knowledge further toward creative applications. . . . Application to identification and solution of practical engineering problems is stressed."[13] In other words, engineering involves putting scientific phenomena and principles to practical use.

Fortunately, the engineer's role can be explained in simpler terms: In any engineering project, the professional engineer is responsible for ensuring that the "factor of safety" (the ratio of capacity to expected load on the system or structure) has been calculated properly, and is adequate. This responsibility for "life, health, property or the public welfare" is required by the Code of Ethics. As several case histories in this text show, disasters can occur whenever engineers neglect this key responsibility.

Definition of Geoscience

Every province and territory (except Prince Edward Island and Yukon) has an Act that defines the term "professional geoscience" (or geology or geophysics). These definitions draw the boundaries between geoscience and engineering, and between both of these and other professions. The definitions vary slightly among the provinces and territories; however, the Canadian Council of Professional Geoscientists (CCPG) has proposed what it calls the "minimum content" that should be included in a definition of Professional Geoscience. (CCPG is a federation of ten of the provincial and territorial Associations that regulate geoscience. CCPG is described in more detail later in this chapter.)

The following definition was developed by a task group of the CCPG and has been approved for circulation by the Canadian Geoscience Standards Board and the Board of Directors of CCPG:

> The "practice of professional geoscience" means the performing of any activity that requires application of the principles of the geological sciences, and that concerns the safeguarding of public welfare, life, health, property, or economic interests, including, but not limited to:
>
> (a) investigations, interpretations, evaluations, consultations or management aimed at discovery or development of metallic or non-metallic minerals, rocks, nuclear or fossil fuels, precious stones and water resources;
>
> (b) investigations, interpretations, evaluations, consultations or management relating to geoscientific properties, conditions or processes that may affect the well-being of the general public, including those pertaining to preservation of the natural environment.[14]

On October 18, 1913, the foundation for CP Rail's huge grain elevator in North Transcona, Manitoba, tilted almost 27 degrees from the vertical during its initial loading. The massive structure was eventually righted and still exists today. However, the accident taught a valuable lesson about bearing capacity and the effects of uneven loading.

Source: Archives of Manitoba, Foote 1801 N2793.

PROVINCIAL AND TERRITORIAL ASSOCIATIONS

To administer its provincial or territorial Act, each province or territory has established a self-governing Association of Professional Engineers and/or Geoscientists. The name of the province is included in the Association's name (Quebec has an *Ordre* rather than an Association). In eight jurisdictions, the Associations include geologists, geophysicists, or geoscientists. In Ontario, the Association has adopted the simpler working name of Professional Engineers Ontario (PEO), although its legal name, the Association of Professional Engineers of Ontario, remains unchanged. In Newfoundland and Labrador, the provincial Association recently revised its name to Professional Engineers and Geoscientists of Newfoundland and Labrador (PEG-NL). A list of Association addresses is provided in Appendix A.

As stated above, each Act delegates the responsibility for its implementation to the provincial or territorial Association. Each Association, in turn, has developed regulations or bylaws and a Code of Ethics (or has assisted in their development). These documents are defined as follows:

- **Regulations** are rules that have been established to implement or support the Act. They concern matters such as qualifications for admission to the Association, professional conduct, and disciplinary procedures.
- **Bylaws** are rules established to administer the Association itself. They concern matters such as election procedures for the council, financial matters, committees, and meetings.
- The **Code of Ethics** is a set of rules of personal conduct. Every engineer and geoscientist must be familiar with this code and endeavour to follow it.

The regulations and bylaws and the Code of Ethics have been established under the authority of each Act. This means they govern the profession with the force of law. The engineering profession is referred to as "self-regulating" because its members play a role in governing the profession. Specifically, the members elect the majority of the Association's governing council (the government also appoints some councillors); furthermore, the members must confirm (by ballot) the regulations and bylaws passed by the council.

Obviously, if a profession is to regulate itself effectively, its members should be well informed when voting on changes to regulations and bylaws, and they must be willing to serve in the elected positions in their Association, especially at the council level.

ADMISSION TO THE PROFESSION

Each provincial Association admits applicants to the profession by registering them as members of the Association and granting them licences to practise. The standards for admission are similar, although not identical, across all provinces and territories. Typically, an applicant is admitted to the profession and awarded a P.Eng. or P.Geo. licence after satisfying six conditions:

- **Citizenship**. The applicant must be a citizen of Canada or have the status of a permanent resident, although temporary nonresident licences are usually permitted.
- **Age**. The applicant must have reached the legal age of majority, which is eighteen in most provinces.
- **Education**. The applicant must have adequate academic qualifications. A university degree from an accredited engineering program (or equivalent) is required.
- **Examinations**. Typically, every applicant is expected to write and pass the professional practice exam.
- **Experience**. The applicant must satisfy the experience requirements. Most jurisdictions now require four years of suitable experience.
- **Character**. The applicant must be of good character, as determined mainly from references.

Although these six conditions are encountered in almost every jurisdiction in Canada, some variations do exist. Clearer explanations are provided below for some of the requirements. Readers should check with their provincial or territorial Association regarding these requirements, as admission processes are constantly under review.

Academic Requirements

Academic qualifications are the most important requirement for admission. They are usually evaluated by a board of examiners or an academic requirements committee. Applicants must provide documents to substantiate their academic qualifications. Graduation from a recognized (CCPE-accredited) program at a Canadian university grants full exemption from the examination program, except for the Professional Practice Exam. (University accreditation is a function of the Canadian Council of Professional Engineers, discussed later in this chapter.)

Degrees in engineering from American universities accredited by the U.S. Accreditation Board for Engineering and Technology (ABET) are usually recognized as equivalent to Canadian degrees. However, examinations may be required in some cases, even for those holding degrees from ABET-accredited universities. Some degrees are not recognized, and applicants holding these degrees are required to pass a series of examinations to confirm their engineering knowledge.

Technical Examinations

People who have not completed university-level engineering degrees may apply for admission to the engineering profession. However, they will usually be required to write examinations. For each branch of engineering there are between fourteen and eighteen three-hour examinations. Applicants may be required to write a subset of these exams to make up deficiencies in their academic qualifications. Permission to enter the examination system varies widely. For example, entrance is virtually open in British Columbia, whereas in Quebec it is fairly tightly controlled. In Ontario the examination system is open only to those who hold, as a minimum, one of the following: a three-year engineering technologist diploma from a college of applied arts and technology; a technologist-level certificate from the Ontario Association of Certified Engineering Technicians and Technologists (OACETT); or other acceptable education as determined by the Association's academic requirements committee.

The number and type of examinations assigned will depend on the applicant's academic achievements to the date of evaluation. The possession of one or more postgraduate degrees may reduce the number of exams assigned. However, postgraduate degrees are not always relevant, since the essential engineering knowledge required for licensing is taught at the undergraduate level in accredited engineering programs.

The examination system provides an alternative route into the profession, but it is not an easy route, because it is not an educational system. People applying for Association exams must be qualified to write and pass the exams. The Associations do not offer classes, laboratories, or correspondence courses.

Professional Practice Exam

Most applicants, whatever their academic qualifications, must pass an examination in professional practice, law, contracts, liability, and ethics. Exceptions are made for people who are transferring from elsewhere in Canada and who have written the exam in the previous five years.

Figure 2.1 shows the typical admission process for graduates of CCPE-accredited university engineering programs. For these graduates the admission process is fairly simple, since they need only pass the Professional Practice Exam. Applicants who are not graduates of accredited programs may be required to pass a more extensive set of exams.

Experience Requirement

Satisfactory work experience is another important requirement for obtaining a licence. Applicants must prepare adequate documentation to prove to the Association that they have met the experience requirements. This documentation is usually assessed by a committee of the Association. Standards for evaluating engineering experience have been developed by CCPE; these standards have in turn been adopted by most provincial Associations, with minor modifications. The applicant's experience is evaluated as to its nature, duration, currency, and quality. As an example, the Ontario experience requirements are summarized below.[15]

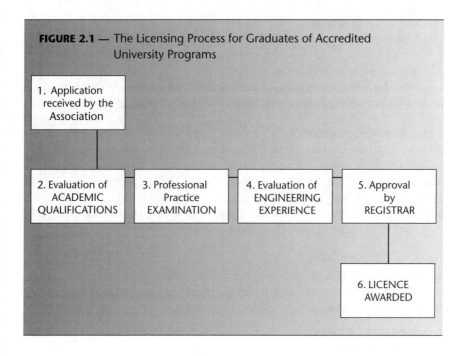

FIGURE 2.1 — The Licensing Process for Graduates of Accredited University Programs

1. Application received by the Association

2. Evaluation of ACADEMIC QUALIFICATIONS

3. Professional Practice EXAMINATION

4. Evaluation of ENGINEERING EXPERIENCE

5. Approval by REGISTRAR

6. LICENCE AWARDED

NATURE OF EXPERIENCE

Engineering experience is expected to be in the area of the applicant's academic study. Thus, if the experience is incompatible with the academic study—for example, if a mechanical engineering graduate is working in an electrical engineering job—the Association may ask why this incompatibility exists and may require additional experience or additional studies before granting a licence. Also, some jobs may have similar descriptions but, depending on the activities performed, may be given different credit as engineering experience. An applicant should not presume that his or her area of employment will automatically be accepted (or rejected) as valid engineering experience.

In particular, an applicant whose employment falls into one of the following categories should consult the provincial Association for more specific advice:

- teaching (at any level)
- sales and marketing
- military service
- project management
- operations and maintenance
- computer engineering

DURATION OF EXPERIENCE

The length of experience required has increased over the past decade. Before 1993, only two years of direct engineering work were required once an engineering degree had been obtained. Since 1993, every jurisdiction has implemented a four-year experience requirement. The exception is Quebec, which in 2002 raised its experience requirement from two years to three.

CURRENCY OF EXPERIENCE

The Associations consider it important that the applicant's experience be recent; generally speaking, more recent experience is preferred over experience acquired in the distant past. Engineering theory evolves fairly slowly. Procedures and standards change more quickly than this, and computer applications change very rapidly. An applicant must show that experience obtained at the start of his or her career is still relevant.

QUALITY OF EXPERIENCE

Each applicant must prepare an experience résumé and explain how that experience satisfies the requirements. Experience is usually judged in relation to five quality criteria:

- application of theory
- practical experience
- management of engineering
- communication skills
- social implications of engineering

The application of theory is a mandatory requirement, and it must be demonstrated over a substantial part of the experience period (though not necessarily all of it). It must be supplemented by exposure to, or experience in, the remaining four criteria. The different Associations may place a slightly different emphasis on each of these criteria, so the applicant should consult the Association's guidelines when preparing experience documentation.

MISCELLANEOUS EXPERIENCE CREDITS

Some Associations evaluate experience differently:

- Most provinces require all applicants to obtain twelve months' experience in Canada under the direction of a professional engineer who is licensed in Canada.
- Most provinces permit twelve months' experience for completing a postgraduate degree in engineering. In fact, some provinces permit even more credit (up to the total time spent in postgraduate studies), depending on how well the postgraduate experience satisfies the five quality criteria described earlier.
- In many provinces where the four-year experience requirement is in place, the Association will grant up to twelve months' credit for experience obtained prior to university graduation. The experience must satisfy all of

the standards noted earlier and usually must be obtained after the mid-point of the academic program. (Similarly, Quebec permits the three-year experience requirement to be reduced by as much as eight months if the candidate completes an optional sponsorship program, which involves a series of meetings with senior engineers to discuss specified topics, including skills, responsibility, ethics, and social commitment.)

Graduates of Foreign Engineering Schools

An applicant who received his or her engineering education from a foreign college or university must provide the Association with originals (or certified copies) of all transcripts and diplomas. The academic requirements committee (which in some provinces is called the board of examiners) will assess those documents and determine the applicant's admissibility. Each case is evaluated individually. However, since foreign universities are not accredited by the Canadian Engineering Accreditation Board, admission is not generally awarded solely on the basis of foreign educational documents; some further evidence of engineering competence is usually required.

This evidence can be provided in several ways. For example, senior foreign-educated engineers may be able to document their engineering achievements. Other applicants may validate a foreign bachelor's degree by completing a master's degree (or even a few advanced undergraduate engineering courses) at an accredited Canadian university.

In most provinces an applicant who is not a graduate of a CEAB-accredited engineering program, but who has completed engineering studies judged by the Association to be equivalent to the Syllabus of Examinations published by the CCPE, is normally assigned to write a set of four Confirmatory Examinations. These examinations, which are usually set by the Association, cover the advanced topics in a small portion of the engineering program. (Such applicants should contact their provincial or territorial Association for more information on validating credentials.)

Applicants from foreign universities should understand that the request for corroborating evidence of academic ability does not imply any lack of respect for the foreign university or for the individual. The Association is required by law to assess the qualifications of applicants. In the absence of an accreditation process for the foreign university, the Association must obtain and evaluate other evidence in order to justify admission.

Nonresident, Temporary, and Provisional Licences

Most provinces and territories offer other forms of licences, depending on local needs and practices. For example, Associations typically offer full membership for residents and temporary licences for nonresidents. The procedures for obtaining a nonresident or temporary licence vary slightly from province to province. Typically, the applicant must be a member of an Association in another province or territory, qualified to work on a specific project, and familiar with the applicable codes, standards, and laws. Usually the nonresident must be collaborating with a member of the Association on the project specified for the temporary licence. However, this last requirement may be waived if the applicant is highly qualified.

Ontario recently introduced provisional licences for applicants who have satisfied all of the application requirements (including experience), except the one stating that the applicant must obtain twelve months of experience in Canada. Provisional licences entitle the holder to practise professional engineering, but only under the supervision of a licensed professional engineer; furthermore, all final drawings, specifications, plans, reports, or other documents must be signed by the supervising professional engineer. This licence is intended to assist mature applicants.

LICENSING OF ENGINEERING CORPORATIONS

There are unique problems inherent in regulating engineering corporations, because the purpose of licensing is to protect the public against incompetence, negligence, and professional misconduct, and these qualities can be evaluated accurately only for human beings. In almost every province and territory, the Act solves these licensing problems by requiring corporations to identify the individual(s) responsible for the engineering services provided by the corporation, and by requiring corporations to carry liability insurance.

A corporation must satisfy these two criteria in order to obtain a Permit to Practise (also called a Certificate of Authorization). That is, the corporation must employ a professional engineer who assumes personal responsibility for the engineering services provided by the corporation and who acts in a supervisory capacity. The corporation must also obtain liability insurance; however, the rules for insurance vary significantly from province to province (see Chapter 3).

An evolving question for corporations is quality assurance: How can the public be assured of the continuing competence of the corporation's engineers? Most provinces have developed methods to deal with continuing competence (as discussed in Chapter 15).

A professional engineer working for a corporation that has a Permit to Practise (or Certificate of Authorization) does not have to apply for an individual permit. Nor is a permit required if the corporation does not offer engineering services to the public, obviously.

However, an engineer who "moonlights" at night or on weekends may require a permit. In some provinces, such as Ontario, every entity—be it an individual, a partnership, or a corporation—that offers or provides professional engineering services to the public requires a Certificate of Authorization.[16] An engineer who plans to provide services to the general public must make it a point to explore this issue with the provincial Association.

CONSULTING ENGINEERS

At present, Ontario is the only jurisdiction in Canada that regulates the designation of Consulting Engineer. That designation is regulated by Ontario Regulation 941 (section 56), made under the authority of the Professional Engineers Act. To qualify as a consulting engineer, a member

- must have been continuously engaged for at least two years in private practice,
- must have at least five years of satisfactory experience since becoming a member, and
- must pass (or be exempted from) examinations that may be prescribed by the Association Council.

Since applicants for the Consulting Engineer designation must be engaged in private practice, and therefore offering their services to the public, they must also be holders of a Certificate of Authorization in Ontario; or they must be associated with a partnership or corporation that is a holder of a certificate.

THE ENGINEER'S SEAL

In every province, the professional engineering Act provides for each engineer to have a seal denoting that he or she is licensed. All final drawings, specifications, plans, reports, and other documents involving the practice of professional engineering, when issued in final form for action by others, should bear the signature and seal of the professional engineer who prepared and approved them. This is especially important for services provided to the general public. The seal has important legal significance: it certifies that the documents have been competently prepared, and it identifies clearly the person responsible for them. The seal should not, therefore, be used casually or indiscriminately. In particular, preliminary documents should not be sealed; instead, they should be marked "preliminary" or "not for construction."

The seal denotes that the documents have been prepared or approved by the person who sealed them. This implies an intimate knowledge of and control over the documents. An engineer who knowingly signs or seals documents that have not been prepared by himself or herself or by technical assistants under his or her direct supervision may be guilty of professional misconduct, and may also be liable for fraud or negligence if this misrepresentation results in someone suffering damages. When two or more engineers have assumed responsibility for different areas of a project, the areas of responsibility must always be specified unambiguously.

A fairly common problem involves engineers who are asked to "check" documents and then sign and seal them. This is usually not ethical. Such documents should be sealed by the engineer who prepared them or who supervised their preparation. If they were prepared by a nonengineer, then perhaps he or she should have been under the supervision of an engineer. The extent of the work required to check a document is not clearly defined. Many disciplinary cases have arisen involving engineers who checked and sealed documents that later turned out to be seriously flawed.

In some projects, a proper review would require complete duplication of the analysis. Of course, if you were to repeat all of the work, then obviously it would be appropriate for you to assume responsibility for it. However, never assume responsibility for work that you have not thoroughly and independently reviewed. (This topic is discussed in more detail in Chapter 9.)

THE CODE OF ETHICS

Every provincial and territorial Act requires professional engineers and professional geoscientists to subscribe to a Code of Ethics—that is, a standard of professional conduct. The Code of Ethics for each Act is provided in Appendix CD-B. Each code defines, in general terms, the duties of the engineer to the public, to the employer (or client), to fellow engineers, to the engineering profession, and to himself or herself. The code's main purpose is to protect the general public from unscrupulous practitioners. By instilling public confidence in engineering, the code also increases the prestige of the entire profession.

In most jurisdictions, the Code of Ethics is specifically mentioned in the Act and therefore carries the full force of law. The second half of this text discusses the Code of Ethics in more detail as a guide to personal professional conduct.

PROFESSIONAL MISCONDUCT AND DISCIPLINE

The main purpose of the provincial Associations is to protect the public welfare. So it is sometimes necessary to discipline engineers. Each provincial and territorial Act therefore grants the Association the authority to reprimand, suspend, or expel members who are guilty of professional misconduct, which is usually defined as negligence, incompetence, or corruption.

Misconduct, negligence, incompetence, and corruption are well-known concepts used in common speech. However, these terms must be defined legally if they are to be enforceable. Therefore, most of the Acts and/or regulations provide formal legal definitions of these terms. These definitions are included in Appendix CD-B to this text; in addition, many specific examples of professional misconduct are discussed in later chapters.

To enforce the Act and regulations, each provincial Association has a staff that receives complaints, prosecutes those who are practising under false pretences, and arranges disciplinary hearings for members charged with misconduct. Disciplinary decisions are not made by the staff, but rather by the disciplinary committee, which is appointed by the council. Most members of the council have been elected by the Association's members; thus, the self-regulating aspect of the profession is carried through to disciplinary actions. The results of disciplinary hearings are usually published and circulated to all members. (The disciplinary process is discussed in more detail in Chapter 14.)

The most frequent complaints relate to violations of the Code of Ethics. In most provinces this code is specifically included in the Act and is therefore enforceable under the Act. The terms of the Code of Ethics are based on common sense, natural justice, and basic ethical concepts. Although everyone should read and understand the code, it is not usually necessary to memorize it. Most engineers find that they follow it intuitively and never need fear charges of professional misconduct.

The Code of Ethics enjoins every engineer to act in an honest and conscientious manner. In addition to the code, there is a much older voluntary oath, written by Rudyard Kipling and first used in 1925, called the Obligation of the Engineer. Those who have taken this oath can usually be identified by the Iron Ring worn on the working hand. In recent years an Earth Science Ring has been awarded to geoscientists, in a ceremony comparable to that of the Iron Ring. The Iron Ring does not indicate that a degree has been awarded, nor does the Earth Ring; each, however, indicates that the wearer has participated in the ceremony and has voluntarily made a commitment to high standards of performance and thought. A detailed discussion of both the Iron Ring and the Earth Ring is included in Chapter 16.

CANADIAN COUNCIL OF PROFESSIONAL ENGINEERS (CCPE)

The Canadian Council of Professional Engineers (CCPE) was established in 1936 as a federation of the twelve provincial and territorial Associations that license engineers and regulate the profession across Canada. As a federation of Associations, CCPE does not have individual members; however, every licensed engineer is indirectly a member of CCPE. As an umbrella organization for the engineering profession, CCPE is the legal entity that holds the Canadian trademarks on engineering titles such as P.Eng., Professional Engineer, Engineer, Engineering, and Consulting Engineer, as well as the French equivalents and several related terms and logos.

The CCPE coordinates the engineering profession on a national scale by promoting consistency in licensing and regulation. It develops detailed policies, guidelines, and position statements at the national level. Although these are not binding, the member Associations are encouraged to review and adopt the documents when appropriate. CCPE has several important boards or committees, which are described below.

Canadian Engineering Accreditation Board (CEAB)

In 1965 the CCPE established the Canadian Accreditation Board (CAB), now known as the Canadian Engineering Accreditation Board (CEAB). CEAB's role is to evaluate undergraduate engineering degree programs offered at Canadian universities. With the consent of the engineering Associations, CEAB develops academic criteria for evaluating the undergraduate engineering degree programs, arranges accreditation team visits to engineering universities, compares universities' engineering programs against these criteria, and eventually recommends (for or against) the accreditation. This accreditation ensures that graduates receive an education that meets the necessary academic standards for professional engineering registration across Canada.[17]

An accreditation visit is undertaken at the invitation of the engineering university and with the agreement of the provincial Association. A team of senior engineers is assembled, composed of specialist engineers for the subjects involved and at least one engineer from the provincial Association. Armed with documents—including a detailed questionnaire completed by the institution beforehand—the team consults with administrators, faculty, students, and department personnel. The team examines the academic and professional quality of faculty members and the adequacy of laboratories, equipment, computer facilities, and so forth. It also evaluates the quality of the students' work on the basis of face-to-face interviews with senior students and assessments of recent examination papers, laboratory work, reports and theses, records, models or equipment constructed by students, and other evidence of the scope of their education.

Accreditation teams also perform a qualitative analysis of the curriculum content to ensure that it meets the criteria. Finally, the team reports its findings to CEAB, which then makes an accreditation decision. Based on the information provided, CEAB may grant or extend accreditation of a program for up to six years, or may deny accreditation altogether. CEAB publishes an annual listing of the accreditation history of all programs that are presently or have ever been accredited.

Engineering accreditation is extremely important to universities and to graduates of their programs, because accreditation is a measure of quality. A program is accredited only when it can demonstrate academic and professional competence. In recognition of this quality assurance, every provincial Association grants graduates from these programs exemption from all the technical examinations required for licensing as a professional engineer. Although many engineering students—and even graduate engineers—may not be aware of CEAB's role in ensuring quality in programs of study, CEAB nevertheless has a significant influence on their lives.

Canadian Engineering Qualifications Board (CEQB)

CEQB's primary role is to develop national guidelines for professional engineering qualifications, standards of practice, ethics, and professional conduct. These guidelines assist CCPE in its efforts to promote consistent licensing, registration, and other regulations in the provincial and territorial Associations.

CEQB also maintains the CCPE Examination Syllabus. This syllabus is especially important for evaluating the credentials of candidates who have studied in foreign countries. It describes exam programs for seventeen engineering disciplines as well as programs for basic and complementary studies. When applicants cannot show that they have satisfied the syllabus, they are assigned exams to ensure that they meet appropriate standards for admission into engineering in Canada.

In June 1999, CEQB facilitated the negotiation of the profession's Inter-Association Mobility Agreement (IAMA). This was a major achievement, since the agreement simplifies the process for practising engineering in another province or territory of Canada.

CANADIAN COUNCIL OF PROFESSIONAL GEOSCIENTISTS (CCPG)

The Canadian Council of Professional Geoscientists (CCPG), incorporated in 1996, serves the same function for professional geoscientists as CCPE plays for professional engineers. However, CCPG was created more recently, so its structure is still being defined. CCPG's mandate is to coordinate the geoscience profession on a national scale by promoting consistency in licensing and regulation. It is an umbrella organization that provides unifying advice and guidance. Its policies, guidelines, and position statements are not binding; however, the provincial and territorial Associations are encouraged to review and adopt the documents as appropriate. Several geoscience guidelines have been developed and are available on the CCPG website. One standing committee has been formed: the Canadian Geoscience Standards Board (CGSB). CGSB develops national guidelines and examination syllabi for professional registration in the geosciences, assesses geoscience education across Canada, makes this information available to constituent Associations, and provides advice regarding the national mobility of geoscientists.[18]

SUMMARY

Figure 2.2 illustrates how professional engineers and geoscientists interact with the various organizations mentioned in this chapter. As a professional engineer or geoscientist, you are licensed by your provincial Association and you are likely a member of at least one engineering society (as recommended in Chapter 16). All provincial engineering Associations are federated members of CCPE (the umbrella body for engineering). In most provinces, engineers and geoscientists are licensed in the same Association, which is also linked to CCPG (the umbrella body for geoscience). The members of the Association (including you) elect the council, which appoints the staff and committees. The staff confer with the province's professional engineers and geoscientists regarding admission, professional practice, disciplines, professional development, and so forth. Though both are important bodies, CCPE and CCPG affect you only indirectly. They probably assessed and accredited your university program, and they monitor the profession across Canada; however, they advise the Associations and rarely interact directly with members.

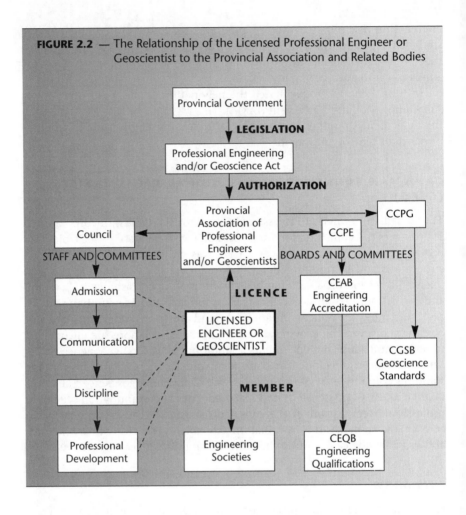

FIGURE 2.2 — The Relationship of the Licensed Professional Engineer or Geoscientist to the Provincial Association and Related Bodies

CASE HISTORY 2.1

THE BRE-X MINING FRAUD

Introduction

In the spring of 1997, Bre-X Minerals Limited, a mining company based in Calgary, became the focus of a spectacular mining fraud. The Bre-X company claimed to have made a gold strike that was richer than any gold discovery in history, and Bre-X stock prices soared. However, after a mysterious death, the fraud began to unravel. The resulting scandal ruined the reputations of almost everyone involved. More seriously, the fraud caused financial calamity for thousands of investors, some of whom had staked their life savings on Bre-X geologists' reports. An independent team of investigators from Strathcona Mineral Services later stated: "The magnitude of tampering with core samples . . . is of a scale and over a period of time and with a precision that, to our knowledge, is without precedent in the history of mining anywhere in the world."[19]

Sequence of Events

For our purposes, the story begins in May 1993, when David Walsh, the founder, chairman, and CEO of Bre-X Minerals, announced the discovery of a gold deposit in Busang, Indonesia. One site, drilled previously by an Australian company, was reported to contain an estimated one million ounces of recoverable gold.[20] This modest estimate was to escalate as the months passed, generating an investment frenzy that pushed Bre-X stock prices from pennies in March 1994 to the equivalent of over $200 per share in September 1996.[21]

Sequence of Events

The roles played by various Bre-X geological staff are not completely clear, although they may be clarified once the many pending lawsuits come to trial. John Felderhof, a 1962 geology graduate of Dalhousie University, is often referred to as the chief geologist[22] of Bre-X Minerals, although he has since declared that his role was that of an administrator or commercial manager. Michael de Guzman, a geologist from the Philippines, was "Bre-X's No. 2 geologist." De Guzman was running four Bre-X camps in Indonesia, so much of the work at the Busang site was reportedly supervised by a fellow Filipino geologist, Cesar Puspos.[23]

Walsh was actively involved in raising funds in Calgary. Meanwhile, Felderhof was in Jakarta and de Guzman and Puspos were in Busang. In addition, about twenty others worked as geologists or project managers for Bre-X in Indonesia.

A reputable Australian drilling company was hired to drill core samples to evaluate the gold content of the Busang site. In March 1996, Bre-X reported estimates of 30 million ounces of gold at the Busang site; this was soon raised to 70 million ounces, with the potential for 100 million ounces. In early 1997, Felderhof increased the "official" reserve estimate at Busang to 200 million ounces of gold.[24]

However, the golden glow began to tarnish in January 1997, when a storage building containing the core samples at the Busang site burned down, allegedly destroying the records of the drilling results. World attention was drawn to the unravelling fraud on March 19, 1997, when de Guzman committed suicide by jumping from a helicopter. In his suicide note he explained that he had been driven to the act by poor health. A body was recovered from the Indonesian jungle and confirmed to be his. At the time, de Guzman had been en route to a meeting with a geological team to discuss discrepancies in the test results. Additional test holes had been drilled next to the Bre-X drill holes by Freeport-McMoRan Copper & Gold, Inc., a company in partnership with Bre-X, and the results had been quite different from the glowing results quoted by Bre-X.

The key events in the "discovery" of the gold and the uncovering of the fraud are summarized concisely as follows:

De Guzman worked on the [Busang] site with a team of Filipino technicians who processed the samples—two-metre-long cylinders of rock—which were then sent to an independent laboratory in Samarinda, the nearest big town, for assaying. (For a single evaluation commissioned by Bre-X last January [1997], more than 16 000 samples were submitted.) Felderhof visited the mine monthly. In public announcements, the chief geologist steadily increased his estimate of the lode's size: from 2.5 million ounces to 30 million, to 70 million to an eventual 200 million, which would be worth nearly $70 billion at current prices. As a relatively small operator, Bre-X lacked the capital and expertise to exploit the discovery by itself, so it took on Freeport as a partner—or rather, the relationship was forced upon Bre-X by the Indonesian government.

After Freeport raised questions about the extent of the find in March [1997], Bre-X commissioned an independent firm, Strathcona Mineral Services of Toronto, to make a thorough analysis of the rock taken from Busang. Strathcona found that the samples had been "salted"—the industry term for adding minerals like gold where none exists. Salting is not especially difficult to detect if it is suspected. Feasibility tests carried out on Bre-X samples by several companies from as early as 1995 showed that the gold found in Busang samples was not the type that comes out of the ground, but rather the kind that is found in rivers—and that Dayak tribesmen had been finding in small amounts for years. Since the tests were never made public and the companies that carried them out say they weren't hunting for fraud, Bre-X never got caught on that point, or on several other breaches of normal sampling methods. Core-sample cylinders are normally sliced up the middle, with one part kept intact and the other crushed, marked with the precise location of its extraction, and sent to a laboratory for evaluation. Bre-X never maintained the uncrushed half for later verification. So it was impossible to say with precision where each sample had come from.

Analysts say the pattern of salting at Busang was remarkably skilled: samples have to have varying quantities of gold to be consistent with the pattern of how the metal is distributed in an underground seam. "This was not just a matter of tapping the salt shaker," says Freeport-McMoRan's [CEO Jim Bob] Moffett. "The results had to give a very specific three-dimensional picture of a plausible deposit. The whole picture had to make sense. It had to be very well-planned and well-executed."

The chief candidate for that role is de Guzman, who had the knowledge: he frequently gave speeches to geological societies on his theory about gold in earthquake zones.[25] (© 1997 Time Inc. Reprinted by permission.)

Aftermath

After the fraud was discovered, Bre-X hired an investigative team, Forensic Investigative Associates (FIA), to perform an independent audit. The FIA report, published in October 1997, exonerated Bre-X's senior staff and stated that FIA had "reasonable and probable grounds" to conclude that de Guzman and others at the Busang site were responsible for the ore salting.[26]

Many Bre-X employees had profited personally by selling shares they had purchased with stock options. The FIA report estimated that de Guzman received $4.5 million in stock sales, Puspos $2.2 million, and Walsh about $36 million.[27] Felderhof reportedly sold about $30 million of his shares.

In March 1997, just before the fraud was discovered, Felderhof was named Prospector of the Year at the annual meeting of the Prospectors and Developers Association of Canada in honour of the Busang discovery, which was believed at that time to be the world's largest single gold deposit. A few months later, he agreed to return the award. He was also asked to resign from Bre-X, and he now resides in the Cayman Islands. Bre-X has been delisted from the stock exchange, and its shares are essentially worthless. Lawsuits are being pursued by many investors, who believe that the corporation and the individuals who controlled its geological activities should have shown greater diligence in controlling the assay samples and verifying the gold estimates. The lawsuits and related problems will persist for decades.

Estimates of the total loss to investors as a result of the Bre-X fraud run as high as $6 billion. The Bre-X scandal did serious damage to the Canadian mining industry. Junior mining companies—even those with no links to Bre-X—have since found it difficult to raise the necessary capital to develop their discoveries.

Conclusion

The Bre-X scandal is a case of skilled geological fraud, apparently perpetrated by de Guzman. If he were still alive, de Guzman would be facing criminal charges for fraud, as well as discipline for professional misconduct. Although he was working for a Canadian company, the criminal activities were carried out in a foreign country, so the laws of that country would have to be followed (although Canadian authorities could act if any of his work was actually conducted in Canada).

In the face of such deliberate fraud, it may seem trivial to refer to the Code of Ethics, since the individual(s) involved clearly were not concerned about breaking the code. Nevertheless, these examples demonstrate the classical results of ignoring the Code of Ethics: the perpetrator has come to a bad end, but in the process, thousands of people have suffered serious financial harm, and an entire industry has been held up to contempt and ridicule. Many lawsuits have been launched, and there will undoubtedly be more ruined lives before the scandal plays itself out.

The vital importance of clear and unambiguous duties and titles is evident, after the fact. The chief geologist was reportedly Felderhof. He certainly seemed to consider himself qualified for this title when he made estimates of the gold content in Busang and when he accepted the Prospector of the Year award, although he later claimed to be merely an administrator. In any case, the chief geologist—whether Felderhof or some other person—had a responsibility to show due diligence in safeguarding the core samples and ensuring the following: that the gold assay was conducted properly; that the gold content, based on the samples, was accurately calculated; and that double-checks were made to confirm the results. As noted earlier, core samples are usually split before testing, and half of each sample is retained for further confirmation as necessary. It is clear that in the Bre-X case none of this was done. Security was loose, and all of the Busang samples were crushed and tested under the control of a single individual. This was why the salting could proceed undetected for many months.

Stock promoters who unwittingly encouraged the investment of billions of dollars in a fraud may need to question the standards of due diligence in their own profession as well. The Bre-X case is a forceful reminder that the mining and resource industries depend massively on professional attitudes and high ethical standards.

DISCUSSION TOPICS AND ASSIGNMENTS

(Additional assignments can be found in Appendix CD-E on the CD-ROM included with this text.)

1. Should the professional person be more concerned about the welfare of the general public than the average person? Should people in positions of great trust, whose actions could cause great harm to the general public, be required to obey a higher Code of Ethics than the average person? Apply your answer to engineering and geoscience. Can negligent or unscrupulous professionals really cause great harm? (Hint: Read the case histories.)

2. The definitions of engineering and engineering practice vary from province to province (to territory), as described in this chapter and in Appendix CD-B. Consult Appendix CD-B and review the definitions for your province or territory. Do they agree with the CCPE definition and the U.S. Model Law definitions given in this chapter? Can you think of any activities that are clearly engineering activities but would not be covered by the definition for your province or territory? From the various definitions available in the text or Appendix CD-B, select the definition you consider the most accurate, and explain why.

3. Repeat question 2 for geoscience: The definitions of geoscience and the practice of geoscience also vary across Canada. Consult Appendix CD-B and review the definitions for your province or territory. Do they agree with the CCPG definition given in this chapter? Can you think of any activities that are clearly geoscience activities but that would not be covered by the definition for your province or territory? From the various definitions available in this text, select the definition you consider the most accurate, and explain why.

4. Most professions in Canada have a two-pronged structure whereby one organization regulates the members of the profession, and an independent organization works on behalf of the members by setting fees, organizing pension plans, and providing other benefits. A good example is law: the Law Society regulates members, and the Bar Association works on their behalf. Similarly in medicine, the College of Physicians and Surgeons regulates the profession, and the Medical Association works on behalf of the members. In engineering, a similar two-pronged structure does not fully exist. In some provinces the provincial Association is empowered to act on behalf of engineers, but in other provinces the Association is not legally permitted to act on behalf of engineers as directly as the Bar Association and the Medical Association can. Discuss the advantages and disadvantages of the two-pronged structure. Should it be implemented in engineering? Explain why or why not. (Hint: See the discussion of the Canadian Society of Professional Engineers [CSPE] and the Ontario Society of Professional Engineers [OSPE] in the next chapter, and using the Internet, examine their websites.)

5. To regulate the engineering profession effectively, the provincial Associations require a definition of engineering (or the practice of engineering) that is sufficiently specific that it does not encroach on other professions (such as science and architecture) yet sufficiently general that it can include new areas of engineering practice (such as computer engineering and software engineering) as they are established. Which of the various provincial definitions best satisfies these conflicting criteria? In your opinion, what are the differences between computer engineering and computer science? What are the differences between software engineering and software development?

NOTES

1. J.R. Millard, *The Master Spirit of the Age: Canadian Engineers and the Politics of Professionalism, 1887–1922*, University of Toronto Press, Toronto, 1988.
2. H.A. MacKenzie [opening address for the debate on the Professional Engineers Act, 1968–69], Ontario, Legislature, Debates.
3. Millard, *The Master Spirit of the Age*.
4. Ibid.

5. W.N. Pearson and G.D. Williams, *Professional Registration of Geoscientists in Canada*, Canadian Council of Professional Geoscientists, <www.ccpg.ca/news/professional_registration_geoscientists.html> (August 28, 2003).

6. D.R. O'Connor, *Report of the Walkerton Inquiry: The Events of May 2000 and Related Issues* (Part One: A Summary), published by Ontario Ministry of the Attorney General, Copyright © Queen's Printer for Ontario, 2002.

7. Canadian Securities Administrators (CSA), *Standards of Disclosure for Mineral Projects, National Instrument 43-101*. Document NI 43-101can be found on the CSA web sites in B.C. (www.bcsc.bc.ca), Ontario (www.osc.gov.on.ca), Quebec (www.cvmq.com), and Alberta (www.albertasecurities.com) (September 15, 2003).

8. Professional Engineers Ontario (PEO), *Professional Engineers Providing Reports on Mineral Properties*, PEO Guideline, Toronto, September, 2002. <www.peo.on.ca> (August 28, 2003).

9. Canadian Council of Professional Engineers (CCPE), *Guideline on the Definition of the Practice of Professional Engineering*, CCPE, Ottawa. <www.ccpe.ca/e/guide_guidelines.cfm> (September 22, 2003). Reprinted with permission of CCPE.

10. Ibid.

11. *Professional Engineers Act*, Revised Statutes of Ontario, 1990, c. P28, s.1.

12. National Council of Examiners for Engineering and Surveying (NCEES), *Model Law*, NCEES, Clemson, S.C. © 1996 National Council of Examiners for Engineering and Surveying.

13. Canadian Engineering Accreditation Board (CEAB), *2002 Accreditation Criteria and Procedures*, CCPE, Ottawa, 2002, p. 12. <www.ccpe.ca/e/prog_publica-tions_3.cfm> (September 22, 2003).

14. Canadian Council of Professional Geoscientists (CCPG), *Definition of the Practice of Professional Geoscience*, Canadian Council of Professional Geoscientists, approved January 11, 1999. <www.ccpg.ca/guidelines/definition_practice_geo-science.html> (August 28, 2003). Reprinted with permission of CCPG.

15. Professional Engineers Ontario (PEO), *Guide to Required Experience for Licensing as a Professional Engineer in Ontario*, PEO, Toronto, 1998. <www.peo.on.ca> (August 28, 2003).

16. Professional Engineers Ontario (PEO), *Guideline to Professional Practice*, PEO, Toronto,1998, p. 16. <www.peo.on.ca> (August 28, 2003).

17. CEAB, *2002 Accreditation Criteria and Procedures*, p. 4.

18. Canadian Council of Professional Geoscientists (CCPG), *Canadian Geoscience Standards Board: Terms of Reference*, CCPG website, www.ccpg.ca/boards_and_committees/cgsb_terms_reference.html (December 4, 2003).

19. A. Willis and D. Goold, "Bre-X: The One-Man Scam," *Globe and Mail* (July 22, 1997), A1.

20. J. Stackhouse, P. Waldie, and J. McFarland (with files from C. Donnelly), "Bre-X: The Untold Story," *Globe and Mail* (May 3, 1997), B1.

21. A. Spaeth, "The Scam of the Century," *Time*, Canadian ed. (May 3, 1997), p. 34.

22. J. Wells, "The Bre-X Bust," *Maclean's* (April 7, 1997), p. 50.

23. Stackhouse et al., "Bre-X: The Untold Story," B1.

24. Wells, "The Bre-X Bust," p. 50.

25. Spaeth, "The Scam of the Century," p. 34. © 1997 Time Inc. Reprinted by permission.

26. P. Waldie, "De Guzman Led Tampering at Gold Site," *Globe and Mail* (October 8, 1997), A1.

27. Ibid.

Part Two
Professional Practice

Chapter 3
Professional Employment and Management

Everyone appreciates good advice that helps them succeed. This chapter answers questions typically posed by graduate engineers and geoscientists entering full-time employment—questions such as: How do I obtain a licence? How do I document my experience? What are reasonable salary expectations? How do I remedy poor working conditions? This chapter also discusses a key decision that almost all professionals must make: whether to become a specialist in a discipline, or to develop the "people skills" needed for management. In 2002, about 82 percent of the professional engineers who responded to a CCPE survey on working conditions in Canada were permanent employees of corporations or private companies.[1]

ENTERING THE PROFESSIONAL WORKFORCE

Graduating from university and entering the professional workplace is usually exhilarating: you move to a new location, meet new colleagues, and participate in new projects that create wealth and influence people's lives. You apply your knowledge to real problems and—especially if you are lucky enough to be involved in creative work such as design—you see your ideas taking tangible form in the workshop, on the construction site, in the test laboratory, or on a production line. But unless you had very good jobs during your university work terms, your first full-time professional job may prove to be a challenge.

A formal training period is best, but is also rare. If possible, ask your supervisor for a brief orientation. You need to meet your new co-workers, learn the building layout, and find out where you fit in the organization. You also need to learn important passwords, as well as basic information relating to the company structure, sources of technical information, file organization, computer hardware and software systems, and so on. By the time you take full responsibility for your job, it may be too late to discuss such basic details.

Your first surprise may be how little supervision you receive and how much responsibility is assigned to you because you are "a recent graduate and familiar with the theory." You may have to work more carefully and aggressively to justify this confidence in your ability, and you should accept any help or advice offered by co-workers.

A second surprise may be the apparent lack of order or structure. University courses usually have well-defined durations; in contrast, real engineering projects can change drastically at any time. If a crisis should arise, you may suddenly be asked to move to a new project because you have least seniority. If this happens, don't be offended—new projects are usually chaotic, but they are also challenging and interesting. And anyway, as a professional, it's your job to create order from chaos.

A third surprise may well be the strong emphasis on obtaining useful results. Products must perform as promised, and when they don't work properly, the engineer must "find out why and correct it." If you need information, you must be aggressive in getting it and applying it properly. The focus has changed—although your goal in university was to get an education, your employer wants results.

Finally, you may be surprised at the importance of meeting deadlines. Late deliveries, whether of products or information, cost money—especially when contracts have penalty clauses for late completion, or when "just-in-time" assembly lines have narrow windows for delivery of components. Time management is much more important in the engineering workplace.

Most engineering graduates easily overcome these challenges. Your university courses and past work experiences are intended to equip you to succeed. However, the following advice may help you get a good start on your first permanent job:

- **Begin your employment on a professional note.** As soon as you receive your degree, apply to your provincial Association to begin the licensing process.
- **Start to document your engineering experience.** Follow your provincial Association's criteria for documenting experience, and begin immediately. It is very easy to document experience as you go, but very tricky to remember details later.
- **Project a professional attitude.** Examine your work style to improve your efficiency, productivity, and self-confidence.
- **Prepare yourself for advancement and promotion.** Eventually you will want to be promoted, so think about your next step. Do you have the knowledge and ambition to become an engineering specialist? Do you have the management skills—especially the "people skills"—necessary to lead the organization? Are you independent enough, and sufficiently self-confident, to succeed in private practice or engineering entrepreneurship? It is never too early to identify and develop the skills and knowledge you will need.

These topics are discussed in more detail in the following sections.

APPLYING FOR A PROFESSIONAL LICENCE

It is important for you to apply to your provincial or territorial Association for licensing (sometimes called membership or registration) as soon as possible. Some graduates mistakenly believe that they must wait until they satisfy all of the admission requirements (including the experience requirement) before applying for a licence. This is not true; applicants can apply as soon as they graduate from university. A simple letter, e-mail, fax, or phone call will get the process started. (You will find the address of your Association in Appendix A of this text.)

As explained in the previous chapter, graduates of university engineering or geoscience programs must be licensed before they can practise, or use any title that would lead people to conclude that they are licensed. In particular, you should avoid using titles such as Junior Engineer, Assistant Engineer, or Assistant Geoscientist. These titles may suggest to people that you are licensed; for that reason they are not permitted by your provincial or territorial Act. However, every Association admits recent graduates to a temporary internship or training category, and this simplifies and expedites licensing. During the internship, applicants are typically working under the guidance of a licensed engineer or geoscientist.

Before you can be admitted to an internship, you must satisfy all educational requirements. Typically, this simply means completing an accredited degree program. Once your education has been evaluated and accepted, you are extended internship status and permitted to use an appropriate title (and its initials). Your internship title will vary according to your province and discipline, but it may well be one of the following: Engineering-Internship-Training (EIT), Engineer-in-Training (EIT), Member-in-Training (MIT), Geologist-in-Training (Geol.IT), Geophysicist-in-Training (Geoph.IT), *ingénieur junior* (ing. jr.), Junior Engineer (Jr. Eng.), or Geoscientist-in-Training (GIT). These titles and initials may be used in formal correspondence, on business cards and desk plaques, and so forth.

As an intern, you will be in contact with the provincial Association, which will provide you with information and guidance to help you complete the licensing process; for example, it will advise you on the experience requirements. You may find this guidance very useful. Also, you will be able to attend Association meetings, and you may also be able to participate in group insurance, investment plans, and similar benefit programs.

At one time, internship programs were most attractive to applicants who had graduated from CEAB-accredited universities, perhaps because they were more aware of the benefits of contact with the Associations. However, foreign-educated applicants should also participate in the internship program, even if they have many years of experience, to benefit from any advice that may speed the licensing process.

DOCUMENTING YOUR EXPERIENCE

As you begin your new employment, you should begin to document your experience. As explained in the previous chapter, you will need four years of documented engineering experience to satisfy the licensing requirements (the exception is Quebec, where the requirement is three years). You must submit your engineering experience to the provincial Association in the form of a personal résumé. Most Associations require that your engineering experience satisfy the following five quality criteria, so you should classify your experience in these categories:[2]

- **Application of theory.** This is the best form of experience. It includes analysis, design and synthesis, testing methods, and project implementation. Most professional Associations will expect a sizable portion (typically 20 percent) of your experience to be in this category.

- **Practical experience.** Practical experience helps you appreciate the capabilities and limitations of the theory, equipment, systems, procedures, and standards that are typically used in your discipline. For example, you will be a more competent professional if you have personal experience of the capabilities and limitations of manufacturing methods and tolerances, operating procedures, maintenance schedules, equipment reliability, computer software, safety codes, design standards, and so forth, which are commonly used in your discipline. Fortunately, many activities that do not fall under the other headings would likely qualify as practical experience.

- **Management of engineering.** Management experience includes planning, scheduling, budgeting, supervision, project control, and risk assessment. New graduates are not usually assigned management duties, so you may have to wait for these opportunities.

- **Communication skills.** Engineering is critically dependent on effective communication. Your experience should include evidence of preparing important written work (such as formal reports, design specifications or standards, and contracts or similar documents), drawings or sketches (where appropriate), and oral reports or presentations to supervisors, management, clients or the general public.

- **Social implications of engineering.** This area typically includes any experience that heightens the engineer's awareness of his or her professional responsibility to guard against conditions that are dangerous to life, health, property, or the environment, and to call any such conditions to the attention of those responsible. Most of the topics in this book concern the social implications of technical decisions.

PROJECTING A PROFESSIONAL ATTITUDE

As a professional, you will be busy—typically solving problems—so if you hope to find time for sports, leisure, and other personal activities, you will have to be effective on the job. The following suggestions are intended to help you develop your professional skills and professional image:

- **Time management.** Get a good digital organizer or pocket calendar, and schedule your time efficiently. You probably did this in university, or you likely would not have graduated. Good scheduling is even more important in the workplace. To balance conflicting time commitments, you may have to say "no" to some people and tasks.

- **Accuracy**. Double-check your work to make sure it's correct. Your work may influence expensive engineering decisions, and you won't have a chance to correct and resubmit it, as you could at university.
- **Clarity.** Learn to describe what you are doing in clear, concise terms. Don't use long words or technical jargon when writing reports, letters, or memoranda. Buy a dictionary and use it.
- **Courtesy.** Be tactful and positive. Don't be afraid to disagree, but do so constructively. Older employees may be apprehensive, initially, about new employees. If you maintain a friendly attitude and speak courteously, the workplace will be friendlier and you will make allies out of colleagues who might otherwise be more aloof.
- **Challenges.** Once you are acclimatized to the new environment, accept tough jobs. The challenges inherent in unusual assignments will help you develop your skills and knowledge. Select competent, energetic engineering colleagues as friends and role models.

LEVELS OF ENGINEERING RESPONSIBILITY

As you begin your career as an engineer, you will want to know what lies ahead. The following list shows the typical levels of engineering responsibility. Each provincial Association uses these levels (or similar criteria) to compile salary statistics.[3]

A word of caution: Engineering managers and technical specialists have separate—albeit equivalent—career paths. The levels are therefore based on either management responsibility or engineering responsibility. Levels often overlap, and some companies may recognize more (or fewer) levels, depending on company size. Moreover, a typical company (especially in manufacturing) usually employs many more managers than specialists.

Level A—Entry Level. A bachelor's degree in engineering or applied science, or its equivalent, is usually required for entry into engineering-oriented jobs. Recent university graduates—usually with little practical experience—receive on-the-job training in office, plant, field, or laboratory engineering work, or (rarely) in classrooms. Level A employees work under close supervision.

Level B—Junior Engineer Level (or Engineer-Internship-Training). During the first two or three years of work experience after graduation, an engineering employee will be assigned tasks of limited range and complexity; most of these will involve completing smaller parts of larger projects. Such assignments provide more advanced training and development. During this period the employee is usually registered with the provincial Association in an internship training program (EIT, MIT, Geol.IT, Geoph.IT, *ingénieur junior*, or GIT, depending on the province and discipline). Level B employees may give technical advice to technicians or to level A engineering graduates.

Level C—Professional Engineer. Level C is the first fully qualified professional engineering level. The engineer carries out responsible and varied engineering assignments in a broad field of engineering, and is also expected to understand the effects of decisions on related fields. Typically, this stage requires a minimum of three to five years of related work experience after graduation. A professional engineer must be licensed by a provincial or territorial Association. Level C engineers make independent analyses and interpret results without supervision.

Level D—First Supervisory (or First Specialist Level). Job titles at this level have many variations, such as project engineer, team leader, lead engineer, or engineering specialist. This is the first level that involves direct and sustained supervision of other professional engineers, or the first level of full specialization. This level requires engineers to apply mature knowledge in planning and conducting projects or in coordinating difficult and responsible assignments. To reach this level, engineers typically require a minimum of five and as many as eight years of experience in the field of specialization.

Level E—Middle Management (or Senior Specialist Level). Job titles at this level include chief project engineer, chief industrial engineer, group head, and senior specialist. This level usually requires knowledge of more than one field of engineering, or it requires performance as an engineering specialist in a particular field of engineering. The incumbent participates in short- and long-range planning and makes independent decisions on work methods and procedures within a general program. Originality and ingenuity are required for devising practical and economical solutions to problems. The engineer may supervise large groups that include both professional and nonprofessional staff, or may direct a small group of highly qualified professionals in complex technical applications. This level normally requires at least nine and as many as twelve years of engineering and/or administrative experience.

Level F—Senior Management (or Senior Consultant Level). Job titles at this level include chief engineer, director of engineering, plant manager, and senior consultant. Levels F and F+ may overlap, depending on company size (that is, a chief engineer in a large corporation may have essentially the same duties as the vice-president of engineering in a medium-size corporation). The incumbent is usually responsible for an engineering administrative function, directing several professional and other groups engaged in interrelated engineering responsibilities; or may be an engineering consultant, recognized as an authority in a field important to the organization. The level F engineer independently conceives programs and problems to be investigated, determines basic operating policies, and devises ways to reach program objectives economically and to overcome problems. The job requires extensive engineering experience, including responsible administrative duties.

Level F+—Senior Executive Level. Job titles at this level include president, vice-president of engineering, vice-president of manufacturing, general manager, and partner (in a consulting firm). At this level, the person receives general strategic guidance but conceives independent programs and problems to be investigated. He or she plans or approves projects that require considerable amounts of human financial resources. This level requires many years of author-itative engineering and administrative experience. The incumbent is expected to possess a high degree of originality, skill, and proficiency in the various broad phases of engineering.

SALARY EXPECTATIONS FOR PROFESSIONAL ENGINEERS

For a professional, salary is not the only motivator, but it is still important. Several Associations conduct annual surveys of their members' salaries and post the results on their websites. A word of caution is appropriate here: salary data may vary, depending whether employees or employers provide the num-bers, because data provided by employees may include income from other sources. Moreover, salary surveys show past history; in other words, they may not be good predictors of future trends.

Some recent salary survey data are summarized in Tables 3.1 and 3.2. Table 3.1 is an excerpt from *Ontario Engineers' Salaries: Survey of Employers, Summary Report, 2003*, published by Professional Engineers Ontario (PEO).[4] Table 3.2 is an excerpt from the *June 2002 Salary Survey* (survey of employers) published by the Association of Professional Engineers, Geologists and Geophysicists of Alberta (APEGGA).[5] Note the following:

- The PEO and APEGGA salary surveys were taken at different times in the year and were the most recent surveys available as this text went to press.
- Both surveys were conducted by soliciting salary data from employers. The data were then listed using the responsibility levels outlined earlier in this chapter.
- The complete salary surveys were far more comprehensive than the brief excerpts shown here. Also, they typically classified income by responsi-bility level, year of graduation, type of industry, city or region, and so on.
- Contact your Association for copies of its most current salary survey.

TABLE 3.1 — Ontario Engineering Salaries by Responsibility Level, 2003

Level of responsibility	Median salary	50% spread	80% spread
Level A	$50,000	$45,000–55,000	$41,000–57,553
Level B	$57,977	$53,458–65,069	$48,483–72,000
Level C	$68,436	$62,122–76,749	$57,000–84,960
Level D	$83,000	$74,127–92,405	$67,476–106,000
Level E	$94,920	$86,800–107,000	$79,704–125,020
Level F	$110,000	$100,000–126,588	$92,376–147,000
Level F+	Not reported		

SOURCE: Professional Engineers Ontario (PEO), *Ontario Engineers' Salaries: Survey of Employers, Summary Report, 2003*. Available from the PEO website at <www.peo.on.ca/> (September 28, 2003). Reprinted with permission of PEO.

TABLE 3.2 — Alberta Engineering Salaries by Responsibility Level, 2002

Level	Mean	Low decile (D1)	Low quartile (Q1)	Median	High quartile (Q3)	High decile (D9)
Level A	$48,122	$40,912	$44,357	$48,600	$51,960	$55,200
Level B	$56,101	$48,000	$52,152	$55,900	$59,976	$63,900
Level C	$67,795	$57,720	$63,000	$68,100	$72,800	$75,755
Level D	$82,849	$70,500	$75,765	$82,438	$89,000	$96,000
Level E	$98,336	$81,952	$89,772	$100,000	$106,600	$111,900
Level F	$117,936	$95,992	$108,072	$118,800	$126,200	$137,000
Level F+	$138,814	$115,000	$126,000	$133,800	$145,600	$165,000

SOURCE: Association of Professional Engineers, Geologists and Geophysicists of Alberta (APEGGA), *June 2002 Salary Survey* (survey of employers). Available from the APEGGA website at <www.apegga.ca> (September 28, 2003). Reprinted with permission of APEGGA.

The surveys are intended to help employees and employers compare compensation. All of the salary surveys are accompanied by a checklist so that each engineer can determine the responsibility level that corresponds to his or her duties, years of experience, supervision received, supervisory (or leadership) authority exercised, and so forth.

The salary survey conducted by the Association of Professional Engineers and Geoscientists of British Columbia (APEGBC) includes an especially detailed calculation of responsibility levels. These levels are similar to those described earlier in this chapter, but are more specific, and are determined using an Employment Evaluation Guide.[6]

WORKING CONDITIONS FOR PROFESSIONAL ENGINEERS

Professional engineers are not interested only in salaries, of course. Typically, they also want challenging technical work, and they want opportunities for advancement that arise from jobs well done. In addition, all professional employees typically expect basic benefits such as extended health care and disability and pension plans; as well as some degree of job security, especially for times of economic recession. Engineers usually care less about fringe benefits such as a good office environment and secretarial or technical help, and care more about having high-quality computer hardware and software, high-speed Internet connections, friendly colleagues, and clear communication with management. All of these increase productivity and job satisfaction.

In summary, professional employees expect professional working conditions. Fortunately, most engineers get them. A CCPE survey of more than 26,000 practising engineers showed that the majority *agreed* or *agreed strongly* with the following statements:

- They were satisfied with the job (37% agreed; 49% agreed strongly).
- They had freedom to decide on work issues (35% agreed; 47% agreed strongly).
- They were satisfied with their career prospects (37% agreed; 39% agreed strongly).
- They had opportunities for advancement in the job (38% agreed; 23% agreed strongly).[7]

Ideally, working conditions should be specified in an employment contract (which is discussed further in a later chapter). Unfortunately, many engineers (especially newly hired graduates) do not have personal employment contracts; instead, they are hired on a simple letter of appointment and their working conditions are set by company policy.

Enlightened employers review working conditions on a regular basis (at least annually) as employees gain experience and assume new responsibilities. Failure to review working conditions regularly can lead to declining morale, reduced productivity, and turnover of key personnel. There must be some method for negotiating and amending employment contracts (or company policies). This text cannot offer specific advice; however, help is available to professional engineers who have problems with working conditions.

The Canadian Society of Professional Engineers (CSPE). CSPE is an advocacy group for professional engineers. It is not a union and does not face the legal and bureaucratic problems typically encountered by unions; it is modelled after the medical associations and bar associations, which work collectively for their respective members, who are also professionals. CSPE has recently been restructured as a national advocacy group, whose purpose is to coordinate the activities of provincial advocacy groups. As this text goes to press, the Ontario Society of Professional Engineers (OSPE) is the only provincial group that has been created under the CSPE umbrella. Both CSPE and OSPE are proposing advocacy initiatives for their members; however, they have yet to provide the type of extensive information that is offered by a corresponding American organization, the National Society of Professional Engineers (NSPE).

The NSPE Guidelines. The National Society of Professional Engineers has developed professional employment guidelines that may be of interest to Canadian engineers, since comparable Canadian guidelines do not yet exist. The NSPE Guidelines are discussed in more detail in Chapter 7 (and are included in their entirety in Appendix CD-C). These guidelines contain more than sixty detailed clauses that discuss general rules for engineering recruitment, employment, professional development, and termination. Most clauses discuss very specific, practical employment problems and are directed at both employers and employees.

The Avro Arrow, a Canadian-designed all-weather fighter-interceptor, was a techno-logical marvel and decades ahead of its time. The Arrow, which first flew in March 1958, could exceed Mach 2 at 50,000 feet in normal flight. The Arrow was can-celled by the government of Prime Minister John Diefenbaker on September 28, 1958, in a scandalously abrupt manner, on the basis that manned fighters were obsolete and too costly. Avro's 14,000 employees were immediately laid off. The two prototypes were ordered to be destroyed.

Source: CP Photo/Toronto Star.

PREPARING FOR ADVANCEMENT AND PROMOTION

It is human nature to seek increased responsibility as time passes; with this comes promotion and, of course, a higher salary. It is never too early in your career to prepare for promotion. The first step is to accumulate a wide range of experience by volunteering for challenging tasks. However, you must also acquire advanced knowledge through personal study (and possibly an advanced degree).

The ideal career path is different for each individual, but you are more likely to achieve success if you choose a career path deliberately, prepare for it, and make a success of it. Conversely, simply drifting along without a goal leads to mediocrity—neither a specialist nor a good manager, and lacking the skills to enter private practice or business.

For a typical engineer, the first promotion is usually to project leader, group leader, or team leader (level D in the levels of responsibility discussed earlier). The promotion paths diverge at level D, leading either to engineering *specialist* or to engineering *manager*. This divergence is sometimes subtle.

Specialization. All engineering companies require experts who can solve difficult engineering problems and design new products that bring technical success to the company. At present, the provincial and territorial Associations neither examine nor recognize specialized engineering knowledge beyond the basic P.Eng. qualification. (The one exception is the structural engineering specialization, which is recognized by several Associations.) Even so, whether they are designated properly or not, these specialist engineers are essential—they create and analyze new products, which generate profits. So employers must provide adequate training and experience to engineers who have the ability and determination to develop into specialists. In particular, the salaries for engineering specialists (especially for those experts who achieve world-class ability) must keep pace with (or exceed!) the salaries of managers, to ensure that high-level engineering expertise will flourish within the company.

If specialization is your goal, you must learn every aspect of your specialty. This means seeking out technical problems in order to expand your problem-solving experience. Also, you must know your specialty's codes, standards, and specifications, and you should join the relevant engineering society and attend its conferences. (Engineering societies are discussed in Chapter 16.)

Specialization implies further study. Self-directed study is sometimes enough; however, an advanced degree in engineering is a definite asset, especially in the more mathematical, high-technology industries. Engineering master's degree programs prepare graduates to advance as specialists in high-tech companies.

Management. Every corporation also requires good managers, and many chief executive officers (CEOs) received their basic degree in engineering. Engineers make good managers because they are trained in problem-solving, which is sound preparation for management. Moreover, Canadian industry is becoming increasingly automated and technically sophisticated, and engineers have an advantage when it comes to understanding the problems of these high-tech companies. However, although engineering managers are selected from the most competent engineers, they must usually improve their leadership, organizational, and financial skills if they hope to rise to the highest levels of management. They can obtain these skills through personal study and practice; however, in recent years formal education has become a faster route to personal advancement.

A master's degree in business administration (MBA) is usually the fast track to senior management, because an MBA teaches the skills that engineering graduates are usually lacking. Several universities offer "modular" MBA programs that students can complete while continuing to work. However, MBA graduates are sometimes looked on with suspicion. For example, in the 1990s the American corporation Enron, a giant energy trader, hired about 250 recent MBA graduates every year.[8] Enron declared bankruptcy in December 2002; as a result, thousands of employees were thrown out of work and the assets of Enron shareholders were demolished. At the time this was the largest bankruptcy in American history (until the crash of Worldcom). Dozens of senior Enron executives were charged with fraud, and many were convicted amid accusations of money laundering and conspiracy. The lesson to be learned from this is that while an MBA may be a useful entry into management, true success is based on accomplishments, not credentials.

Note well that it is dangerous for a recent engineering graduate to move directly into management before acquiring basic engineering experience, because that experience is essential for licensing as a professional engineer. An aspiring manager may need better leadership, organizational, and financial skills, but not at the expense of solid, basic engineering experience.

In closing, remember that no one is ever promoted as quickly as he or she would like. The best advice for a new engineering graduate is to develop both technical knowledge and management skills, and delay the career divergence into specialization or management until your career goals are clear in your own mind. As you gain skills and experience, your employment prospects will improve, and you may decide to change employers or perhaps enter private practice or business (see the following chapter).

ENTERING ENGINEERING MANAGEMENT

If you decide to enter engineering management, be prepared for constant challenges! For those who have the energy and enthusiasm to make the commitment, the challenges of leadership can be exhilarating. Most engineering managers thrive on competition and enjoy the constant race for new contracts, new products, and new opportunities. This competitive spirit is what drives our free enterprise system. Whether it is fair or not, salaries are generally higher for engineering managers, so the people who can meet the challenges are usually rewarded generously. However, the first step is to ensure that you have adequate management skills. Even engineers who want to specialize must be able to manage small engineering projects.

Developing Your Management Skills

Few people are born leaders, but leadership skills can be learned. In an article in *Engineering Dimensions*,[9] John Farrow defined three needs that an effective manager must be able to satisfy: *task* needs (that is, the needs set by the company or the project), *group maintenance* needs, and the needs of *individuals*. To satisfy these needs effectively, a good manager must develop six key leadership attributes:

Vision. To be an effective leader, the engineering manager must be more than a competent engineer; he or she must be able to rise above the immediate task and visualize the desired goal. This insight will not come in a flash of inspiration; it will require hard work—such as a thorough study of the organization or the product (or perhaps the competition). Once the manager has a vision of what is to be achieved, the task can be subdivided into smaller goals.

Planning. Planning involves building a bridge from the present capabilities to the desired end point—achieving the vision. The plan defines what is to be done and how to do it.

Organizing. Organizing is closely related to planning, but defines how the team members must interact with one another. Organizing means imposing a structure on the engineering team and coordinating the group's activities to implement the plan.

Communicating. Team members must be given specific tasks. However, goals will be easier to achieve if the members of the team understand the vision. An effective manager conveys the vision and the goals to team members, and does so clearly, and then checks to make sure the message was received. *People* skills such as courtesy, tact, sympathy, and compassion are essential to this function.

Monitoring, evaluating, and controlling. The manager must monitor (check the progress against the plan), evaluate whether the progress is adequate, and then control (or intervene) to bring the progress into line when it is deviating. Engineers recognize these tasks as the typical functions of a *feedback* control system. These are the most visible tasks for a manager, but they are not the only tasks. Moreover, monitoring and controlling organizations is not as simple as the theory might imply, because people often react unpredictably.

Role modelling. The manager also has a symbolic role to play—one that can influence the project's success. An effective manager typically projects an attitude of competence, trustworthiness, openness, and consistency. In other words, he or she must exhibit the characteristics that the manager wants the team members to develop.

In summary, management requires skill. An engineer, however competent, cannot simply assume the title of manager and automatically become effective. Fortunately, management skills can be learned.

Advantages and Disadvantages of Management Careers

Managers take on responsibilities that most engineers would usually avoid. For example, managers face the typical engineering problems of meeting design deadlines, following codes and standards, and ensuring that engineering analyses are thorough and accurate. However, managers also must follow up on production and assembly schedules to make sure projects are on time. Managers must also worry about the future: What new products or projects will have to be developed to make a profit and keep everyone employed? What are the competitors doing? How will new projects be funded? At the same time, managers must monitor the work environment: Is new computer software or hardware needed? Do I need legal help to negotiate the contracts? Do we have enough office space? And finally, managers must worry about people problems: Do we have enough staff? How can personality conflicts among employees be avoided? How am I to deal with lay-offs and severance packages? These issues show the breadth of responsibilities for the typical engineering manager; they also indicate why engineers who hope to be effective managers must develop their leadership, organizational, and decision-making abilities.

Managers usually enjoy higher salaries, greater autonomy, and challenging tasks; but they also face greater personal stress because they are usually "in the middle" when conflicts arise. Management increases the demands on your time, which means less personal and family time. When the good of the organization conflicts with the good of society as a whole, managers may also be faced with agonizing conflict-of-interest decisions. If you are considering engineering management as a career, you must ask yourself whether these drawbacks constitute major obstacles or minor inconveniences. It can be argued that, as a manager, the engineer is in a better position to influence company policy and to prevent problems from arising.

To summarize: Engineering managers who are competent and confident are in strong demand, but most engineers hesitate to move into management positions. Management has its rewards, but management responsibilities take time that could be spent developing engineering, analytical, and consulting skills. How will a management position help you achieve your long-term career goals? Consider this question before you decide to become an engineering manager. Then, if you take on the challenge, make a success of it!

CHOOSING A PERSONAL LEADERSHIP STYLE

Management involves personal interactions. These interactions take many different forms, which depend on the style set by the manager. Douglas McGregor was the first researcher in human relations to explain management theory in a popular way. In his classic book he described the two extremes of management style: Theory X and Theory Y. These theories, which relate the management style to the manager's view of human nature, can be summarized in a very simple way:

Theory X. Theory X states that work is basically distasteful to most people and that people will avoid it whenever possible. Therefore, employees must be closely monitored and controlled. Furthermore, they must be made to work by threatening or penalties or by luring them with rewards.

Theory Y. Theory Y states that people are naturally inclined to work and merely need favourable working conditions in order to be productive. Furthermore, psychological factors such as perceived control over one's activities and opportunities for creative work are important for proper motivation. If properly motivated, employees will produce beyond expectations.[10]

McGregor recommended that managers adopt Theory Y. Most engineering managers today would say they try to follow this theory. In fact, although Theory X may be effective in the armed forces, it is rarely encountered in today's engineering organizations.

The best leadership style depends on many factors, such as the personality, maturity, and competence of the manager; the type of corporation; the degree of initiative, creativity, and independence required from employees; and, to a

TABLE 3.3 — A Comparison of Leadership Styles

Leadership style	Typical example
Collegial	Manager treats employees as colleagues, permits them to function independently within agreed terms of reference. (Similar to Theory Y)
Team-oriented	Manager defines goals, but asks employees to suggest solutions and make decisions.
Interactive	Manager presents problem, obtains ideas and suggestions from employees, then makes decision.
Responsive	Manager presents tentative decision to employees, subject to change after discussion.
Paternal	Manager presents decision, invites questions, and may change decision.
Authoritarian	Manager makes decision, explains it to employees.
Military	Manager makes decision, instructs employees. (Similar to Theory X)

lesser degree, the knowledge or skill of the employees and their willingness or motivation to achieve the corporation's goals. The spectrum of leadership styles ranges from the collegial (similar to Theory Y) to the military (similar to Theory X), as shown in Table 3.3. Most people would find that this table is arranged in order of desirability, with the collegial style most preferred and the military style least preferred. Where does your usual management style (or what you expect your style to be someday) fit in this rough spectrum?

If we consider a typical engineering environment, such as a design office, and if we assume that we have a motivated, mature, and competent manager and professional and well-motivated personnel engaged in creative activity, then it should be possible for the engineering manager to adopt a collegial or team-oriented leadership style. However, different situations may require different techniques. Even when dealing with the same employees, it may be necessary to be more authoritarian in some matters—for example, in specifying engineering goals such as incorporating safety features into a design. However, a good manager will try to give the maximum latitude as to how the goals are achieved. Consistency is important, but flexibility is necessary to deal with exceptional issues. Leading highly skilled and well-motivated personnel is difficult, because it is the manager's task to ensure that they are sufficiently challenged and rewarded to stay that way.

DISCUSSION TOPICS AND ASSIGNMENTS

(Additional assignments can be found in Appendix CD-E on the CD-ROM included with this text.)

1. The acronyms CCPE (Canadian Council of Professional Engineers) and CSPE (Canadian Society for Professional Engineers) both refer to "umbrella" organizations that are intended to assist engineers across Canada. However, although these organizations have similar names, they have totally different functions. Explain briefly the similarities and differences between these organizations. In particular, discuss the purpose (or mission) of each, the way that each is organized, the number of members, and how a licensed professional engineer becomes a member of each organization.

2. The Canadian CSPE and the OSPE (in Ontario) have similar aims as the NSPE in the United States. However, NSPE has been in operation for more than thirty years and is much larger than CSPE (or OSPE). By gathering information from their websites, compare and contrast these three organizations in terms of membership size, access to resources, and the assistance that they can extend to practising engineers.

3. The Codes of Ethics for provincial Associations all contain the clause that the engineer must consider the health, safety, and welfare of the public to be paramount; yet another clause states that the professional engineer should be loyal to the employer. (The Codes of Ethics are included in Appendix CD-B and are available on the Internet.) In some situations these two clauses could come into conflict. For example, an employer might ask the engineer to design piping that would discharge untreated wastewater into a public creek. In such situations, what is the proper course of action for the professional engineer? (This type of ethical conflict is discussed in more detail later in this text.)

4. The discussion of leadership styles in this chapter implies that authoritarian managers are unable to motivate their workers effectively. Discuss this point. Is this necessarily true? Machiavelli would have disagreed. History shows that many authoritarian managers have successfully motivated their workers (or followers) in the past and some will likely do so in the future. While one may disagree with using authority (that is, fear) to motivate workers, it does sometimes work. Discuss the benefits and disadvantages of an authoritarian management style, and give one or two examples of where it would be (or might be) most effective. As a contrast, give a few examples of situations where it would be least effective.

5. The following questions concern management ideas:
(a) Two well-known "laws" in management are Parkinson's Law and Peter's Principle. Parkinson's law states: "Work expands to fill the time available." Peter's Principle states: "People are promoted within an organization until they reach their level of incompetence." Using the Internet or your library, find the sources of each of these "laws" and explain them in more detail. Do you agree that they are universally (or even widely) true? Can you find any examples where they may apply in public life, in your personal life, or in your employment?

(b) The president of a famous computer manufacturing company is alleged to have said: "We try to promote people without making them into managers." Discuss this concept, briefly. Do you think this is a positive statement about the importance of having the dual paths of specialization and management to promotion, or is it a negative statement about managers?

NOTES

1. Canadian Council of Professional Engineers (CCPE), *National Survey of the Canadian Engineering Profession, 2002*, CCPE, Ottawa.
2. Professional Engineers Ontario (PEO), *Guide to the Required Experience for Licensing as a Professional Engineer in Ontario*, PEO, Toronto, January 2002. Available from <www.peo.on.ca> (September 28, 2003).
3. Professional Engineers Ontario (PEO), *Ontario Engineers' Salaries: Survey of Employers, Summary Report*, 2003, PEO, Toronto, 2003. Available from www.peo.on.ca> (September 28, 2003).
4. PEO, *Ontario Engineers' Salaries*, ibid. Excerpts used with permission.
5. Association of Professional Engineers, Geologists and Geophysicists of Alberta (APEGGA), *June 2002 Salary Survey*, (survey of employers), APEGGA, Edmonton, 2003. Available from <www.apegga.ca> (September 28, 2003). Excerpts used with permission.
6. Association of Professional Engineers and Geoscientists of British Columbia (APEGBC), *Salary Survey*. Available from <www.apeg.bc.ca> (September 28, 2003).
7. CCPE), *National Survey of the Canadian Engineering Profession*, 2002, ibid.
8. M. Gladwell, "The Talent Myth," *The New Yorker*, July 22, 2002, p. 28.
9. J. Farrow, "Getting Rid of Management," *Engineering Dimensions* 10, no. 5, September–October 1989, pp. 22–24.
10. Douglas McGregor, "Theory X and Theory Y," *Business: The Ultimate Resource*, Perseus Books Group, Cambridge, UK, 2002, p. 1022–23.

Chapter 4
Private Practice, Consulting, and Business

Engineers and geoscientists who offer services to the general public are said to be in "private practice" and are usually called "consulting engineers," "consulting geoscientists," or simply "consultants." Private practice or consulting is a productive example of free enterprise in the provision of engineering and geoscience services. Rewarding and enterprising careers are also to be found in starting technical businesses or in similar entrepreneurial activities. This chapter is an overview of private practice, consulting, and entrepreneurship, and encourages readers to consider careers in these areas. (The term "engineer" is used throughout the chapter, but almost all of the discussions apply equally to geoscientists.)

CONSULTING OPPORTUNITIES

Most engineering graduates see private practice or consulting as the apogee of a career. The variety and challenge are very attractive, and salary surveys indicate that the financial rewards are very good. Moreover, engineers in these areas tend to be extremely busy—they must maintain high professional standards while seeking new clients, developing new products, and/or competing for new projects. The field is not crowded. In spite of the attractions of private practice and consulting, only a minority of engineers practise in these areas, as the statistics show:

- A survey of practising engineers carried out by the Canadian Council of Professional Engineers (CCPE) in December 2002 revealed that 82 percent of the 27,108 respondents were permanent employees. Only 9 percent were self-employed. (However, 3.8 percent were employed on short-term or project contracts.)[1]
- In 2003, when over 160,000 professional engineers were licensed in Canada, the Association of Consulting Engineers of Canada (ACEC) claimed a membership of only about 600 consulting engineering firms (including sole proprietorships, partnerships, and corporations). Most of the member firms were privately owned and employed fewer than 25 people.[2]

Thus, engineering is unlike other professions, in that only a handful of practitioners are self-employed or in private practice. This is disappointing, because consulting and entrepreneurial activities generate new products, services, and jobs—and by extension, prosperity for society as a whole. Many engineers may take a contrary opinion, and may see the present situation as an attractive opportunity.

CONSULTING ACTIVITIES

The engineer in private practice or consulting is usually an advisor to a client. That is, the engineer is not usually a risk-taking entrepreneur, but is advising some individual or company that has made a financial investment in an engineering project and needs advice on design, development, management, or construction. The client assumes the risk for the enterprise; the advice and skills of the consulting engineer increase the likelihood that the enterprise will succeed.

Engineers provide a very broad range of services to the public, and the list of ACEC's member firms reflects this. Civil engineering is probably the largest single specialty, but ACEC's directory lists engineers who can provide advice and assistance for almost any type of project. The tasks most often required of consulting engineers are:

- **Engineering advice.** Most consulting engineers work on specific projects, in design, development, inspection, testing, quality control, management, and so on. Sometimes a client may require a forensic analysis of a device or structure that has failed.
- **Expert testimony.** An independent consultant may be called on to provide engineering opinions or advice to a court, commission, board, hearing, or similar government or judicial body. An expert witness provides an independent opinion to the court. Even when paid by one side in a dispute, such a witness is not a spokesperson for either side.
- **Feasibility studies.** Consultants are especially useful in the preliminary stages of a project, when a client requires guidance as to the feasibility, financial justification, duration, and cost of the project. In such situations the consultant helps the client decide whether to go ahead with the project.
- **Detail design.** Consulting engineers have the expertise to carry out detailed designs. This includes preparing drawings, specifications, and contract documents.
- **Specialized design.** Custom design and development are available, especially for manufacturing processes, machine design, mining, and other specialized areas. Consulting engineers may work independently or alongside the client's in-house staff. Consultants may also help develop inventions or prepare patents.
- **Project management.** Supervision of part or all of a project is commonly carried out by consulting engineers. This could include the design, manufacturing, construction, or assembly phases of a project, or the commissioning (initial start-up) of a large plant.

In summary, engineers in private practice can be found performing any task that requires professional engineering knowledge. They usually do so on behalf of a client who lacks the personnel or the expertise to conduct the work.

ADDITIONAL LICENSING AND INSURANCE REQUIREMENTS

All provinces and territories impose licensing or registration requirements on partnerships or corporations that offer engineering services to the general public. Also, some jurisdictions require individual engineers offering services to the public to satisfy additional licensing standards. This scrutiny is necessary in order to protect the public, but each province and territory interprets this responsibility differently. No two Acts are identical on all issues.

Practice by Partnerships or Corporations

All provinces and territories (except Quebec) require partnerships and corporations that offer services to the public to be licensed or registered in some way. Typically, the licence is called a "permit" or a "certificate of authorization," and it has two main purposes: to identify the people who are assuming responsibility for the corporation's engineering work, and to ensure that their qualifications are acceptable. The provincial Association usually requires the holder of the permit (or certificate) to identify at least one full-time member of the firm who directs and accepts responsibility for the professional practice. The people so identified must be licensed by the Association and must, of course, be qualified in the field of engineering or geoscience for which the firm is providing services.

EXPERIENCE REQUIREMENTS

Many Acts—for example, those of Alberta and British Columbia—allow professional engineers to provide services to the general public without additional experience requirements. Many provinces require more experience. Ontario requires individuals or corporations providing services to the public to obtain a certificate of authorization. This certificate requires a minimum of five years of relevant experience.

LIABILITY INSURANCE REQUIREMENTS

Liability—in other words, being sued—is a greater risk for engineers in private practice than it is for engineers who are employees. Because of this, almost every engineering Act requires engineers offering services to the public to maintain professional liability insurance (also called "errors and omissions" insurance). Some Acts allow an exemption (especially for sole proprietorships), provided that the clients are informed in writing that the engineering services are not covered by liability insurance. Typically, the client must acknowledge and accept this condition. It would seem that in several jurisdictions, liability insurance is voluntary, since it is not mentioned in the Act.

However, practising without liability insurance is not advisable; moreover, before engaging the engineer's services, some clients routinely insist that he or she take out insurance.

EXEMPTION FOR HAZARDS

A holder of a permit or certificate of authorization is not usually required to be insured against professional liability for pollution hazards, nuclear hazards, aviation hazards, or shipping hazards. This is not because these areas are safer; rather, when accidents happen in these areas, they may be so disastrous that insurance is unavailable.

SECONDARY LIABILITY INSURANCE

The Associations in Alberta, British Columbia, Manitoba, New Brunswick, Newfoundland, Northwest Territories, Nova Scotia, Prince Edward Island, Quebec, and Saskatchewan require all individual members in good standing to participate in a group plan of secondary liability insurance. This group policy is referred to as "secondary" because it is not the main protection against liability claims. Consulting engineers and design, manufacturing, and similar companies must maintain separate "primary" liability insurance to cover their activities, projects, and products and to protect their employees. Secondary insurance typically covers professional engineers and geoscientists who are employees. However, these people are not covered for "moonlighting" activities (small jobs they may undertake outside of their regular employment), nor are they covered if they work for nonengineering companies that lack primary protection. The secondary coverage is fairly narrow, and many exclusions are attached to it. Even so, it advances the profession by providing extra protection to the public. Group coverage increases participation and reduces premium costs. If you are covered by such a policy, you should review the contract agreement, coverage, restrictions, and exclusions. This information is available from your Association and on the Internet.[3]

If you want to offer services to the general public, you should first consult your provincial or territorial Association and check the rules for experience, licensing, and liability insurance. Table 4.1 compares some of these rules. You should also see a lawyer, and contact your provincial consulting engineering organization (listed later in this chapter) for specific guidance.

SELECTING A BUSINESS FORMAT

A private practice, consulting firm, or similar enterprise must follow a recognized business structure. It is wise to discuss this with a lawyer, to ensure that you make the right choice. The common business structures are discussed below:

Sole proprietorship. A sole proprietorship is an easily established one-person business. The proprietor (owner) simply registers the name of the business with the provincial or territorial government. The sole proprietor is

TABLE 4.1 — Comparison of Licensing and Liability Insurance Requirements

Province/ territory	Licensing rules for offering services to the public	Liability insurance requirements
Alberta*	• The Act requires partnerships and corporations (and similar entities) that practise engineering, geology, and geophysics to have a permit. A member who practises under his/her own name does not require a permit.	• Coverage is not mandatory. (The Engineering, Geological and Geophysical Professions Act does not mention liability insurance.)
British Columbia*	• Corporations, partnerships, and other legal entities must hold a certificate of authorization.	• Members, licensees, and certificate holders must notify clients in writing (and receive their acknowledgment) whether professional liability insurance is held and applies to the services offered. (BC Act, s. 17)
Manitoba*	• Corporations, partnerships, and other legal entities (including sole practitioner corporations) must hold a certificate of authorization. Sole proprietors are not required (or allowed) to hold a certificate.	• Each professional member shall either have professional liability insurance under the member's (or employer's) policy, or shall notify the client that no coverage applies. The client must acknowledge and accept. (Manitoba By-Laws, s. 14.2)
New Brunswick*	• A partnership, association of persons, or corporation must be licensed under a certificate of authorization.	• Coverage is not mandatory. The Act authorizes bylaws to be passed to require liability insurance, but the bylaws are silent.
Newfoundland*	• A person, partnership (or other association of persons), or corporation is licensed as a permit holder.	• Coverage is not mandatory. The Act authorizes bylaws to be passed to require liability insurance, but the bylaws are silent.
Northwest Territories*	• A firm, partnership, corporation, or association must be licensed as a permit holder. A member who practises under his/her own name does not require a permit.	• Coverage is not mandatory. Neither the Engineering, Geological and Geophysical Professions Act nor the bylaws mention liability.
Nova Scotia (engineering)	• S. 13(b) of the bylaws requires a partnership or corporation providing engineering services to the public to hold a certificate of compliance.	• Voluntary. Neither the Nova Scotia Engineering Profession Act nor the bylaws mention liability insurance.

Continued

TABLE 4.1 — Comparison of Licensing and Liability Insurance Requirements

Province/territory	Licensing rules for offering services to the public	Liability insurance requirements
Nova Scotia (geoscience)	• S. 14 of the Geoscience Profession Act requires a partnership or corporation providing geoscience services to the public to hold a certificate of authorization.	• Voluntary. The Geoscience Profession Act does not mention liability insurance.
Nunavut	• Administered by Northwest Territories.	• As for Northwest Territories.
Ontario (engineering)	• A person, partnership, or corporation that offers engineering services to the public must have a certificate of authorization.	• A holder of a certificate of authorization must be insured against professional liability, but may be exempted if each client is notified that the certificate holder is not insured and if the client acknowledges this fact in writing.
Ontario (geoscience)	• A partnership, corporation, or individual that offers geoscience services to the public must have a certificate of authorization.	• Mandatory. Every certificate of authorization holder must be insured against professional liability. (Ontario Reg. 59/01)
Prince Edward Island*	• A partnership, association of persons, or corporation must be licenced under a certificate of authorization.	• Voluntary. The PEI Engineering Profession Act does not mention liability insurance.
Quebec*	• Separate licenses are not required for corporations.	• Mandatory for all members.
Saskatchewan*	• A partnership, association of persons, or corporation must obtain a certificate of authorization.	• Voluntary, but prior to providing professional services, a member or holder of a certificate of authorization shall notify the client, in writing, whether professional liability insurance applies to those services. (Saskatchewan Reg. By-Laws, s. 28)
Yukon	• A partnership (or other association of persons) or a corporation or other entity must be licensed as a permit holder.	• Voluntary. (The Yukon Engineering Profession Act does not mention liability insurance.)

Note: All individual members in good standing of the provincial/territorial Associations in Alberta, British Columbia, Manitoba, New Brunswick, Newfoundland, Northwest Territories, Nova Scotia, Prince Edward Island, Quebec, and Saskatchewan participate in a group plan of secondary liability insurance.

responsible for all of the business's financial and professional problems. That is, if the business should develop large debts, creditors may seek to seize both the proprietor's business and his or her personal property.

General partnership. A partnership enables the sharing of knowledge, experience, and friendship. Two or more individuals can agree to form a partnership (preferably in writing, but not necessarily). Most often, the partners contribute the capital to start the business and agree to manage it together; they then share the profits (and losses). The partnership must be registered as a business with the provincial government (as for sole proprietorships). However, the people involved must have absolute confidence in one another's skills, competencies, ethical standards, and personalities before entering a general partnership, because each partner is liable for the business debts or obligations incurred by all the other partners. Errors, omissions, and fraudulent acts by one partner in the name of the partnership create a liability for the innocent partners. In particular, if charges of professional misconduct are made against the partnership under the provincial Act, all partners may be required to answer.

Limited partnership. Limited partnerships are formed by investors when one (or more) of them does not want to participate in running the business. The general partners carry on the business; the limited partners contribute assets for use in the business (most typically money, but sometimes property such as a building or a patent). The liability of the limited partners is restricted to the value of the assets contributed. However, limited partners are not permitted to intervene in the business's operations. Limited partnerships must be registered provincially if they are to enjoy this liability protection. If charges of professional misconduct are made against a limited partnership under the provincial Act, only the general partners are required to answer. Limited partners are not involved in the day-to-day running of the partnership and typically are not at risk.

Incorporation. A corporation can be formed by a person or by a group of people and has many of the rights of a real person. For example, a corporation can enter into contracts, own property, and conduct business. To establish a corporation, you have to apply to the government, which involves some paperwork and fees. Also, you may want to consult a lawyer to help you decide whether to incorporate under federal or provincial law, and for assistance in processing the application. Forming a corporation is an effective way to protect personal assets. When a practice is incorporated, creditors (and judgment awards) can seize only the corporation's assets; the personal assets of shareholders are not at risk. A corporation that performs engineering services must usually obtain a Permit to Practise (or Certificate of Authorization) or must register with the provincial or territorial Association. The corporation must designate a licensed engineer, within the corporation, who will take responsibility for the corporation's engineering work. The designated engineer must possess the competence and the initiative to ensure that the corporation's work meets professional standards; it is that person who must answer if any charge of professional misconduct is made against the corporation under the provincial Act.

Obviously, each business structure has advantages and disadvantages. When the goal of the engineer in private practice is to protect personal assets, the best actions are as follows:

- Incorporate the practice in order to limit liability against demands from creditors and against civil judgments such as breach of contract.
- Obtain professional liability (or errors and omissions) insurance, regardless of the form of business organization.
- Obtain comprehensive general liability insurance to cover the risk of accidents within the business premises, as well as product liability insurance to cover the risk of damage claims for dangerous products. (More specific forms of liability, accident, and corporate insurance are also available.)

Engineers cannot shield themselves from disciplinary actions for negligence, incompetence, or corruption by taking out insurance or by incorporating. Negligence, incompetence, and corruption can be avoided only by careful, competent, ethical practice.

CONSULTING ENGINEERING ORGANIZATIONS

Designation as a Consulting Engineer

Engineers in private practice typically use the title Consulting Engineer. In Ontario this title (or any variation with the same meaning) is regulated under the Professional Engineering Act (Ontario) and may not be used without specific authorization from Professional Engineers Ontario (PEO). To be designated as a consulting engineer in Ontario, a licensed professional engineer must apply to PEO and satisfy several requirements:

- **Authorization.** The engineer must obtain a Certificate of Authorization from PEO.
- **Experience.** Consulting engineers must have five years' experience in addition to that required for registration. At least two years of this experience must be in private practice.
- **Professional liability insurance.** Proof of liability insurance must be filed with PEO, although permission to offer services to the public is awarded even in the absence of liability insurance (subject to certain requirements, such as notifying all clients of this fact using a specified written format).

Consulting Engineering Organizations

Organizations devoted to assisting consulting engineers have been established in every Canadian province and territory except Prince Edward Island and Nunavut. The names, locations, and websites (where they exist, as of February 2004) of these consulting organizations are provided in Table 4.2.

TABLE 4.2 — Consulting Engineering Organizations

Association of Consulting Engineers of Canada (ACEC)
Ottawa, ON

International Federation of Consulting Engineers (FIDIC)
Geneva, Switzerland

Consulting Engineers of Alberta
Edmonton, AB

Consulting Engineers of British Columbia (CEBC)
Vancouver, BC

Consulting Engineers of Manitoba
Winnipeg, MB

Consulting Engineers of New Brunswick
Moncton, NB
Contact through ACEC website <www.acec.ca>

Consulting Engineers of Northwest Territories
Yellowknife, NT
Contact through ACEC website <www.acec.ca>

Consulting Engineers of Nova Scotia
Halifax, NS

Consulting Engineers of Ontario
Toronto, ON

Association des Ingénieurs-Conseils du Québec
Montreal, PQ

Consulting Engineers of Saskatchewan
Regina, SK

Consulting Engineers of Yukon
Whitehorse, YT
Contact through ACEC website <www.acec.ca>

The organizations listed in Table 4.2 promote the interests of their members (that is, the consulting engineers) in several ways:

- The organizations publish directories of their members and the services they provide. Other publications on various topics—how to select a consulting engineer, how to operate a professional practice, how to write contracts and agreements, and so on—are also available at a nominal charge, usually through each organization's website.

- The organizations communicate with their members on issues that affect the profession. They also provide a central source of information on consulting engineers for the public, the industry, and governments.
- The organizations represent their members before municipal, regional, and provincial governments when requested, or when an issue affects consulting engineers as a group.

Association of Consulting Engineers of Canada (ACEC)

Membership in one of the provincial consulting engineering organizations automatically includes membership in the Association of Consulting Engineers of Canada (ACEC), a Canadian nonprofit organization founded in 1925. ACEC comprises about 600 independent consulting engineering firms, who belong to 11 provincial and territorial member organizations, all of which can be contacted through the ACEC website.[4] ACEC promotes good business relations between its member firms and their clients. It also fosters the exchange of professional, management, and business information. ACEC safeguards the interests of consulting engineers when necessary, raises the high professional standards in consulting, and provides liaison with the federal government. The American equivalent of ACEC is the American Consulting Engineers Council (also known as ACEC).

International Federation of Consulting Engineers (FIDIC)

Both North American ACECs belong to the International Federation of Consulting Engineers/*Fédération Internationale des Ingénieurs-Conseils* (FIDIC), founded in 1913, an umbrella group for more than fifty national consulting engineers' associations. The member associations must comply with FIDIC's code as it relates to professional status, independence, and competence. FIDIC publishes an international directory and works on behalf of consulting engineers at the international level. FIDIC also publishes a wide range of documents to help members draft contracts and agreements, manage projects, and operate consulting firms.[5]

THE CONTRACTING PROCEDURE

Selecting a consulting engineer is one of the most important decisions a client makes, and experience shows that selections based on qualifications ultimately produce the best value.

Procuring engineering services is not the same as purchasing materials. Material properties are more easily standardized, and the purchaser's goal is simply to obtain the lowest price. It is certainly not illegal or unethical to select a consultant based solely on lowest bid, and provincial Associations cannot and would not discourage this practice. However, it is wiser for clients to separate the evaluation of qualifications from the fee negotiation.

The fee is negotiated after the consulting engineer has been selected. In fact, the fee can be set only after the scope of the project is fully understood, and often this point is not reached until after the client and the consulting engineer have begun working together on the project. In most projects the most significant cost savings are achieved in the early stages, and employing the best-qualified engineers will usually generate more cost savings. Penny-pinching at the design stage may result in much higher capital, operating, and maintenance costs.

Quality-Based Selection (QBS)

Guidelines for the contracting process are available from the provincial Associations, the provincial consulting organizations, ACEC, and FIDIC. The quality-based selection (QBS) process recommended by ACEC is reproduced here with the permission of ACEC.[6]

QUALITY-BASED SELECTION (QBS)

Many clients rely on an engineering firm with which they have a long-term relationship (this is called "sole sourcing"). However, clients may want to assess the merits of several engineering companies before choosing. The global trend is for clients to select consulting engineers in accordance with the Quality-Based Selection (QBS) system. QBS means that the client chooses a consultant according to:

• Technical competence

• Managerial ability

• Experience on similar projects

• Dedicated personnel available for the project's duration

• Proven performance

• Location and/or local knowledge

• Professional independence and integrity.

Quality-Based Selection is recognized and used around the world. Since 1972, the United States Federal Government has applied it to all federal work. More than 30 US state departments also use it. The Interamerican Development Bank and the Asian Development Bank are only two of the many international financing institutions advocating applications for QBS for their projects. In Canada, QBS is used by the Province of Quebec and the Municipality of Metropolitan Toronto. The system is strongly endorsed by the International Federation of Consulting Engineers (FIDIC) and by the Association of Consulting Engineers of Canada (ACEC).

ADVANTAGES FOR THE CLIENT

A good client-consultant relationship is assured from the beginning of the process. Adversarial relationships are avoided. By first agreeing on the scope of the project, the client can make clear the required emphasis on factors such as environmental impact, cost, schedules, and social implications before fees are negotiated. Fees are fairer to both client and consultant because they are negotiated after the parameters of the work are established. Provincial and territorial fee guidelines published by associations of professional engineers should be used as a basis for such negotiations.

EVALUATION AND SELECTION

The recommended steps in the selection of a consulting engineer are:

- Prepare a list of qualified consultants (long list). For small projects, ACEC recommends selecting a single candidate based on quality of earlier work rather than going through the entire process;

- Contact consulting engineers so selected, outlining the nature of the project and enquiring about the engineer's interest;

- Identify three to five candidate consultants from the long list to develop a short list;

- Request proposals [from the candidate consultants on the short list];

- Interview candidate consultants separately to examine their qualifications and discuss the project and scope of services required. Clients may also wish to check carefully with recent clients of each consulting engineering firm;

- Select the consulting engineer who appears best suited for the project;

- Negotiate fees and execute a contract with the selected consulting engineer. If negotiations are not successful, negotiate with second choice;

- Notify all those interviewed when a contract has been awarded.

FACILITATING THE SELECTION PROCESS

[ACEC recommends the following actions on the part of the client and the consulting firm:]

The Client:

- Describes in general terms the needs of the proposed project and its purposes and objectives;

- Designates the various phases into which the work is to be divided;

- Sets out a desired timetable for the work;

- Identifies problems that are likely to arise;

- Determines budgets and estimates for all phases of the total project;

- Selects firms that offer the required services either from personal knowledge or from an appropriate directory such as those published by ACEC and its member organizations; and

- Gives the selected firms the project information set out above and invites them to offer their services.

The Consulting Firm:

- Responds with a letter of interest;

- Demonstrates an understanding of the project;

- Provides evidence of the firm's ability to perform the work;

- Submits profiles of the firm's principals and staff who will be assigned to the project;

- Gives references, including previous clients for whom similar projects have been carried out.

The Client Then:

- Evaluates the responses and selects a firm from the short list with which to begin negotiations;

- Makes reference to the fee schedules and list of alternative methods of remuneration published by the Association of Professional Engineers and by the Association of Consulting Engineers for each province or territory;

- Negotiates a fair fee and contract based on the most appropriate method of remuneration;

- Makes reference to the [ACEC Contract Documents or FIDIC Client/Consultant Model Services Agreement].

Provincial fee guidelines (or recommended schedules) and performance standards may be obtained from the provincial engineering Associations.

COMPENSATION ARRANGEMENTS FOR CONSULTING ENGINEERS

In the past, engineer's fees were often calculated as a percentage of the construction costs. This procedure is appropriate in some cases (mainly civil engineering); the main drawback to it is that it penalizes the engineer for creating an economical design. A good compensation process should reward efficiency and innovation. The following paragraphs describe most of the common methods currently employed for establishing consulting fees, as adapted from the FIDIC directory (with permission).[7]

Per diem. Fixed daily rates are known as per diem payments. They are the most basic method of calculating remuneration for consulting engineering services and are generally used for assignments for which the scope of work cannot be accurately determined. Studies, investigations, field services, and report preparation all fall within this category. For example, some consulting engineering companies specialize in short-term overload assistance, sometimes referred to as contract engineering. These consulting firms will, on request, place their personnel within a client's firm to work side-by-side with the client's engineers. In this way the client can absorb peak workloads without hiring and training new engineers, who may become surplus when the peak has passed. The per diem payment is obviously the best compensation method in these cases. Direct out-of-pocket expenses, such as travel costs, are reimbursed in addition to the per diem payments.

Payroll costs times a multiplier. Payroll costs, multiplied by a factor to cover overhead and profit, are most often used for site investigations, preliminary design, process studies, plant layout, and detailed design. The multiplier is usually in the range of 2 to 3. Under this method the client essentially pays the engineering costs as they arise, including a sufficient amount to cover overhead and profit. Direct out-of-pocket expenses are also reimbursed in addition to the payroll costs.

Lump sum. With this method the consultant determines in advance a unit or lump sum fee that will cover costs, overhead, and profit. Many clients prefer this method of compensation because it tells them in advance how much they will be paying for engineering services. When the services to be performed are fairly obvious, this is a simple approach. The disadvantage is that the consultant may incur a serious loss if costs have been underestimated.

Fee as a percentage of estimated or actual costs of construction. As noted earlier, the percentage of construction costs approach is now used less frequently. It has generally been applied to consulting engineering services that mainly involve the preparation of drawings, specifications, and construction contract documents. This method is becoming less and less popular with consultants because of the difficulties inherent in relating design costs to rapidly changing construction technologies and to unpredictable market conditions in the construction industry.

The Alberta oil sands contain more petroleum than Saudi Arabia, but these immense reserves face two challenges for engineers and geoscientists to overcome: the oil sands must be mined (as shown in the photo) and processed to extract the oil. The processing releases carbon dioxide, a greenhouse gas that causes global warming. (To get a sense of scale, the truck tires in the photo are about 3 metres in diameter.)

Source: CP Photo/Jeff McIntosh.

The FIDIC directory also suggests some variations on these basic compensation procedures. Consult this directory for further details.

STARTING A PRIVATE PRACTICE OR BUSINESS

Starting a Private Practice

Many engineers enter private practice (or other engineering-related businesses) when they are offered partnerships or share options, after many years of diligent service to the firm. This route to ownership carries the least risk. For example, after working with an older colleague for several years, you may be offered an opportunity to buy out the practice or business upon the colleague's retirement. However, very few engineers enjoy such ideal situations. Besides, most entrepreneurial engineers want to move into leadership roles as soon as they can—they don't want to wait.

Private practice (or consulting engineering) is an attractive option, especially for an engineer who has mastered the engineering knowledge, has good self-discipline, and wants a challenge. Yet even an engineer who is well prepared for a plunge into private practice may require two or three years to get fully established, and those early years will test the new consultant's determination. Extra hours of work must be expected, of course. Also, most engineers will need to acquire more business management skills. Working overtime cannot offset poor business sense or inadequate financial reserves.

Also, to succeed in private practice, you must have an entrepreneurial spirit! Private practice makes you both boss and employee, and it is impossible to be the boss if you object to working long hours or if you cannot cope with stress and uncertainty. To succeed in private practice, you need a special set of entrepreneurial skills and attitudes. The following section lists the entrepreneurial attitudes, technical skills, and personal characteristics required to succeed in private practice or in an engineering-related business.

Starting an Entrepreneurial Business

A recent survey shows that young people today are more likely to start a business than previous generations were.[8] In fact, young people today welcome change, are more independent, and seek opportunities and challenges more than money and security. Among this generation, the idea of staying with one company for a lifetime is obsolete; they know they must adapt to constant change and triumph over it. Most university students have held serious jobs prior to (or during) their university years, and they know far more about the business world than their parents did at the same age. In fact, many university students are already entrepreneurs, some with companies created in high school. Most of today's graduates are well-prepared to be the business leaders of tomorrow.

The most recent engineering graduates are strongly motivated, as well as technologically precocious. Even so, a leap directly into business could result in disaster for many of them. So the best advice to engineering graduates is basically the same as it has always been: get some work experience, evaluate the marketplace, and learn how to avoid mistakes, before you start your own business. Also, before you strike out on your own with a new enterprise, evaluate your skills and opportunities carefully. The following paragraphs may help you do this.

EVALUATING YOUR POTENTIAL AS A CONSULTANT

Not everyone is meant to enter private practice or to become an entrepreneur. Just as we need inspired researchers to develop new theories, and competent engineers to turn theories into practical ideas, we need a class of fearless people who will gamble their time and money to get projects started and bring good ideas to the marketplace. The successful entrepreneur displays certain special skills and personality traits. The following list was assembled from various sources and applies to engineers considering a private practice or an engineering-related business.[9] Do you have these skills, attitudes, and aptitudes?

- **Education and licensing.** Obviously, the first two requirements before you can start any engineering-related enterprise are a degree and a licence. The rare exception might involve entering a business or partnership with a person who is already licensed.
- **Adequate experience and technical knowledge.** You should know the business or practice very well. Leaving a paying job to start a totally new and unknown business is very risky. (For example, if a construction engineer wants to be a machine design consultant, some experience in machine design is obviously essential.) If you don't have adequate knowledge, you must get it, even if you have to change jobs to do so.
- **A network of contacts.** A network of contacts, who will become potential customers, suppliers, or investors, is extremely valuable. Government contacts may also be useful. You will need these people during the start-up phase of your business—a critical phase. They will give you advice on new products, standards, and regulations.
- **Determination.** Determination is perhaps the most important personal trait you will need, because entrepreneurs face many obstacles and rejections before their efforts yield success.
- **Confidence and independence.** Entrepreneurship is often a lonely business, so you must like being your own boss. However, you must also be able to learn from your mistakes.
- **Business skills.** Running a business requires discipline and good management skills. Can you manage budgets, business operations, and employees? A basic knowledge of accounting is also essential. If you're not familiar with terms such as "cash flow," "balance sheet," and "profit and loss statement," you will need to upgrade your business skills.
- **People skills.** You may have innovative engineering ideas, but you will still have to sell your ideas to people. You are well prepared if you can say,

honestly, that you have a positive personality, enjoy working with people, and can communicate well.

- **Good health.** Most successful entrepreneurs enjoy getting up early in the morning to attack the day's problems. Good health is important, especially in the early years, when the stress and physical demands are greatest.
- **Intelligence.** Regardless of their education, successful entrepreneurs are intelligent, think quickly, and enjoy working with new ideas.

GETTING YOUR ENTERPRISE STARTED

Obviously, the above traits will not guarantee success. You will also need good advice, good planning, and enough money to survive until the business starts to pay for itself. Before you risk your time and money, you must conduct a market survey—that is, you must gather as much information as possible about the potential marketplace. When you have solid market research showing that your service, product, or idea is needed, you must then prepare a business plan. A business plan describes what you learned from the market research; it also defines your business objectives and outlines a strategy for reaching those objectives. Your business plan should describe every aspect of your proposal, including business structure, manufacturing, advertising and marketing, and whatever other topics are appropriate.[10] A business plan is extremely important, as you are going to rely on it to raise money—the essential ingredient. In fact, the main obstacle to entrepreneurs is the shortage of investors willing to gamble on a new venture—especially in Canada.

The details of starting a business are beyond the scope of this text, but they are discussed extensively elsewhere. Some advice and references are provided in Appendix CD-G on the CD-ROM included with the text.

DISCUSSION TOPICS AND ASSIGNMENTS

(Additional assignments can be found in Appendix CD-E on the CD-ROM included with this text.)

1. Consider the list of nine personal characteristics explained in this chapter under "Evaluating Your Potential as a Consultant." On a scale of one to ten, rate your own ability under those nine headings:

   - Zero means you have serious doubts about your ability in the area.
   - 5 means you have reasonable confidence in your ability in the area.
   - 10 means you have absolute confidence in your ability in the area.

   Sum your scores to get a rating out of 90. Give yourself an additional 5 points for each further qualification listed below.

   - You have adequate, current experience in computer software related to your field.

- You own computer hardware that runs the software mentioned above.
- You have published three or more technical papers.
- You have already been involved in consulting.
- You have a master's degree.
- You have contacts in three or more local companies that might need your services.
- You have previous experience in making group presentations and writing technical proposals.
- You enjoy making important (and expensive) decisions under pressure.

Total your points; they should not exceed 130. If your total is 100 points or more and you have been scrupulously honest in your personal assessment, then you are probably ready to move into private practice or business. If your total is less than 80 points, then you probably need more experience, education, or determination.

2. Assume that you have carried out the quiz in question 1 and have evaluated your qualifications for entering private practice. The quiz should give you a score between zero and 130.

   - If your score was higher than 100, prepare a business plan for establishing your private practice or business. When the plan is complete, discuss it with your professor (or your banker).
   - If your score was lower than 100, consider whether you see private practice as a career objective. If not, what other career paths (management, specialization, etc.) appeal to you? If private practice is your objective, do you need to improve your qualifications, or do you simply need to obtain more experience in your field? Make a list of steps that would better enable you to move into private practice or business.

3. Imagine that you have decided to enter private practice and that you are trying to become better known in your local area so that you can attract more clients and contracts. Advertising is a sensitive issue, since it must be consistent with the Code of Ethics. Read the sections in this text on advertising (consult the Index). Then devise at least five methods for becoming better known as a competent and ethical professional engineer that are clearly consistent with this text and with your provincial or territorial Code of Ethics.

4. Assume that you are a member of a community service club (such as Rotary, Lions, or Kinsmen). As a fundraising activity for a local children's hospital, your club takes part in a fall fair by building and running a "dunk tank." People pay the club to throw baseballs at a small target. When a ball hits the target, the impact activates a mechanism that tips a perch (or seat) on which a well-known local politician has volunteered to sit, and the person falls into a large tank of (warm) water. Since you are a professional engineer, the club asks you to design the mechanism, and you do so. The mechanism works fine for several days and is a profitable

fundraising activity. However, on the last day of the fair, the politician sitting on the perch inadvertently gets a finger caught in the mechanism. When it is activated by a baseball, the mechanism accidentally severs most of the politician's index finger. The politician is taken to the hospital by ambulance, but surgeons are unable to reattach the finger. The question of liability arises. Note that you would contact your lawyer for proper legal advice, but the goal of good engineering practice is to avoid situations requiring such advice. Answer the following:

- What could and/or should you have done to prevent this situation from occurring?
- What can be done at this stage to minimize damage or alleviate suffering?
- From your perspective, who is liable?
- If your design were to be judged unsafe, what liability insurance would you have (or should you have) to cover this liability?

5. Assume that you are a consulting engineer in private practice. You are contacted by a colleague who once practised as a consulting engineer but who has since lost his licence to practise engineering as a result of a disciplinary action by your provincial Association. The former engineer was convicted of fraudulent misappropriation of a client's property, and as a result of subsequent actions by the Association, he lost his licence. The former engineer has extensive expertise in an area of value to you, and he is willing to use his knowledge to assist you. However, most Codes of Ethics forbid practising engineers from assisting nonlicensed people to practise engineering, so would it be ethical and legal for you to hire him? If so, on what basis could he legally and ethically work for you? Would it be necessary to inform your clients of the employee's conviction for fraud?

NOTES

1. Canadian Council of Professional Engineers (CCPE), *National Survey of the Canadian Engineering Profession, 2002*, Ottawa.
2. Association of Consulting Engineers of Canada (ACEC), *Directory of Member Firms*, ACEC, Ottawa, 2003 <www.acec.ca/> (December 6, 2003).
3. Association of Professional Engineers and Geoscientists of British Columbia (APEGBC), *National Secondary Professional Liability Insurance Program: Questions & Answers* (website), APEGBC, <www.apeg.bc.ca/members/sec-liability-ins.html> (December 6, 2003).
4. The Association of Consulting Engineers of Canada (ACEC) is located at 130 Albert Street, Suite 616, Ottawa, ON K1P 5G4. ACEC's website is <www.acec.ca/> (February 2, 2004).
5. International Federation of Consulting Engineers (FIDIC), *FIDIC International Directory of Consulting Engineers* <www.fidicdirect.com/> (December 6, 2003).
6. The Association of Consulting Engineers of Canada (ACEC), Ingénieurs – Conseils CANADA Consulting Engineers, *Quality-Based Selection*, (Ottawa: ACEC, 2003), pp. x, xi. Reprinted with permission of ACEC.

7. International Federation of Consulting Engineers (FIDIC), *FIDIC International Directory of Consulting Engineers, 1997–1998*. Reprinted with permission of RhysJones Publishing Limited.
8. "Know future," *The Economist*, December 21, 2000.
9. "Assessing Your Entrepreneurial Profile," *Business Encyclopedia, Practical Guides*, Bloomsbury Publishing Plc., <www.economist.com/encyclopedia/> (July 25, 2003).
10. Perseus Publishing, "Writing a Business Plan," *Business: The Ultimate Resource*, Perseus Books Group, Cambridge, UK, 2002, p. 486.

Chapter 5
Hazards, Liability, Standards, and Safety

The Code of Ethics obliges professional engineers and geoscientists to protect the health, safety, and welfare of the public. Most codes also include an obligation to protect the environment. Hazards and risks can never be totally eliminated from a project, so your task is to reduce them to reasonable levels. But how do you recognize hazards, and what is "reasonable"? And if you ignore public safety, what legal liability can result?

This chapter defines basic liability concepts and explains the importance of reducing hazards. It emphasizes a new (but obvious) step in the design process—using the Internet to obtain the codes, standards, and safety regulations required to design a safe product. The case history of the Westray mine, where twenty-six miners lost their lives as a result of unsafe practices, illustrates the vital importance of these standards.

BASIS FOR LEGAL LIABILITY

A professional engineer or geoscientist (or their employer) may be ordered to pay damages if someone suffers loss or damage because of negligent or incompetent advice provided by the professional. Defective products may create similar liability. A lawsuit may be based on three sources of legal liability: contract law, tort law, and consumer legislation.

Contract law. Breach of contract is a fairly common risk in private practice. For example, an engineer who agrees to carry out a specified design project and who then performs the work incorrectly or incompletely may be liable for breach of contract. That is, the engineer may be sued for damages by the other participants in the contract. Engineers can avoid this liability by making more careful contract commitments. Put another way, the engineer must undertake only work for which he or she is competent and must seek help with problems outside this area of competence. It is good legal advice (and simple common sense) to consider all reasonable ways that a contract could go wrong, to foresee any damages that could result, and to include clauses in the contract that specify bonuses for good results, payments for damages, and/or limits to liability.

Tort law. Yet even in the absence of a contract or legislation, a professional may be sued for negligence under tort law. Tort law, which is separate from contract or criminal law, entitles a person who has suffered a loss as a result of someone's negligence to seek damages from the negligent person.[1]

Consumer legislation. If a defective product is manufactured and sold, the designer, manufacturer, dealer, or seller may be liable for damages under provincial legislation, such as the Sale of Goods Act or the Consumers Protection Act. Canadian and American consumer laws differ on a basic premise. Under the "strict liability" concept in American law, the manufacturer is presumed to be at fault unless it can be proved otherwise. In Canada, the injured party must usually prove negligence in the design or manufacture of the product.

In addition to the legal liability noted above, an engineer who has been negligent, incompetent, or indifferent to public safety may be subject to disciplinary action by the provincial Association.

The preceding legal discussion may sound threatening. The good news is that safe products and ethical decisions are rarely challenged in the courts and are easy to defend if they do reach the courts. That is, when you act ethically, you are automatically protecting public safety, avoiding most legal problems, and eliminating the need for most legal advice. If this chapter helps you avoid legal hassles, you will be glad you read it.

PROFESSIONAL LIABILITY—TORT LAW

The word *tort* means *injury* or *damage*. If an injury or damage is caused by wrongful behaviour or by defective merchandise, the *plaintiff* (the injured party) may sue the *defendant* (the accused person). The Canadian law of tort is inherently fair, in that a lawsuit cannot be based on bad luck or a truly random accident. Rather, the nature of the defendant's conduct is central. This conduct is classified as *intentional, negligent,* or *accidental.*[2]

In general, torts must be intentional or negligent to result in liability: "There is no liability without fault."[3] Truly random or accidental events may be tragic but they cannot be the basis for a tort lawsuit. To succeed in a tort action, the plaintiff must prove:

- that the defendant owed a *duty of care* to the plaintiff, *and*
- that the defendant *breached* that duty (intentionally or negligently), *and*
- that the plaintiff suffered *damage* as a result.[4]

If any of these criteria are absent, the tort lawsuit will fail. For example, the plaintiff must have suffered some actual damage in order to claim recompense; under tort law a negligent act does not in itself constitute a sufficient basis for a claim. The terms *duty of care* and *negligence* in the preceding discussion perhaps require further definition.

Defining the Duty of Care

Certain actions create a duty of care between people, even in the absence of a contract between them, and even if the people have never met. The most common example is on our highways: all drivers have a duty of care to avoid injury or damage to other drivers. A driver who, through negligence, causes an automobile accident has breached the duty of care and is liable for the resulting damage. A primitive definition of *duty of care* requires positive ("yes") answers to the following two questions:

- Was there a reasonably foreseeable risk of injury or damage to others, as a result of the action?
- Was anyone close enough to be affected by the activity?[5]

In other words, a duty of care is not owed to everyone, but only to those who are likely to be injured by a potentially dangerous act. However, the scope of the duty of care has expanded over time. For example, legal precedents have extended the duty of care to include people who were not very close during a dangerous act, but who suffered damage later (or the danger did not exist or was not apparent until later). For example, poor building design, poor-quality construction, pollution, defective products, and so on may be considered dangerous even if they do not become dangerous or cause damage until many years later. Other professionals have observed a similar expansion of the duty of care. For example, psychiatrists who fail to warn others of dangerous psychiatric patients have been found liable for damage caused by those patients. Similarly, physicians have been found liable for failing to prevent people with disabilities from driving automobiles. All provinces now have laws requiring medical doctors to report incapacitated patients to the vehicle licensing bureau.[6]

Over the years, the law of tort has repeatedly found that engineers and geoscientists have a duty of care for their actions and decisions. When engineers design a device or structure, or when geoscientists plan a mine or drill for oil, they have a duty of care to anyone who suffers from the harmful effects of these activities, even if the harm occurs years later. The following example demonstrates the duty of care expected from a professional person:

> A house in England was built on a garbage dump, and should have had deep foundations to avoid settling. The foundations were supposed to be inspected by the municipal inspector, but the inspector approved the foundations without inspecting them. Over a period of time, the foundations settled, and could not carry the weight of the building, which collapsed. The municipal authority who employed the inspector was held liable to a later purchaser of the house. In his 1972 judgement, the judge said: " . . . in the case of a professional man who gives advice on the safety of buildings, or machines, or material, his duty is to all those who may suffer injury, in case his advice is bad." (*Dutton v. Bognor-Regis*)[7]

Defining the Standard of Care

After it is determined that a duty of care existed, the next question in tort deliberations involves determining what *standard of care* was owed. In other words, was the person negligent? The courts will apply a *reasonable person* test, and ask: "What would a reasonable person do, under the circumstances?" In the provincial and territorial engineering Acts, the term reasonable appears often in the definition of *negligence*. For example, Ontario's Professional Engineering Act defines negligence as an act or omission that "constitutes a failure to maintain the standards that a reasonable and prudent practitioner would maintain, in the circumstances."[8] Obviously, the precise interpretation of this definition depends on the accepted standards of the engineering discipline concerned. Perfection is not required; however, professionals are expected to use reasonable care, established practices, and well-tested engineering principles.

The best protection against negligence is careful, thorough, accurate work. Professionals can also obtain liability insurance to protect against some of the costs of negligent conduct. This insurance, typically called "errors and omissions" insurance, is a wise investment for engineers in private practice; in fact, in most provinces it is compulsory. Of course, a negligent engineer is always subject to disciplinary action by the provincial Association (as discussed in Chapter 14).

Proving Negligence

Although Canadian tort lawsuits require evidence of negligence, courts increasingly are accepting circumstantial evidence which indicates that negligence must have occurred, even though the exact cause cannot be defined precisely. The courts will usually assume that negligence has been proved, even if no explanation can be presented, when:

- whatever caused the harm was under the sole control of the defendant, *and*
- the event that caused the harm ordinarily would not occur without negligence.

For example, when a surgeon leaves a sponge inside a patient during a surgical operation, there is no need to discover precisely how the sponge was left there; the surgeon was in control of the surgery, and had the surgeon not been negligent, the sponge would not have been left behind. This type of circumstantial evidence is described by the Latin phrase *res ipsa loquitur* ("the thing speaks for itself").[9]

The precedent for *res ipsa loquitur* dates back to 1863 and the lawsuit *Byrne v. Boadle*. A barrel of flour fell from a warehouse window above a shop, striking and injuring a passerby, who sued for damages. Although no evidence was produced to explain how the barrel came to fall from the window, the judge (on appeal) stated: "A barrel could not roll out of a warehouse without some negligence, and to say that a plaintiff who is injured by it must call witnesses from the warehouse to prove negligence seems to be preposterous." The judge ruled that the barrel was "in the custody" of the warehouse owner (or his employees) and that he was therefore responsible for the control of it.[10]

This reliance on circumstantial evidence, in the form of *res ipsa loquitur*, is important to engineers, because it has been applied to defective products. When an injury has resulted from a product, and when the cause is so obvious that "the thing speaks for itself," the designers must prove that negligence did not occur. In effect, in situations where *res ipsa loquitur* can be applied, a requirement is imposed on defendants in Canadian tort cases that is similar to the requirement imposed by strict liability in American courts.

In summary, regardless of the law's basis, the engineer has an obligation (and an incentive) to act ethically, to avoid negligence and incompetence, and to eliminate hazards before products reach the marketplace.

PRODUCTS LIABILITY

We are all consumers, and we know that sellers are expected to provide goods of acceptable quality. When a consumer purchases a defective product from a manufacturer (either directly or through the manufacturer's agent), the contract (written or implicit) in this action is the basis for demanding reparation. There are three basic reasons that support a claim for damages:

- defective manufacturing, *or*
- negligence in design, *or*
- failure to warn of dangers associated with the product.[11]

Contract Conditions and Warranties

The contract clauses must be examined closely. Contract clauses are typically divided into conditions and warranties. Usually these terms are identified in the contract:

- *Conditions* are key clauses that must be satisfied, or the contract may be terminated. Obviously, conditions are important, since they have the potential to end the contract.[12]
- *Warranties* are clauses that permit the consumer to demand repairs, replacement, or damages. However, a warranty clause does not permit a contract to be terminated. (The term *warranty* is usually applied to goods and products, whereas the term *guarantee* is usually applied to services or agreements.) Warranties are promises that a manufacturer makes about a product. If the product fails to meet the terms in the warranty, the manufacturer may be liable for any resulting damage.

Sale of Goods Act

Every province and territory has a law—typically called a Sale of Goods Act—that may be invoked when a sales contract lacks specific wording. The Sale of Goods Act defines certain conditions and warranties in order to protect the general public. For example, typical sale conditions require a basic quality (or *merchantability*) as well as fitness for use. These terms imply that the product must not be defective and must be usable as intended. Thus, a lawn mower must be able to cut grass, and a refrigerator must be able to keep food cold.[13] Regulations for the sale of goods typically apply to quality. Recently, however, the courts have been interpreting Sale of Goods Acts to encompass safety as well.[14]

Consumer Protection Act

In addition, every province now has a Consumer Protection Act that imposes further provisions on consumer sales, or that prevents certain basic rights from being waived. However, this legislation is intended to protect individuals; typically, it does not apply to goods purchased for commercial use or resale.

U.S. Products Law—Strict Liability

The concept of strict liability applies mainly in the United States; however, Canadian engineers and manufacturers must be aware of it, because the North American Free Trade Agreement (NAFTA) now permits a freer flow of products across the border into the United States. Moreover, since Canadian and American laws are both based on the British concept of established precedents, legal decisions made in one country may, over time, be applied in the other country.

Strict liability covers product defects and consumer safety. The focus is on the product itself, and no questions of negligence arise. D.L. Marston, in his text *Law for Professional Engineers*, states:

> In products liability cases in the United States, a manufacturer may be strictly liable for any damage that results from the use of his product, even though the manufacturer was not negligent in producing it. Canadian products-liability law has not yet adopted this "strict liability" concept, but the law appears to be developing in that direction.[15]

In both Canada and the United States, the requirements for care were set by a 1932 British case, *Donoghue v. Stevenson*. The judgment in this case stated in part:

> A manufacturer of products which he sells in such a form as to show that he intends them to reach the ultimate customer in the form in which they left him, and with no reasonable possibility of intermediate examination, and with the knowledge that the absence of reasonable care in the preparation or putting up of the products will result in injury to the consumer's life or property, owes a duty of care to the consumer to take that reasonable care.[16]

In the United States, the American Law Institute has published the following two-part rule for products liability, which contains the idea from the British precedent but also adds a new concept:

1. One who sells any product in a defective condition, unreasonably dangerous to the user or consumer or to his property, is subject to liability for physical harm thereby caused to the ultimate user or consumer, or to his property, if
   (a) the seller is engaged in the business of selling such a product, and
   (b) it is expected to and does reach the user or consumer without substantial change in the condition in which it is sold.
2. The rule stated in Subsection (1) applies although
   (a) the seller has exercised all possible care in the preparation and sale of his product, and
   (b) the user or consumer has not bought the product from or entered into any contractual relation with the seller.[17]

The key difference between Canadian and American law is that the American definition specifically states that the rule applies even when the seller has taken all possible care; in other words, it applies even when no negligence can be shown. Instead of the plaintiff being required to prove negligence, the defendants must prove their innocence. Moreover, the American law applies to sellers; as a result, everyone in the design/manufacturing/sales chain is included.

Canadian Products Law—Risk–Utility Analysis

Canadian products liability law differs somewhat from American law; even so, the obligation to design a safe product is well established in Canada. American law focuses on the *reasonable expectations* of the consumer. Osborne makes the following comments regarding this test, and illustrates the *risk–utility* approach that is more commonly followed in Canada:

> [In the United States:] If the product is not as safe as a consumer might reasonably expect, the design is defective. This test has, however, proved to be very difficult to apply in a fair and predictable manner. The reasonable expectations of consumers in respect to some products may be unreasonably high and in respect of others it may be unduly low. The test also fails to consider if there is an alternative and safer design available. [A different] test, which is favoured in Canada, is a *risk–utility* analysis that seeks to determine if the utility of the product's design outweighs the foreseeable risks of the design.

> The risk-utility test was applied in *Rentway Canada Ltd. v. Laidlaw Transport Ltd.* [1989] The case involved a head-on collision between two trucks, when both of the headlights of one of the trucks failed. The defendant had designed the lighting system of that vehicle. Flying rubber from a tread separation of a tire knocked out one headlight and, because both headlights were on the same circuit, the other one also failed. The plaintiff claimed that the two headlights should have been on independent circuits, in which case, one headlight would have remained on. The trial judge assessed the safety of the

design on a risk-benefit analysis and decided that the design was defective. Consideration was given to the degree of danger arising from the design, the availability of a safer design, and the functionality, the costs, and the risks of that alternative design. The ultimate question was whether or not, in the light of these factors, the product was reasonably safe. The Court held that the danger of having both headlights on a single circuit and the availability of a functional and affordable alternative design outweighed the utility of the single circuit system used by the defendant.[18]

Generally, most lawsuits involving product liability are brought against the manufacturers and sellers of the products, and are usually based on breach of warranty or on strict liability. A lawsuit is usually brought against the design engineer only in the case of alleged negligence. However, some of these lawsuits have been enormously costly, so safety is a good investment. When an engineer makes a product safer, this protects the public from harm and simultaneously protects the manufacturer from financial loss.

HAZARDS AND THE DUTY OF THE DESIGN ENGINEER

Hazards, Flaws, and Failures

Hazards exist in normal life, and flaws exist in materials, and even the most painstaking attempts by the most diligent engineers will not prevent all failures. Nevertheless, engineers have an obligation to seek and eliminate design hazards and material flaws wherever reasonably possible. It is reasonable to expect an engineer to:

- conduct a hazard analysis of a new design to seek unexpected or potential dangers,
- require inspection or testing to detect flaws in materials and parts, *and*
- conduct failure analysis of a failed device to learn the cause of the failure and how to avoid it in future.

Designing for Safety

The design engineer is responsible for taking action to eliminate a hazard in the following circumstances:

When a clear and evident danger exists. The Code of Ethics requires practising engineers to take action to eliminate obvious hazards. For example, a high walkway clearly needs a railing for safety. Remedial action is especially important when the hazard is concealed from the user (that is, when the hazard is not obvious). The engineer has an obligation to remove concealed hazards. When such a hazard cannot be removed, the engineer must make it obvious to the user and/or warn the user about it.

To follow established design standards. The design engineer must know and follow the appropriate standards—whether they are embodied in law or are voluntary design guides—unless there is a convincing engineering analysis to justify deviating from them. Thanks to the Internet, design standards are now easier to find and retrieve than ever before.

To follow legal requirements and government regulations. Engineers must follow provincial and federal safety laws and regulations. For example, every province has laws and regulations on occupational health and safety to protect workers, and environmental laws and regulations to protect the environment. The designer must be familiar with these laws and regulations and follow them. Government regulations are also very easy to find through the Internet, and may usually be copied or printed immediately, at no charge.

To follow good engineering practice. In the absence of design standards or government regulations, the engineer must simply use good engineering practice. A *hazard analysis* should be conducted for every design, and the design engineer should keep a permanent record of the analysis. The hazard analysis is a systematic review of the design and typically has at least four steps: to identify hazards; to eliminate them where possible; to shield them where they cannot be eliminated; and to provide remedial action where shielding is not possible (by warning users, by recalling products for repair, by providing escape routes, and so forth).

The hazard analysis is intended to remedy dangers that might occur during normal operation, but if the system is complex, a *failure analysis* should also be carried out. A failure analysis examines the consequences if a single system component should fail. Failure analysis is especially important in large systems, to determine whether a component failure could lead to a disastrous failure of the whole system. In some cases, the risks (or probabilities of failure) can be estimated mathematically using reliability theory, and the design can be changed until the probability of failure is reduced to an acceptable value.

Formal failure analysis techniques are complex and are typically applied only to large systems, such as electrical power plants, aircraft, or computer control systems. The two best-known methods are failure modes and effects analysis (FMEA) and fault tree analysis (FTA). However, such computations are not feasible in most design situations, and the engineer's judgment must suffice. When in doubt, the engineer's bias should always lean toward increasing safety. A more detailed discussion of hazard analysis, FMEA, and FTA is included in Appendix CD-GI.

HAZARD REDUCTION CHECKLISTS

Design engineers are creative people. They put great effort into their designs, and they want their products to be appreciated. However, consumers often misuse or abuse products; damage, injury, or death results; and the product designer (and/or the manufacturer) faces a lawsuit. In defending against the lawsuit, it becomes obvious that, if even a fraction of the cost of the lawyer's bills had been spent on design safety, the legal problems might have been avoided. An old adage says that a dollar spent on design safety may save ten dollars on defence lawyers. The following discussion of design safety is addressed to both engineers and manufacturers.

A New Design Step—Checking the Internet

Many textbooks discuss methods for making products safer and for avoiding product liability lawsuits. However, few texts emphasize the critical importance of design standards and safety regulations. The first step in the design process includes gathering information (as discussed briefly in Chapter 6), so design texts should specifically recommend an Internet search for appropriate design standards and safety regulations.

The almost instant availability of this wealth of standards and regulations was not known by earlier generations of engineers. Designers must take advantage of this new resource, which is certain to continue expanding. The best advice to designers is to be aware of, and adhere to, established industry standards and government safety regulations. More advice follows, arranged as a checklist:

Designer's Checklist

A design project will run more smoothly, and the result will be safer and of better quality, if you take the following steps:

- Find and apply standards and regulations.
- Use state-of-the-art design methods.
- Conduct formal design reviews.
- Carry out a formal hazard analysis.
- Carry out a formal failure analysis.
- Warn consumers and/or clients of hazards.
- Prepare and distribute instruction manuals.
- Maintain complete design records.

Each of these points is explained at length in design textbooks[19] and is summarized in the file "Reducing Hazards: A Checklist for Designers and Manufacturers," included in Appendix CD-G1.

Manufacturer's Checklist

Manufacturers want to create products that people will buy and use. They also want to avoid wasting the time, money, and effort (not to mention distress and hardship) involved in having to deal with products that fail. After an accident, analysts present a damage case in court and the entire history of the product is discussed. Unless a safety and/or hazard analysis was carried out (and thoroughly documented) before the product was sold, the product will likely be judged unsafe, and the manufacturer found liable. So manufacturers have several incentives to help their engineers design safe products. The manufacturer's typical responsibilities are as follows:

- To establish safety as a company policy.
- To conduct adequate quality assurance and testing.
- To review warranties, disclaimers, and other published material.
- To act promptly on consumer complaints.
- To warn owners immediately of hazards.[20]

Each of these points is explained concisely in the file "Reducing Hazards: A Checklist for Designers and Manufacturers," included in Appendix CD-G1.

THE STANDARDS COUNCIL OF CANADA (SCC)—NATIONAL STANDARDS

As noted earlier, the best method to improve product quality is simply to follow accepted design standards. Never before in history have design and safety standards been so plentiful or so easy to obtain. The Internet has made it easy to find, search, and purchase safety standards. Often, the Internet permits you to download standards immediately. There is no reason to neglect this important design step.

An excellent starting point in such a standards search is the Standards Council of Canada (SCC), a federal Crown corporation with a mandate to promote standardization in Canada. The following description of SCC's role is excerpted from SCC publications, with permission.

Standards: Standards are publications that establish accepted practices, technical requirements and terminologies for diverse fields of human endeavour. There are standards for almost everything, from the simplest screw to the most complex information technology equipment.

By applying standards, organizations can help to ensure that their products and services are consistent, compatible, safe and effective. Today, products are assembled from components made in different countries, and are then sold around the world, so standards are more important than ever. Standardization is an essential element of technology, innovation and trade.

The Standards Council of Canada: The Standards Council of Canada (SCC) is a federal Crown corporation with the mandate to promote efficient and effective standardization. Located in Ottawa, the Standards Council has a 15-member governing Council and a staff of approximately 70. The organization reports to Parliament through the Minister of Industry.

The National Standards System: The Standards Council has the mandate to coordinate and oversee the efforts of the National Standards System, which includes organizations and individuals involved in voluntary standards development, promotion and implementation in Canada. More than 14,000 Canadian volunteers contribute to committees that develop national or international standards.

As well, some 250 organizations have been accredited by the Standards Council. Some of these develop standards, others are conformity assessment bodies which determine the compliance of products or services to a standard's requirements. The list of accredited organizations includes:

- standards development organizations;

- certification organizations;

- testing and calibration laboratories;

- environmental management systems (EMS) registration organizations that perform registrations to the ISO 14000 series standards; and

- auditor certifiers and trainers that provide QMS and EMS auditors with their training and credentials.

International Standardization: The Standards Council co-ordinates the contribution of Canadians to the two most prominent voluntary international standards development forums:

- The International Organization for Standardization (ISO) and

- The International Electrotechnical Commission (IEC).

These two bodies publish standards in a wide variety of fields, including information technology, medical technology, the environment and quality management. ISO and IEC standards are often adopted by countries as voluntary standards, or included in national rules and regulations. Many trade agreements, including the World Trade Organization (WTO), call upon signatories to adopt international standards wherever possible. The Standards Council encourages the adoption and application of international standards.

The Standards Council is also involved in efforts to establish mutual recognition of its accreditation programs for conformity assessment organizations with similar bodies in other countries. This effort facilitates the movement of goods and services across borders.

National Standards of Canada: Accredited standards-development organizations may submit standards to the Standards Council for approval as National Standards of Canada. This designation indicates that the given document meets criteria that are important to many standards users.

For example, a National Standard of Canada must be developed by consensus of a balanced committee representing producers, consumers and other relevant interests. It must undergo a public review process, be available in both official languages and must not be framed in such a way that it will act as a restraint to trade. Further, the standard should be consistent with or incorporate appropriate international standards as well as pertinent national standards.

Standards Information: The Standards Council provides customers with the latest information on standards and technical regulations used in Canada and around the world—including such important markets as the European Union, the United States and Asia-Pacific. Much of this information is available through SCC.CA, the Standards Council's new Internet-based information service.[21]

SCC's sales centre greatly simplifies the purchase of non-Canadian standards by eliminating the need to deal with foreign distributors and currencies. More information on SCC's programs and services is available from this address: Standards Council of Canada, Communications Division, Suite 1200—45 O'Connor Street, Ottawa, ON, KIP 6N7 (www.scc.ca).

THE INTERNATIONAL ORGANIZATION FOR STANDARDIZATION—ISO STANDARDS

A standard exists for almost every manufactured object (and a simple Internet search will find it). Design standards are extremely useful to engineers and have had a strong impact on maintaining and improving our quality of life. The International Organization for Standards (ISO) was founded in 1947 with the mandate to make standards more widely available and to "standardize the standards" among countries. By 1996, more than one hundred member countries were participating in its standards activities, including Canada, Britain, and the United States. In its present form, ISO is a network formed by the national standards institutes of 147 countries (with one member per country). SCC is Canada's representative in ISO. The Central Secretariat, located in Geneva, Switzerland, coordinates the system. However, ISO is a nongovernmental organization; in other words, its members do not specifically represent national governments.

Whenever a new standard is proposed, ISO brings together a technical committee comprising experts from the various member countries. Typically, each nation that participates on the technical committee sets up an advisory group composed of experts from within its own borders; this group then generates a national consensus regarding the proposed standard.

The proposed standard must pass through three drafts. At each draft, differing opinions and alternative wordings are proposed by members of the technical committee. The standard is then voted on by the member countries. If the final draft standard receives a two-thirds positive vote, it becomes an ISO standard and is translated into the three official ISO languages: English, French, and Russian. Each country can take a further step and adopt the ISO standard as a national standard, and publish the standard in the language of that country.

ISO has developed more than 13,000 International Standards on a wide range of topics. The ISO standards titles may be searched through the Internet, and the listings (in many cases) include brief abstracts. The ISO website is <www.iso.ch/iso/en/ISOOnline.frontpage> (February 3, 2004).

ISO 9000 and ISO 14000 Standards

ISO standards for many different products have been in use for many years. Two recent standards have generated a great deal of interest among engineers, engineering managers, and manufacturers: ISO 9000 and ISO 14000. Each of these standards is described below.

ISO 9000—QUALITY MANAGEMENT AND QUALITY ASSURANCE STANDARDS

The ISO 9000 standard is very different from most ISO standards for products, because it is a standard for managing a manufacturing corporation in order to maximize the quality of its manufactured products. The standard is very effective and has been widely adopted. Between 1987 (when the first version of ISO 9000 was released) and 1996, more than 100,000 corporations obtained ISO 9000 certification. All 13,000 first-tier suppliers to the "big three" automakers (Chrysler, Ford, and General Motors) were required to adopt an extended form of ISO 9000 by the end of 1997. Every supplier to the automotive industry is now expected to be certified to ISO 9000 standards. By the end of 2002, more than 560,000 ISO 9000 certificates had been issued to organizations worldwide, making ISO 9000 certification the dominant quality management certification system in the world.[22] It is estimated that an investment in ISO 9000 certification usually pays for itself within three years through increased productivity and reduced scrap.[23]

The ISO 9000 standard is very comprehensive. It requires a corporation to examine almost every aspect of its management, design, purchasing, inspection, testing, handling, storage, packaging, preservation, delivery, and documentation systems. Improving the quality of these systems enables effective evaluation of the manufacturing process and shows where quality improvements are required. An important part of ISO 9000 involves developing a "quality manual" to document the four key aspects of the certification process. This manual documents the following:

- Quality policies for every aspect of the corporation's operations.
- Quality assurance procedures, which involve twenty clauses in the ISO 9000 standard.
- Quality process procedures (or practices, or instructions), which include all of the company's production processes.
- Quality proof: a repository for all of the forms, records, and other documentation that give objective evidence—or proof—that the quality system is operating properly.

An important aspect of the ISO 9000 quality management process is that it is arranged to permit internal and external audits. This process is very similar to a financial control system, for which audits are an established, accepted, and routine procedure. A customer who purchases the company's products may be invited to examine the company's operations for verification that the quality management system complies with ISO 9000 standards. Typically, the audit is carried out by independent quality auditors, or "registrars," to ensure impartiality. These audits should be conducted every six months or so, with a complete recertification audit carried out every third year.[24]

ISO 9000 certification is a long and detailed process. Every aspect of a company's operations must be examined in detail; fifteen-step processes, which may take more than a year to implement, are common. However, ISO 9000 is clearly becoming the accepted world standard for quality management, so many companies may find it necessary in order to survive in business.

ISO 14000—ENVIRONMENTAL MANAGEMENT SYSTEMS

Companies around the world are registering to the ISO 9000 standard for quality management. Many of these companies are also adopting another new ISO standard, ISO 14000, for environmental management.

ISO 14000 was developed using the international consensus procedure (as for all ISO standards) and fits well with the ISO 9000 standard. In fact, companies that have registered to ISO 9000 will find that ISO 14000 certification is very similar.

The ISO 14000 process requires the company to examine every function of its operations with the goal of identifying activities with a significant environmental impact; it then commits that company to preventing pollution in all of its forms. The standard does not set acceptable environmental levels; that is left to regulatory agencies. However, the standard does require that these environmental levels be determined and followed. Monitoring and measurement are, of course, essential; procedures for corrective action and emergency response are also required. This may require that procedures and performance criteria be set, that responsibilities be defined and assigned, that training be provided, and that adequate communication be ensured. ISO 14000 does not require the writing of an environmental management system manual; however, most companies would probably want to develop such a document.

ISO 14000 was released in 1996, and many major companies immediately committed themselves to implement the standard.[25] By the end of 2002, more than 49,000 organizations had certified under ISO 14000, in 118 countries.[26]

GOVERNMENT STANDARDS AND SAFETY REGULATIONS

Government standards and regulations are very important; they affect your professional work, and they must be followed. There isn't enough room in this text to discuss all of these standards and regulations, but the advice below tells readers how to start the search.

As noted earlier, professional engineers and geoscientists have, at their immediate disposal, the most powerful tool for information retrieval that has ever existed—the Internet. A general search using a few relevant keywords will find the documents almost instantly. When the document is not free, you will usually be able purchase it at little cost by credit card, and download it immediately. The following discussion, while not exhaustive, will give you a sense of what is available and how to start looking for it on the Internet (as of December 2003).

Codes and Standards

As noted earlier in this chapter, technical codes and standards are distributed by SCC and ISO. Many standards are also developed and distributed by engineering societies (as discussed in Chapter 16).

Many construction codes and standards have been developed by the National Research Council's Institute for Research in Construction (IRC). For example, the National Building Code was developed in 1941 to consolidate the patchwork of municipal codes that existed across Canada in the 1930s. The code was so successful that it has been maintained ever since. IRC now publishes several codes, including the following:

- National Building Code of Canada
- National Fire Code of Canada
- National Plumbing Code of Canada

All provincial governments have adopted these codes, or have adopted them with modifications, or have developed provincial codes based on the national codes.

Contact: Standards Council of Canada (<www.scc.ca>)
Contact: Federal Publications Inc. (<www.fedpubs.com/>)

Federal and Provincial Product Safety Legislation

Most provinces have similar legislation governing product safety, usually modelled on a national code. In general, federal or provincial safety acts authorize regulatory powers but provide only general guidance. The detailed safety rules are contained in the regulations established under the act. For example, a typical product safety act might authorize the creation of standards for products to be sold in Canada; however, the precise safety standards will be specified in the act's regulations.

SCC has recently published a comprehensive report summarizing Canadian federal, provincial, and territorial legislation dealing with products purchased through retail outlets to be used in the home by the typical consumer.[27] This publication summarizes the applicable federal and provincial laws for packaging and labelling, textile labelling, hazardous products, energy efficiency, radio and telecommunications products, electrical safety, motor vehicles, components, and accessories. (It does not cover products such as food, drugs, medical products, and products intended for industrial or commercial applications.)

Contact: Standards Council of Canada (<www.scc.ca>)

Workers' Compensation Act

Workers' compensation is a provincial jurisdiction, which means that the acts relating to it vary across Canada. However, a federal agency known as the Canadian Centre for Occupational Health and Safety (CCOHS) serves as an umbrella organization. The CCOHS website provides links to the provincial health and safety websites. It also provides occupational health and safety information in clear language to the general public and connects the public to reference sources. Thus, it is a good place to start gathering information about health and safety.

Generally speaking, the provincial workers' compensation acts authorize a governing board (that is, a Workers' Compensation Board) to require the reporting of any accident that involves a serious injury or death, or a major structural failure, or the release of a hazardous substance. This board has the authority to inspect workplaces, investigate incidents, require employers to improve worker safety, and impose penalties.

Contact: Canadian Centre for Occupational Health and Safety (CCOHS) (<www.ccohs.ca/oshanswers/information/wcb_canada.html>)

Provincial Occupational Health and Safety (OHS) Regulations

All OHS acts and regulations are provincial. OHS regulations exist to promote occupational health and safety and to protect workers from employment-related hazards. OHS regulations are very comprehensive—that is, they regulate many hazards, such as the safe use and storage of chemicals, the guarding of machinery, the use of mobile equipment, building safety, emergency preparedness, violence prevention, adequate lighting and air quality, smoking, blasting, diving, firefighting, first aid, and so on. To review the act or regulations for your province, start with the CCOHS website cited above.

Federal Workplace Safety Regulations

The federal Hazardous Products Act and the Controlled Products Regulation made under the act apply to suppliers. Together they define which materials (or controlled products) are included in the Workplace Hazardous Materials Information System (WHMIS), and what information suppliers must provide to employers for controlled products used in the workplace. Suppliers who sell or import a controlled product for the workplace must provide a material safety data sheet (MSDS) for the controlled product. They must also ensure that either the controlled product or its container is labelled with all required information and hazard symbols. If you are involved with hazardous substances, check these out. These laws are usually administered by workers' compensation boards. For the federal act and its regulations:

Contact: Government of Canada, Hazardous Products Act, ch. H-3 (<laws.justice.gc.ca/en/h-3/text.html>)

Environmental Legislation

Every province and territory is governed by environmental laws and regulations. Some of these regulations are listed in Chapter 11 of this text, which also provides the website addresses where the legislation (or the department that administers the legislation) can be found. For government publications on environmental law and sustainable development:

Contact: Federal Publications Inc. (<www.fedpubs.com/>)

CASE HISTORY 5.1

THE RIVTOW MARINE CASE: FAILURE TO WARN

This case history illustrates how important it is to warn of hazards. Rivtow Marine, a British Columbia logging company, chartered a logging barge from a B.C. dealer named Walkem. The barge was fitted with a crane manufactured by Washington Iron Works, an American manufacturer. Washington had constructed a similar crane, which collapsed, killing the crane operator. When Rivtow Marine learned of that collapse, it stopped operating its crane and inspected it. Serious cracks were found. These indicated that the Rivtow crane might soon collapse, so it was withdrawn from service. While repairs were being made, in the middle of a busy log-harvesting season, the barge and crane stood idle.

It was later learned that the dealer (Walkem) and the manufacturer (Washington Iron Works) had both been aware for some time of cracks on cranes of this type, yet neither had informed Rivtow Marine. So Rivtow sued both, alleging negligence for failing to provide a warning, and claiming damages for the cost of repairing the crane and for economic losses while the barge and crane were idle. The trial was appealed to the B.C. Appeal Court and then to the Supreme Court of Canada.

In a unanimous ruling, the Supreme Court awarded damages on the basis of negligent failure to warn. The defendants (Walkem and Washington) had a "knowledge of the risk": they knew the business that Rivtow operated, they were aware that the crane was inadequate, they knew it was the busy season, and they knew what harm could arise from having the barge and crane out of operation. Furthermore, the loss was a "direct and foreseeable" consequence of the inadequacy of the crane, and not a remote or unforeseeable result, which might not have justified award of damages.

No personal injury occurred; however, the failure to warn caused economic losses. The Rivtow case is a key part of Canadian tort law, because prior to this case, economic losses were generally considered too remote or too indirect to justify an award. The Rivtow case has been cited and applied in many recent judgments.[28]

CASE HISTORY 5.2

THE WESTRAY MINE DISASTER: FAILURE TO FOLLOW SAFETY STANDARDS

This case history shows how a disaster can result when managers sacrifice safety standards to maximize profit. Twenty-six men died when the Westray mine exploded—one of the worst industrial tragedies in Canadian history. As the inquiry later noted: "Westray was an accident waiting to happen."

The Westray coal mine explosion in 1992 killed twenty-six miners. In his report on the ill-fated mine, Justice Peter Richard blamed the coal company and the provincial government for the disaster, saying that the Westray mine operations were a "violation of the basic and fundamental tenets of safe mining practice."

Source: CP Photo/Andrew Vaughan.

Introduction

Mines are dangerous, and coal mines are the most dangerous mines, because the rock is soft and because coal dust and methane can explode unless the mine is properly maintained and ventilated. The first major Canadian mine disaster happened in 1873, and thousands of lives have been lost since then. In Springhill, Nova Scotia, 424 miners were killed in the mines between 1881 and 1969. Canada's worst coal mine disaster occurred in June 1914 in Hillcrest, Alberta, when 189 men were killed by an explosion.[29] As the years passed, people came to believe that the use of advanced ventilating, monitoring, and excavating methods had enhanced mine safety. This belief was shattered on May 9, 1992, when an explosion in the Westray mine in the village of Plymouth, Nova Scotia, killed twenty-six miners. The inquiry into this disaster revealed a "complex mosaic of actions, omissions, mistakes, incompetence, apathy, cynicism, stupidity, and neglect."[30]

Details of the Explosion

The Westray mine explosion occurred at 5:20 a.m. on a Saturday morning. The shaking of the earth was felt by most of the residents of Plymouth.

Within hours, mine rescue experts had assembled from neighbouring towns. With oxygen tanks on their backs, they descended into the destroyed mine. It was soon clear that rescue efforts would be pointless—the explosion had killed everyone below ground. Nor could all the dead be retrieved. Eleven dead miners are permanently entombed behind rock falls in the mine, much of which was flooded to prevent further explosions.

The Westray Inquiry

Within days of the tragedy, anecdotal evidence of unsafe practices was being widely reported. One miner described several infractions of the safety regulations: acetylene torches had been used in areas where methane levels could be dangerous; a supervisor had tampered with a methane level monitor to permit higher methane levels; and potentially explosive coal dust had accumulated so thickly that some machinery could not be operated.[31] For fear of retaliation or intimidation, the miners rarely complained about the safety infractions, especially since management seemed to place production ahead of safety.[32]

In the midst of bitter accusations, the provincial government appointed Justice K. Peter Richard to carry out a far-reaching inquiry into how and why the twenty-six miners died. Shortly afterwards, the Royal Canadian Mounted Police opened a criminal investigation. In October 1992 the Nova Scotia Labour Department laid fifty-two non-criminal charges of unsafe practices against Curragh Resources, Inc., the company that owned the mine. These safety charges were later dropped to avoid jeopardizing the police investigation, which resulted in charges of manslaughter and criminal negligence being laid against Curragh Resources and two of its managers. These charges were later stayed; however, on appeal the Supreme Court of Canada upheld an order for a new trial. The trial, scheduled for 1999, was cancelled when the charges were withdrawn because of uncertainty over the precise cause of the explosion.

Throughout the early (1996) court proceedings, the Westray Inquiry continued. Justice Richard's final report, *The Westray Inquiry: A Predictable Path to Disaster*, was published in December 1997. Justice Richard commented: "Westray is a stark example of an operation where production demands resulted in the violation of the basic and fundamental tenets of safe mining practice."[33] The following paragraphs constitute a synopsis of the key facts that the inquiry brought to light. They are excerpted from the report's Executive Summary:

Prelude to the Tragedy

...

In the rush to reach saleable coal, workers without adequate coal mining experience were promoted to newly created supervisory positions. Workers were not trained by Westray in safe work methods or in recognizing dangerous roof conditions—despite a major roof collapse in August. Basic safety measures were ignored or performed inadequately. Stonedusting, for example, a critical and standard practice that renders coal dust non-explosive, was carried out sporadically by volunteers on overtime following their 12-hour shifts.

...

The regulatory framework in Nova Scotia requires that almost every person employed in underground coal mining hold a certificate of competency issued by an appointed provincial board of examiners. Section 11 of the Coal Mines Regulation Act (1989) sets out the education and work experience required for the various certificates. The administration of certification for mine rescue and for competency as a coal miner was delegated to the Department of Labour. In Nova Scotia, the company is responsible for training miners. The role of the Department of Labour is to ensure that the company complies with the *Coal Mines Regulation Act* and the *Occupational Health and Safety Act*.

It is clear that the company was derelict in carrying out its obligations for training. . . . Quite simply, management did not instill a safety mentality in its workforce. Although it stressed safety in its employee handbook, the policy it laid out there was never promoted or enforced. Indeed, management ignored or encouraged a series of hazardous or illegal practices, including having the miners work 12-hour shifts, improperly storing fuel and refuelling vehicles underground, and using non-flameproof equipment underground in ways that violated conditions set by the Department of Labour—to mention only a few. Equipment fundamental to a safe mine operation—from the cap lamp to the environmental monitoring system—did not function properly.

It was equally clear that the Department of Labour was derelict in its duty to enforce the requirements of the two acts.

The Explosion: an Analysis of Underground Conditions

...

[V]entilation is the most crucial aspect of mine safety in an underground coal mine. Methane fires and explosions cannot happen if the gas is kept from accumulating in flammable and explosive concentrations. A coal mine can be quite "forgiving" with respect to other aspects of safety, as long as the ventilation system is properly planned, efficient, and conscientiously maintained. The other major requirement of coal mine safety is control of coal dust, through strict clean-up procedures and regular stonedusting.

The ventilation system of any underground mine is a network of interconnected passages, many of which are also used as transportation routes for personnel, vehicles, and the products of mining. Fresh air is drawn from the surface atmosphere. As the air passes through the underground passages, its quality deteriorates as a result of pollutants produced from the strata and from the effects of machines and mining procedures. The contaminated air is returned to the surface. A mine ventilation system has to deal with both gaseous and particulate pollutants. Methane is a dangerous pollutant present in coal. Although non-toxic, it is hazardous because of its flammability. It will explode in concentrations of between 5 and about 15 per cent by volume in air, and it reaches maximum explosiveness at about 9.6 per cent.

Methane is a natural component of coal, a by-product of the decomposition of the plant matter from which coal is formed. Methane is released as the coal-cutting machines break coal away from the face. As methane continues to emerge from the coal, it moves through fissures in the coal that remains after mining, and it can escape into the active roadways from abandoned or mined-out sections, depending on the effectiveness of the stoppings constructed at the entrances to abandoned sections. One of the principal functions of a ventilation system is to clear the methane at the working face of the mine and to exhaust it from the mine in non-explosive concentrations. It is clear that the Westray ventilation system was grossly inadequate for this task. It is also clear that the conditions in the mine were conducive to a coal-dust explosion.

...

The consensus of the experts suggests strongly that Westray was an accident waiting to happen.

...

Responsibility

As the evidence emerged during this Inquiry, it became clear that many persons and entities had defaulted in their legislative, business, statutory, and management responsibilities. . . . [T]here is a clear "hierarchy" of responsibility for the environment that set the stage for 9 May 1992, and we ought not to lose sight of this hierarchy.

The fundamental and basic responsibility for the safe operation of an underground coal mine, and indeed of any industrial undertaking, rests clearly with management. The internal responsibility system merely articulates this responsibility and places it in context. Westray management, starting with the chief executive officer, was required by law, by good business practice, and by good conscience to design and operate the Westray mine safely. Westray management failed in this primary responsibility, and the significance of that failure cannot be mitigated or diluted simply because others were derelict in their responsibility.

The Department of Labour through its mine inspectorate must bear a correlative responsibility for its continued failure in its duty to ensure compliance with the *Coal Mines Regulation Act* and the *Occupational Health and Safety Act*. Indeed, the many and varied faults of Westray management and its derelict attitude towards safety should have prompted the Department of Labour inspectorate to adopt a firm and uncompromising position on strict compliance. Instead, the evidence indicates that the demeanour of the inspectorate was one of apathy and complaisance.

With its "hands-off" attitude, its general indifference to the quality of mine planning, and its lassitude about any safety responsibility, the Department of Natural Resources failed to discharge its duties in a creditable manner. The general attitude of wilful blindness pervaded the department's dealings with Westray. Thus, the stage was set for Westray management to maintain an air of arrogance and cynicism, knowing that it was not going to be seriously challenged.

Compliance with the Coal Mines Regulation Act

Much has been said throughout this Inquiry about the inadequacy of the *Coal Mines Regulation Act*. As outdated and archaic as the present act is, it is painfully clear that this disaster would not have occurred if there had been compliance with the act.

- If the "floor, roof and sides of the road and the working places" had been systematically cleared so as to prevent the accumulation of coal dust,

- If the "floor, road and sides of every road" had been treated with stonedust so that the resulting mixture would contain no more than 35 per cent combustible matter (adjusted downward to allow for the presence of methane); and

- If the mine had been "thoroughly ventilated and furnished with an adequate supply of pure air to dilute and render harmless inflammable and noxious gases," then ...

the 9 May 1992 explosion could not have happened, and 26 miners would not have been killed. Compliance with these sections of the *Coal Mines Regulation Act* was the clear duty of Westray management, from the chief executive officer to the first-line supervisor. To ensure that this duty was undertaken and fulfilled by management was the legislated duty of the inspectorate of the Department of Labour. Management failed, the inspectorate failed, and the mine blew up.[34]

DISCUSSION TOPICS AND ASSIGNMENTS

(Additional assignments can be found in Appendix CD-E on the CD-ROM included with this text.)

1. This chapter suggests that the first step in a design project should include a search of the Internet for appropriate technical codes and standards. As an exercise, find at least one design standard for each of the following items: automobile headlights, elevators or escalators, buildings, pressure vessels, snowmobiles, children's toys, and the Canadian flag.

2. Road vehicles, such as cars, trucks, and SUVs, are the most dangerous machines that most people use on a daily basis. These vehicles probably kill more people each year than other modes of transportation (from ox-carts to aeroplanes) have killed since their invention. Moreover, cars, trucks, and SUVs pollute: carbon dioxide emissions contribute to global warming, oil leaks pollute, and carbon monoxide emissions—when not properly vented—can injure or kill. However, no serious campaign has ever been mounted to prevent these vehicles from being manufactured. Conversely, no one has ever been killed as a result of a nuclear incident at any Canadian nuclear generating plant, yet public opinion is occasionally very negative to nuclear power generation. Write a brief essay comparing and contrasting these two perceptions of risk and benefit. Gather statistics and verify the above statements. What lesson is to be drawn?

3. In this chapter (in "Hazards and the Duty of the Design Engineer"), engineers are warned that a hazard analysis is needed for every product, and typically consists of four steps. (This process is described in more detail in Appendix CD–G1.) As an exercise, consider the hazards associated with a simple portable electric heater, composed of a heating coil on a metal stand, with an electrical cord and a plug that is inserted into a 120-volt outlet. Such heaters are commonly found in the home, and their radiant heat is beneficial for local heating in the cold Canadian winter. Conduct a hazard analysis for a typical electric heater. Identify the hazards that may be created by the electrical cord or the heating element. How could these hazards be prevented or eliminated? Obviously, the heating element cannot be eliminated without defeating the purpose of the heater. How can this element be shielded without affecting performance? What conditions or occurrences would justify the use of warnings, recalls, or assisting people injured by the heater? Summarize your answer in one or two pages.

NOTES

1. A.M. Linden, *Canadian Tort Law*, Butterworths Canada, Markham, ON, 1993, p. 536.
2. P.H. Osborne, *The Law of Torts*, Irwin Law, Toronto, 2000, p. 8.
3. Ibid., p. 9.
4. D.L. Marston, *Law for Professional Engineers*: Canadian and International Perspectives, 3rd ed., McGraw-Hill Ryerson, Whitby, ON, 1996, p. 33.
5. M. Kerr, J. Kurtz, and L.M. Olivo, *Canadian Tort Law in a Nutshell*, Carswell, Thomson Canada, 1997, p. 34.
6. J.D. Weir and S.A. Ellis, *Critical Concepts of Canadian Business*, Addison-Wesley, Don Mills, ON, 1997, p. 137.
7. Marston, *Law for Professional Engineers*, p. 46.
8. Government of Ontario, Regulation 941/90 under the *Professional Engineers Act*, RSO 1990, c. P.28, s. 72.
9. Weir and Ellis, *Critical Concepts of Canadian Business*, p. 139.
10. Linden, *Canadian Tort Law*, p. 215.

11. Weir and Ellis, *Critical Concepts of Canadian Business*, p. 161.
12. Ibid., p. 69.
13. Ibid., p. 73.
14. Ibid.
15. Marston, *Law for Professional Engineers*, p. 46.
16. Ibid., p. 51.
17. D.W. Noel and J.J. Philips, *Products Liability*, 2nd ed. West Publishing, Saint Paul, MN, 1981.
18. Osborne, *The Law of Torts*, p. 130.
19. G. Voland, *Engineering by Design*, Addison Wesley, Don Mills, ON, 1999; J. Kolb and S.S. Ross, *Product Safety and Liability: A Desk Reference*, McGraw-Hill, New York, NY, 1980; Consumer Products Protection Commission (CPSC), *Handbook and Standard for Manufacturing Safer Consumer Products*, U.S. Government document, 1977; D.L. Goetsch and S.B. Davis, *Understanding and Implementing ISO 9000 and ISO Standards*, Prentice-Hall, Toronto, 1998; and G.C. Andrews, J.D. Aplevich, R.A. Fraser, and H.C. Ratz, *Introduction to Professional Engineering in Canada*, Prentice-Hall, Toronto, 2002, p. 233.
20. C.O. Smith, "Products Liability: Severe Design Constraint,"in *Structural Failure, Product Liability and Technical Insurance*, Proceedings, 2nd International Conference, 1–3 July 1986, Interscience Enterprises, Geneva 1987, pp. 59–75.
21. Standards Council of Canada (SCC) website: <www.scc.ca/> (October 16, 2003). Excerpt reproduced with permission.
22. International Organization for Standardization (ISO) website: <www.iso.org/iso/en/commcentre/pressreleases/2003/Ref864.html> (October 20, 2003).
23. Goetsch and Davis, *Understanding and Implementing ISO 9000*, p. 150.
24. Ibid., p. 151.
25. S.L. Jackson, *The ISO 14000 Implementation Guide*, John Wiley & Sons, Toronto, 1997, p. 1.
26. ISO website: <www.iso.org/> (October 20, 2003).
27. Standards Council of Canada (SCC), *Consumer Product Safety Legislation in Canada: An Introductory Guide*, SCC, Ottawa, ON, March 2003.
28. Linden, *Canadian Tort Law*, p. 387; Marston, *Law for Professional Engineers*, p. 55.
29. H.A. Halliday and J. Joegg, "Mining Disasters," *The Canadian Encyclopedia Plus*, McClelland & Stewart, Toronto, 1996.
30. Justice K. Peter Richard, "Executive Summary," *The Westray Story: A Predictable Path to Disaster: Report of the Westray Mine Public Inquiry*, published on the authority of the Lieutenant Governor in Council, Province of Nova Scotia (1 December 1997). Copyright by the Province of Nova Scotia, 1997. Permission is given by the copyright holder to reproduce the report or any part thereof. The summary is available on the Internet at: <www.gov.ns.ca/enla/pubs/westray/execsumm.htm> (October 17, 2003).
31. MacIssac, "Miners Testify at Westray," *Maclean's* (January 29, 1996). See also *Canadian Encyclopedia Plus*.
32. Ibid.
33. Justice Richard, "Executive Summary."
34. Ibid.

Part Three
Professional Ethics

Chapter 6
Principles of Engineering Ethics

Ethical problems arise often in engineering and geoscience. For example, a manufacturer may challenge the design engineer concerning the need for safety guards on equipment such as lawn mowers or chain-driven vehicles. At what point does the potential harm to the user outweigh the real cost to the manufacturer? This is a common ethical question and, fortunately, technical codes, standards, and guidelines have been established over the years to help engineers make such decisions. However, ethical guidelines are not as clearly defined in other areas. For example, an engineer in a position of authority may have to decide whether a small gift is an innocent kindness or a conscious attempt at bribery. As we will see in later chapters, many situations in engineering generate ethical problems.

In this chapter we examine four useful and important ethical theories that have evolved over the centuries. We then review the professional Codes of Ethics that are based on those theories and develop a methodical approach for solving more general ethical problems.

ENGINEERS AND ETHICS

Engineers are very competent problem solvers. An engineer faced with a technical problem will derive a solution using the well-known axioms, theorems, and laws of mathematics, science, and engineering. It is therefore reassuring to know that when faced with an ethical problem, various sets of ethical theories exist—developed over the centuries—that are tested and true. These ethical theories are the basis of the laws, regulations, and Codes of Ethics that guide us. Let us begin this overview by defining "ethics."

Ethics is commonly classified as one of the four branches of philosophy. Each branch studies important and fundamental questions. The four branches are:

- **Ethics.** The study of right and wrong, good and evil, obligations and rights, justice, and social and political ideals.
- **Logic.** The study of the rules of reasoning. For example, under what conditions can an argument be proved true?
- **Epistemology.** The study of knowledge itself. What is knowledge? Can we know anything? What can we know? What are the sources of knowledge?

- **Metaphysics.** The study of very basic ideas such as existence, appearance, reality, and determinism. Metaphysics asks questions about the most abstract and basic categories of thought: thing, person, property, relation, event, space, time, action, possibility/actuality, and appearance versus reality.[1]

Ethics has been an important field of study since the dawn of civilization, and ethical writings can be traced back over 2500 years. In fact, many ethical concepts that we commonly apply today are much older than the basic concepts of engineering analysis (calculus, statics, dynamics, stress analysis, etc.), which trace their origins to the seventeenth century.

Epistemology and metaphysics are interesting but highly theoretical fields, with few practical applications; in contrast, logic and ethics have many uses. Logic is essential to engineers and geoscientists because it is the basis of every mathematical derivation. Similarly, ethics has practical applications in our lives. It helps us differentiate between right and wrong, as the following review of some ethical theories reveals.

FOUR ETHICAL THEORIES

Many prominent philosophers have devoted their lives to developing ethical theories, and a thorough discussion of their thought would fill a thousand textbooks. We can hardly hope to condense this treasury of philosophical thought into a single chapter. However, some review of ethical theories is essential for understanding the origin of Codes of Ethics and for dealing with cases that "fall through the cracks"—that is, with ethical problems that are not clearly addressed by Codes of Ethics. The following brief summary will illustrate the basic concepts and perhaps inspire you to investigate the subject more deeply.

At least four of the ethical theories that have evolved over the centuries are relevant to ethics in engineering. All of them have stood the test of time and are useful aids to decision making. They differ significantly, and none of them is clearly universally superior to the others; even so, it is startling to see how much all four agree when applied to common ethical problems. Each theory is identified by the name of its best-known proponent; remember, though, that many earlier philosophers contributed to formulating the theories and that many modern philosophers have suggested improvements to them. These are the four ethical theories:

- Mill's utilitarianism
- Kant's formalism, or duty ethics
- Locke's rights ethics
- Aristotle's virtue ethics

Mill's Utilitarianism

This theory was stated most clearly by John Stuart Mill (1806–1873). Utilitarianism states that the best choice in a ethical dilemma is that which produces the maximum benefit for the greatest number of people. This theory is probably the most common justification for ethical decisions in engineering—indeed, in modern society. Democratic government itself can be justified on utilitarian grounds, since it permits the maximum good (control over government) for the maximum number of people (the majority of voters).

The difficulty of applying the utilitarian principle lies in quantitatively calculating the "maximum benefit." Mill proposed that three key factors should be considered in determining the maximum benefit: the number of people affected and the intensity and duration of the benefit or pleasure (or, conversely, the intensity and duration of the pain to be avoided). For example, when we evaluate the benefit of automobile seat-belt legislation, we find that all drivers and passengers suffer some inconvenience when they buckle up, whereas only a few people obtain the benefit (the avoidance of death or severe injury). However, lawmakers have generally concluded that the deaths and severe injuries caused by accidents more than outweigh the inconvenience; thus, the laws require everyone to wear seat belts.

In evaluating benefits, it is important that we apply certain principles. The benefit to oneself must not be given any greater value or importance than the same benefit to any other individual. And benefits should, of course, be calculated without regard to nationality, creed, colour, language, sex, and so on; put another way, no preference should be given to any particular group. Moreover, the equality of distribution of the benefit is important when choosing a course of action: an equal distribution of benefits is preferable to an unequal distribution. That is, the best course of action in an ethical dilemma is the choice that produces the maximum benefit for the greatest number of people, with the benefit most equally divided among those people.

The utilitarian theory is easily understood and is consistent with the concept of democracy. In simple cases, it is also easy to apply. For example, income tax is easily justified by utilitarian theory: a modest hardship (paying tax) is imposed equally on all residents (as a percentage of income); this yields an immense benefit to society, because the tax dollars build hospitals, schools, and essential infrastructure. We may sometimes disagree with the details of the taxation system, over who should get tax exemptions, how tax dollars should be spent, and so on. Even so, income tax is generally accepted and is rarely challenged as unethical.

Most engineers find utilitarianism to be very valuable in making ethical decisions.

Kant's Formalism, or Duty Ethics

The theory of duty ethics, or "formalism," was put forward by Immanuel Kant (1724–1804), who proposed that every individual has a fundamental duty to act in a correct ethical manner. This theory evolved from Kant's belief or observation that each person's conscience imposes an absolute "categorical imperative" (or unconditional command) on that person to follow those courses of action which would be acceptable as universal principles for everyone to follow. For example, everyone has a duty not to tell lies, because if we tolerated lying, then no promises could be trusted and our social fabric would be at risk of unravelling. Kant believed that the most basic good was "good will," or actively seeking to follow the categorical imperative of one's conscience. This is in marked contrast to Mill, who believed that universal happiness was the ultimate good. In Kant's philosophy, happiness is the result of good will: the desire and intention to do one's duty.

Kant emphasized that it was the intention to do one's duty that was significant, not the actual results or consequences. One should always do one's duty, even if the short-term consequences are unpleasant, since this strengthens one's will. For example, even "white" lies should not be tolerated, since they weaken the resolve to follow one's conscience. The formalist theory contends that in solving an ethical dilemma, one has a duty to follow rules that are generated from the conscience (the categorical imperative), and that if a person strives to develop a good will, happiness will result.

Many of the rules that support this universal concept are well known— "Be honest," "Be fair," "Do not hurt others," "Keep your promises," "Obey the law," and so on—and not surprisingly, our happiness would certainly increase if everyone followed them. Kant also stated that a consequence of following the categorical imperative would be an increased respect for humanity. Life should always be treated as an end or goal, and never as a means of achieving some other goal. It follows that Kant's formalism would condemn as unethical any engineering activity that endangered life by water or air pollution (regardless of the purpose or cause of the pollution). In Kant's formalism, every engineer or engineering manager has an individual duty to prevent harm to human life and to consider the welfare of society to be paramount. (As you will see later in this chapter, this axiom from Kant is the first rule in almost every engineering Code of Ethics.)

In sum, Kant's formalism emphasizes the importance of following universal rules, the importance of humanity, and the significance of the intention of an act or rule rather than the actual outcome in a specific case. The only problem with applying formalism relates to its inflexibility—duties based on the categorical imperative never have exceptions. Fortunately, we can obtain further guidance by considering the other ethical theories.

Locke's Rights Ethics

The rights-based ethical theory comes mainly from the thought and writings of John Locke (1632–1704). Rights-based theory basically states that every individual has rights simply by virtue of his or her existence. The right to life and the right to the maximum possible individual liberty and human dignity are fundamental; all other rights flow out of these most basic ones. Every individual's rights must be recognized by others, who have a duty not to infringe on those rights. This contrasts with duty-based ethical theory (discussed earlier), which contends that duty is fundamental; in the rights-based theory, duties are a consequence of personal rights. Locke's writings had a powerful impact on British political thought in the 1690s; they also strongly influenced the French and American revolutions. Basic human rights are embedded in Canadian law through the Canadian Charter of Rights and Freedoms. The Charter recognizes that everyone has the following:

- Fundamental freedom of conscience, religion, thought, belief, opinion, expression, peaceful assembly, and association. (clause 2)
- Democratic rights to vote in an election (or to stand for election) of the House of Commons or of a legislative assembly. (clause 3)
- Mobility rights to enter, remain in, and leave Canada. (clause 6)
- Legal rights to life, liberty, and security of the person and the right not to be deprived of these rights except in accordance with principles of fundamental justice. (clause 7)
- Equality rights before and under the law and the right to equal benefit and protection of the law. (clause 15)[2]

We must recognize that everyone has these basic rights and that they should not be infringed. However, the Charter does not contain every right that should exist—only the fundamental rights that have been hammered out in Parliament and in the courts of law over the past two centuries. Other rights have evolved from Locke's theory. Some of these rights are enacted in other legislation, and some exist in grey areas of uncertainty, and some "rights" are claimed as a cloak for selfishness. Examples of these three types follow:

- Locke's theory would suggest that everyone has the right to a working environment that is free of sexual harassment or racial discrimination and that the employer has a duty to provide it. This right would appear to be common courtesy. Few would challenge it, and it is generally included in provincial labour laws.
- However, many people claim rights that are not included in the Charter or in other legislation. These rights fall into grey areas and must be examined on a case-by-case basis. For example, it would seem to be common courtesy to extend the right to privacy, the right not to be subjected to loud noise, the right not to be subjected to dangerous pesticides, and so on. These rights are based on human dignity and individual liberty, and they should be respected unless it can be proved that a greater good is satisfied by denying such rights.
- Rights-based ethical theory does have limits. As an example, consider income tax. Even today, some people challenge the concept of income tax, claiming that it infringes on the individual right to retain one's property. Others insist on their right to smoke in public establishments, even though such behaviour has been ruled illegal in many Canadian cities. "Rights" like these conflict with utilitarianism, and in most cases the dispute has been resolved in favour of utilitarianism.

Clearly, rights-based arguments justify a spectrum of claims. Some rights are indisputable and are embedded in law; other rights fall into grey areas, which suggests that they should be respected wherever possible but are not absolute. Furthermore, some people claim rights that amount to selfishness in disguise. In summary, rights-based ethics has an important place in resolving ethical dilemmas, but the theory is not sufficient to deal with every situation.

Aristotle's Virtue Ethics

One of the earliest and most durable ethical theories was proposed by the Greek philosopher Aristotle (384–322 B.C.), who observed that the goodness of an act, object, or person depended on the function or goal concerned. For example, a "good" chair is comfortable, and a "good" knife cuts well.

Similarly, happiness or goodness will result for humans once they allow their specifically human qualities to function fully. Aristotle stated that the one quality humans have that animals do not is the power of thought; therefore, humans would achieve true happiness by developing qualities of character through thought, reason, deduction, and logic. He called these qualities of character "virtues," and he visualized every virtue as a compromise between two extremes or vices. His guide to achieving virtue was to select the "golden mean" between the extremes of excess and deficiency. For example, modesty is the golden mean between the excess of vanity and the deficiency of humility; courage is the golden mean between foolhardiness and cowardice; and generosity is the golden mean between wastefulness and stinginess.

We can apply Aristotle's concept of the golden mean to ethical problems by examining the extremes of excess or deficiency and then seeking the compromise—that is, the golden mean, or the "happy medium"—between the extremes. This approach is often useful in resolving ethical dilemmas. The four ethical theories are summarized in Table 6.1.

AGREEMENT AND CONTRADICTION IN ETHICAL THEORIES

The four theories described above have survived the test of centuries, and all of them are useful in finding fair solutions to ethical problems. Each theory has a wide range of applications; none is superior in every situation. Philosophers have long been seeking the one universal principle on which all ethical thought is founded, but a single unifying concept has yet to emerge. In many applications, all four theories agree completely. Sometimes, however, they contradict one another so that each theory yields its own distinctive or unique answer to the same dilemma. We must not be frustrated by this; we must calmly evaluate the various courses of action and choose the optimum, just as we would in any engineering problem with more than one solution.

As an example of agreement between the theories, consider the Golden Rule: "Do unto others as you would have others do unto you." This is a clear statement of Kant's formalism: it imposes a duty on the individual to view

TABLE 6.1 — Summary of the Four Ethical Theories

Mill's utilitarianism	**Statement**. An action is ethically correct if it produces the greatest benefit for the greatest number of people. The duration, intensity, and equality of distribution of the benefits should be considered.	**Conflict**. A conflict of interest may arise when evaluating the benefits. It is important that a personal benefit be counted as equal to a similar benefit to someone else.
Kant's duty-based ethics	**Statement**. Each person has a duty to follow those courses of action that would be acceptable as universal principles for everyone to follow.	**Conflict**. Conflicts arise when following a universal principle may cause harm. For example, telling a "white" lie is not acceptable, even if telling the truth causes harm.
Locke's rights-based ethics	**Statement**. All individuals are free and equal, and each has a right to life, health, liberty, possessions, and the products of his or her labour.	**Conflict**. It is occasionally difficult to determine when one person's rights infringe on another person's rights.
Aristotle's virtue-based ethics	**Statement**. Happiness is achieved by developing virtues, or qualities of character, through deduction and reason. An act is good if it is in accordance with reason. This usually means a course of action that is the golden mean between extremes of excess and deficiency.	**Conflict**. The definition of virtue is occasionally vague and difficult to apply in specific cases. However, the concept of seeking a golden mean between two extremes is often useful in ethics.

human life as a goal rather than as a means to a goal. On the other hand, it could be considered a utilitarian principle, since it imposes an inconvenience on the individual while benefiting everyone with whom that person comes into contact. The proponents of rights-based ethics would agree with the Golden Rule but would claim that the duty of the individual to act fairly comes from the rights of others to be treated fairly. Finally, the concept of "fairness" would be recognized as a virtue by Aristotle. The four ethical theories are thus in complete agreement that the Golden Rule is a good maxim for guiding human behaviour, as we would expect.

Similarly, the basic ethical precepts of most religions are supported by all four ethical theories. Consider the Ten Commandments from the Book of Exodus, which are the ethical basis of Judaeo-Christian religions. Each of the commandments clearly imposes a duty on the individual and at the same time grants rights to others, requires virtuous behaviour, and creates a stable environment that yields the maximum benefit for all. An investigation of the basic precepts of all the great religions would show similar agreement.

However, although the ethical theories show remarkable agreement on some principles, they sometimes give conflicting or contradictory guidance when applied to specific cases. Consider the following hypothetical case.

An Ethical Dilemma

Professional engineers Smith and Legault are both employed in writing control system software for an electrical power generating plant. They are good friends and often spend time together after work. While they are spending time together after work, Legault learns that Smith has an addiction to alcohol and drugs that often affects Smith's mental stability and technical judgment. Legault believes that Smith's addiction is affecting the software he is writing. Legault knows the software errors will eventually become obvious, but he does not want to intervene, because of their friendship. What should Legault do?

In a real situation, you would have much more detailed information, but it is reasonable to assume that the software being developed will be tested before release and will not be used if it endangers life. However, if the software validation tests find flaws, the project will certainly be delayed (perhaps seriously) while the flaws are investigated and corrected.

- **Duty**. The duty-based theory would say that Legault has a duty to report Smith to management for treatment, reassignment, or disciplinary action. In fact—and as discussed later in this chapter—the Code of Ethics would state that the public interest should be placed ahead of personal feelings. One might also consider that because of their friendship, Legault has a duty to help Smith overcome the addiction.
- **Rights**. Conversely, the rights-based theory would say that Smith's health is a private matter. Thus, Smith has a right to privacy and Legault has no right to investigate it or discuss it with others.

The duty-based and rights-based theories yield simple, clear rules. Yet in this situation those rules contradict each other. We must examine the other theories to hear what they say. Because the utilitarian and virtue-based theories require a subjective judgment, more information is usually needed before we can apply them. In this case, the degree of danger to others, the seriousness of the addiction, and Smith's willingness to seek treatment are relevant factors.

- **Utilitarianism.** The utilitarian theory balances the risk of harm to the software project and to the public (if Legault does not intervene) against the harm to Smith's career (if Legault exposes the addiction). The estimated intensity of such harm is a factor. Here, the facts are not provided in full detail, but if the software fails the validation test—as is likely—then the project will be delayed, the employer will suffer a loss, both Smith and Legault may be accused of negligence or incompetence, and Smith's addiction may become known anyway. The utilitarian theory—even based on such meagre information—would tend to favour intervention.

- **Virtue.** The virtue-based theory would recognize drug and alcohol addiction as an extreme and undesirable abuse. The golden mean between abstinence and addiction is moderate use. The virtue-based theory would condemn Smith's addiction, and thereby imply that some action should be taken to alleviate it.

Conclusion. Even with the limited information provided, three of the four theories seem to recommend intervention. However, knowing the right course of action and finding the courage to implement it are separate matters. The role of Legault, as a friend, is not to conceal the problem, nor is it to be a snitch; rather, it is to find a way to implement the decision with a minimum of personal chaos. An outcome can be influenced drastically by how the decision is implemented. Ideally, Legault would convince Smith to take sick leave and enter a recovery program, thus salvaging Smith's career and finances. Care, concern, and innovation are critically important at this stage. Other solutions may also exist, depending on the circumstances (or other details that are not known to us).

When applying all four theories does not resolve the dilemma, then the ethical theory that is considered *most* appropriate must be selected and followed. This requires a value judgment that may vary from person to person and is therefore not an absolute rule. Nevertheless, if the decision has been made in an orderly fashion, is consistent with at least one recognized ethical theory, and has not been made lightly or wantonly, then the person who has made the decision will have a clear conscience. Later in this chapter, these ideas are expanded into a decision-making strategy.

A word of caution: We should always be alert to subconscious bias when selecting a course of action that benefits ourselves at a cost to someone else. A decision that benefits oneself is a conflict of interest and must be avoided. To avoid accusations of bias, it is useful to discuss ethical decisions with someone who does not have a conflict of interest. Where life, safety, security, or personal reputation is at stake, a decision must not only be ethical, but also must be seen by others to be ethical.

CODES OF ETHICS AS GUIDES TO CONDUCT

To put ethics into practice, most people require clearer rules than general philosophical theories. Over the centuries, many customs, conventions, laws, and ordinances have evolved that are based on the ethical theories but offer more specific guidance. For example, the criminal and civil law are probably the most important guides to ethical conduct. Throughout the world there is remarkably close agreement in these laws, in spite of the different political systems, cultural influences, and moral attitudes that exist in different countries. The similarities are closest in criminal law: every country forbids theft, perjury, assault, and murder, although the severity of punishment may vary from country to country. These similarities are understandable, since laws are

The Confederation Bridge, which opened on May 31, 1997, links Prince Edward Island and New Brunswick. The Canadian-designed bridge is 12.9 km long and has a navigable clearance of 60 m above the water. The central portion (11 km) has forty-four spans, typically 250 m long. The PEI approach bridge has seven spans (580 m), and the NB approach bridge has fourteen spans (1300 m). The Confederation Bridge is the world's longest bridge over ice-covered waters. It carries two lanes of traffic twenty-four hours a day, seven days a week.

Source: CP Photo/Andrew Vaughan.

merely formal statements of the ethical theories discussed earlier and since the four theories are generally in full agreement where basic standards of conduct are concerned.

Under the authority of the provincial and territorial Acts, professional engineering Associations have been established and empowered to write and enforce regulations, bylaws, and Codes of Ethics that prescribe acceptable conduct for professional engineers. Infringements can lead to penalties enforced by the provincial justice system or by the Association, as described elsewhere in this text. The Codes of Ethics for each province and territory are provided in Appendix CD-B. They give specific rules to guide the conduct of individual engineers and geoscientists.

General Principles

Codes of Ethics usually include statements of general principles, followed by instructions for specific conduct. Principles define the duties the professional owes to society, to employers, to clients, to colleagues, to subordinates, to the profession, and to himself or herself. Although the various codes express these duties slightly differently, their intent and the results are very similar. The following paragraphs summarize the principles that most provincial and territorial engineering Codes of Ethics have in common. (The term "engineer" includes geoscientists.)

Duty to society. An engineer is required to consider his or her duty to the public—or to society in general—as the most important of all duties. This is consistent with the utilitarian concept that each provincial Act has given members of the Association the authority to define the profession's standards of admission and to regulate the behaviour of its practitioners. This authority has been granted in order to benefit society as a whole—specifically, in order to protect the average person from physical or financial harm by ensuring that professional engineers are competent, reliable, professional, and ethical. Professional engineers have a particular duty to protect the safety, health, and welfare of society whenever society is affected by their work.

Duty to employers. An engineer has a duty to his or her employer to act fairly and loyally and to keep the employer's business confidential. Furthermore, an engineer is obligated to disclose any conflict of interest. Such a conflict arises when the engineer benefits at the expense of the employer's business.

Duty to clients. An engineer who is in private practice is employed by clients and has the same obligations to those clients as any engineering employee would have to his or her employer. Since contracts with clients are usually shorter than the typical employment contract, avoiding conflicts of interest and cooperating with other personnel involved in a project are particularly important.

Duty to colleagues. An engineer has a duty to act with courtesy and good will toward colleagues. This simple statement of the Golden Rule is supported by all four ethical theories. This duty benefits the people involved, their clients or employers, and the engineering profession in general. Clearly, a person who is awarded professional status should act professionally and should not permit personal or unrelated problems to intrude upon the professional relationship. Most Codes of Ethics specifically state that it is unethical to review the work of a fellow engineer without that engineer's knowledge.

Duty to employees and subordinates. An engineer has a duty to recognize the rights of others, especially if they are employees or subordinates who are obligated to work with the engineer by contracts of employment.

Duty to the profession. An engineer has a duty to maintain the dignity and prestige of the profession and not to bring the profession into disrepute by scandalous, dishonourable, or disgraceful conduct.

Duty to oneself. Finally, an engineer must ensure that the duties to others are balanced by the engineer's own rights. An engineer must insist on adequate payment, a satisfactory work environment, and the rights awarded to everyone through the Charter of Rights and Freedoms. An engineer also has a duty to strive for excellence and to maintain competence in the rapidly changing technical world.

A Comparison of Codes of Ethics

The characteristics of the provincial and territorial Codes of Ethics are compared and discussed below. All of the codes are reproduced in full in Appendix CD-B on the CD-ROM disk included with this text. You are urged to refer to that appendix (or connect with your Association's website) and read the Code of Ethics for your provincial or territorial Association. The seven general duties described above should be clear to see, but may be described in different terms. Besides the seven general duties, individual provinces and territories may have additional requirements in their Codes of Ethics.

Alberta. Alberta's Code of Ethics was revised in 2000 and is now summarized in a brief preamble and five simple rules of conduct. This code does not explicitly include all of the duties described above; however, its preamble observes that "integrity, competence, dignity and devotion to service" should guide the conduct of practitioners. The five rules instruct professional engineers, geologists and geophysicists as follows:

- In their areas of practice, to hold paramount the health, safety and welfare of the public and to have regard for the environment.
- To undertake only work that they are competent to perform by virtue of their training and experience.
- To conduct themselves with integrity, honesty, fairness, and objectivity in their professional activities.
- To comply with applicable statutes, regulations, and bylaws in their professional practices.
- To uphold and enhance the honour, dignity, and reputation of their professions and, thus, the ability of the professions to serve the public interest.

Alberta's code focuses on rules of personal conduct rather than on duties to others; even so, a professional person who follows these rules would easily satisfy the seven general duties noted earlier.

British Columbia. British Columbia's Code of Ethics was revised in 1994. It is clear and succinct, with ten clauses (similar to those of several other provincial codes) which specify the duties that professional engineers and geoscientists owe to society, to clients or employers, to colleagues, to the engineering profession, and to themselves. In contrast to some other codes, however, clause 1 specifically includes protection of the environment and the need for workplace health and safety, and clause 9 instructs professionals to "report to their association or other appropriate agencies any hazardous, illegal or unethical professional decisions or practices by engineers, geoscientists, or others."

Manitoba. Manitoba's Code of Ethics was revised in 2000. It specifically states that breaches of the code may be considered unskilled practice or professional misconduct and thus subject to disciplinary action under the Act. This code is arranged as five basic "Canons of Conduct" to guide professional behaviour. The five canons comprise thirty-eight very specific clauses that offer useful advice on various aspects of professional practice. Some of the clauses in Manitoba's code appear in few other codes, but are good advice for professional practice anywhere. Examples:

- Canon 1 contains three clauses that specifically require practitioners to obey the laws of the land. It also notes some specific areas where compliance is necessary.
- Canon 2 contains nine clauses to help practitioners protect the physical, economic, and environmental well-being of the public—which is defined as the "prime responsibility" of practitioners—and emphasizes the importance of quality in professional work.
- Canon 3 contains thirteen clauses explaining how to apply skill and knowledge to satisfy the needs of a client or employer in a professional manner, and gives good advice on communication and on avoiding conflicts of interest.
- Canon 4 contains five clauses describing how to uphold the honour, integrity, and dignity of the profession. Among other things, these clauses relate to the exchange of information, proper advertising, and the reporting of people who are violating the Act.
- Canon 5 contains eight clauses that give very meaningful advice to practitioners to be fair to colleagues and to support their professional development.

New Brunswick. New Brunswick's Code of Ethics was recently revised and is Section 2 of the 2002 APEGNB By-laws. The code has five parts:

- "Foreword." This concise introduction summarizes the duties of engineers and geoscientists.
- "Professional Life." These nine clauses set some guidelines for professional practice and personal conduct. This part (like a similar one in British Columbia's code) specifically includes protection of the environment and the need for workplace health and safety. A clause, unique to this code, advises professionals to "observe the rules of professional conduct which apply in the country in which they practise and, if there are no such rules, observe those established by this Code of Ethics."
- "Relations with the Public." These seven clauses focus on dealings with the general public. One of these clauses is unique in that it forbids discrimination.
- "Relations with Clients and Employers." These seventeen clauses define many key aspects of professional practice—accepting tasks, ensuring safety, avoiding conflict of interest, considering environmental effects, receiving payment, and so on.
- "Relations with Engineers and Geoscientists." These fourteen clauses concern relations with colleagues. Some of these clauses give very specific guidance; for example, they uphold the principle of adequate compensation,

prohibit engineers from using free engineering designs from suppliers in return for specifying their products, and instruct engineers not to associate with any enterprise that does not conform to ethical practices.

Newfoundland and Labrador. This province's Code of Ethics comprises an introduction and three sections that specify the duties of the professional engineer or geoscientist to the public, to the client or employer, and to the profession. It is fairly clear, concise (twenty-two clauses), and easily understood.

Northwest Territories. NWT's Code of Ethics has eleven clauses that encompass all of the seven duties alluded to in the previous section. It also requires engineers, geologists, and geophysicists to have proper regard for the physical environment. The duties to the profession noted in this code include the requirement to advise the Association's registrar of any practices by other members of the Association that are contrary to the Code of Ethics. The preamble to this code entreats engineers to serve in public affairs when their professional knowledge may benefit the public and to demonstrate understanding for members-in-training under their supervision.

Nova Scotia—Engineering. The Nova Scotia engineering Code of Ethics is fairly long (twenty-eight clauses) and contains clauses typical of the other Acts. This code is unique, however, in that it instructs engineers to refrain from conduct contrary to the public good, even if directed by the employer or client to act in such a manner, and similarly instructs employers not to direct employees to perform acts that are unprofessional or contrary to the public good.

Nova Scotia—Geoscience. The Nova Scotia geoscience Code of Ethics is comprehensive, yet fairly brief (12 clauses), and contains clauses specifying the seven typical duties that professional geoscientists have to society, to clients and employers, to colleagues and subordinates, to the profession, and to themselves. The code also requires geoscientists to report illegal or unethical geoscience decisions or practices; and to be aware of social and environmental consequences of actions or projects, and to inform clients and employers of these consequences.

Nunavut. The Association of Professional Engineers, Geologists and Geophysicists of the Northwest Territories (NAPEGG) regulates the professions of engineering, geology, and geophysics in Nunavut. Therefore, until such time as separate legislation is enacted by Nunavut, readers should consult the Code of Ethics for the Northwest Territories.

Ontario—Engineering. The Ontario Professional Engineers Act differs from other provincial Acts in that the Ontario Code of Ethics (containing about 15 subclauses) is specifically not enforcable under the Act. A separate regulation defines professional misconduct (in about 20 subclauses). Both the Code of Ethics (s.77) and the definition of professional misconduct (s.72) from Ontario Regulation 941/90 are reproduced in Appendix CD-B, enclosed with this text. The Code of Ethics is a guide to desired behaviour, but only conduct that contravenes the definition of professional misconduct is subject to disciplinary action. The code contains clauses specifying the seven typical duties of professional engineers to society, to clients and employers, to colleagues and subordinates, to the profession, and to themselves. Two clauses in the Code of Ethics that are unique to Ontario (but not enforcable under the Act) request the engineer to display his or her licence at the place of business and request moonlighting engineers to inform their clients that they are employed and to state any limitations on service that may result from this status. The definition of professional misconduct is more specific than the Code of Ethics, and includes a clause that seems to be unique to Ontario, which defines "permitting, counselling or assisting a person who is not a practitioner to engage in the practice of professional engineering" as a form of professional misconduct.

Ontario—Geoscience. The Ontario Code of Ethics for geoscientists is very comprehensive. It contains an introduction, eleven sections, and many subsections. The eleven sections, which encompass all of the duties typically found in Codes of Ethics (see the previous section), are titled as follows:

- Service and Human Welfare
- Public Understanding
- Business Ethics
- Duty to Others and the Environment
- Competence and Knowledge
- Signing and Sealing of Documents
- Faithful Agent or Trustee
- Conflict of Interest
- Overruling of Judgment
- Professional Advertising
- Breach of Code

A breach of Ontario's geoscience Code of Ethics is specifically defined as an act of professional misconduct. (The Code of Ethics for Ontario's engineers defines professional misconduct separately.)

Prince Edward Island. The engineering Code of Ethics for PEI is relatively clear and concise, consisting of an introduction and twenty-four clauses. All of these clauses are similar to those of other codes. Since geoscience is unregulated on PEI, that province has no code for geoscientists.

Quebec—Engineering. Quebec's Code of Ethics is the longest of any province (seventy-two clauses) and has a unique arrangement. The basic duties noted earlier are all represented in the code. Besides these, the code has clauses that define additional duties for engineers and that provide useful professional advice. The Quebec code is far too complex to summarize here but is included in its entirety in Appendix CD-B. Engineers from other provinces may well find it of interest and value.

Quebec—Geoscience. The Quebec Geologists Act came into force in August 2001. A Code of Ethics had not been published as this text goes to press.

Saskatchewan. Saskatchewan's Code of Ethics is one of the shortest (two main clauses and nine subclauses). Nevertheless, it includes all of the typical duties specified earlier, including the duty to report any possible illegal practices, professional incompetence, or professional misconduct to the Association, and the duty to be aware of the societal and environmental consequences of projects.

Yukon. Yukon's Code of Ethics for Engineers has twenty-nine clauses, which cover the seven basic duties noted in the earlier section. Since geoscience is unregulated in Yukon, no Code of Ethics exists for geoscientists.

The above overview shows that the codes for each province and territory are very similar, although not identical. They are useful guides to personal conduct. Adherence to a province's Code of Ethics is not voluntary, nor is it a lofty ideal that would be "nice but not essential" to achieve. Except in Ontario, the Code of Ethics is part of the Act that regulates engineering, and clear violations of the code may result in disciplinary action in the form of a reprimand, suspension, or expulsion from the profession (as described in a later chapter on disciplinary powers). In Ontario, the engineering Code of Ethics is a guide to desired behaviour, but only conduct that contravenes the definition of professional misconduct is subject to disciplinary action.

In addition, most engineering societies have developed their own Codes of Ethics. Several of these are provided in Appendix CD-C. The code promulgated by the National Society of Professional Engineers (NSPE), which is included in Appendix CD-D, is very similar to the provincial codes and has been endorsed by many engineering societies. Since engineering societies do not have regulatory powers such as those enjoyed by provincial and territorial Associations, serious infractions of Society codes can be punished only by expulsion from the Society (although this rarely happens).

A STRATEGY FOR SOLVING COMPLEX ETHICAL PROBLEMS

When an engineer is faced with a technical problem, there is usually more than one solution and the goal is to select the best or *optimum* solution. Similarly, in dealing with ethical problems, there may be several possible solutions and the goal is to determine the best solution from an ethical standpoint. However, finding the best solution to an ethical problem is sometimes difficult, because ethical theories that seem similar may generate solutions that are diametrically opposed to one another. An example from daily life is easy to provide: We are all taught to be truthful and not to tell lies. However, we have all, at times, told "white" lies. We might argue that we should consider the outcome, not the rule. A greater good may result by lying or denying an unpleasant truth rather than following the ethical rule: "Do not tell lies." This conflict of ethical theories is called an "ethical dilemma," which may require breaking one ethical code in order to satisfy another.

Ethical problems in engineering almost always require competing theories to be considered before making a decision. Ethical dilemmas often result, and the best course of action must then be chosen.

Most of the ethical problems that arise in everyday life are clear and simple and are solved by intuitive use of the ethical theories. Complex ethical problems can be much more challenging, because intuitive methods usually do not work for them. Thus, many people find themselves in a quandary when trying to decide where to start and how to proceed to an acceptable solution. Engineers should have an advantage in resolving ethical dilemmas, since problem-solving and decision-making techniques are a routine part of engineering. In fact, a reassuring similarity exists between ethical problem-solving and engineering design methods. Although a formal strategy is rarely needed for solving ethical problems, it is reassuring to know it exists and that it can be applied to complex cases.

The Engineering Design Process

Engineering designs do not spring fully developed into the mind of the designer. Creation requires a series of steps, including inspiration and deduction in the right order and at the right time. The process usually begins with a vaguely perceived need or problem and ends with a plan that satisfies the need. The steps in the design process, defined below, may have to be repeated many times to reach an optimum design:

1. Recognizing that a problem or need exists and gathering information about it.
2. Defining the problem to be solved or goal to be achieved.
3. Generating or proposing solutions or methods to achieve the goal (synthesis).
4. Evaluating benefits and costs of alternative solutions (analysis).
5. Selecting the best solution and, if possible, improving it (optimization).
6. Implementing the best solution.

The design process begins by recognizing a need in the marketplace, the community, or the factory. For example, people living in the suburbs may want a new expressway so that they can drive into the city more conveniently. It is important to gather data and define the problem precisely before committing resources to designing such an expressway. For example, in this case, once the need has been investigated thoroughly, it may become evident that a new subway would be more useful than an expressway.

The detail design work then begins for this new goal. Alternative routes for the subway are proposed. Alternative sources for materials are considered, as well as alternative approaches to buying or manufacturing the necessary subway cars. Each alternative is then analyzed to obtain the cost estimates for acquiring property and constructing the line. Each route must be analyzed for safety, convenience to riders, and inconvenience to nearby residents. Finally, the optimum or best solution—one that gives the maximum benefit at minimum cost—is selected. The design is approved for implementation, materials are ordered, and construction starts.

Applying the Design Process to Ethical Problems

The design process described above is actually a methodical problem-solving technique. We can develop a strategy for solving ethical problems based on this design process, as follows.

1. RECOGNIZE THE ETHICAL PROBLEM AND GATHER BASIC INFORMATION

Especially in the early stages, ethical problems may be poorly defined and difficult to recognize. Moreover, some engineers may simply be unaware that they are responsible, under the provincial Act and other laws, for ensuring the safety of their work and for minimizing any adverse impact on society and the environment. Recognizing that an ethical problem exists, and collecting the essential basic information, is therefore an important first step. At this stage we typically ask the questions that all newspaper reporters are taught to ask: Who, What, Where, When, Why, How? For example:

- *Who is involved?* (Identify the participants or "stakeholders.")
- *What type of harm or damage has occurred?* (or may potentially occur?)
- *Where, when, why, and how has this harm occurred?* (or may it potentially occur?)

For example, a manufacturing process, once thought to be harmless, may be suspected as the cause of cancer or toxicity in the workplace. This is an ethical (and perhaps a legal) problem, because workers cannot be placed at risk in this way. Information about the process and the toxic agent must be gathered and verified to confirm this suspicion. Once the information has been collected and examined, the proper course of action may become obvious. The remedial action, however, may be quite different from what was initially expected.

2. DEFINE THE ETHICAL PROBLEM

Once you have all the facts, you must define the ethical root of the problem. Sometimes this will be more difficult than other times. You must ask yourself, "What exactly is wrong?" Do any actions contravene the law or the Association's Code of Ethics? What is unfair about the situation? Once the problem is clearly defined, the solution may be obvious. The necessary action may be dictated by law, Codes of Ethics, or ethical theory. If so, you can skip directly to the implementation step. Often, however, the solution will not be obvious, and it will be your job to generate possible solutions.

3. GENERATE ALTERNATIVE SOLUTIONS (SYNTHESIS)

Having defined the ethical problem, you must now find a solution. This phase of problem solving requires creative thought and is usually difficult. Many solutions may be possible; that being so, the goal is to select the best or optimum solution. This optimum may be a compromise or it may be a modification of other solutions. The important thing is to avoid confronting an ethical dilemma in which you must choose between two undesirable or unacceptable courses of action. Many of the creative methods used for solving design problems (such as brainstorming) can also be applied to solving ethical problems.

For example, consider an engineer who receives a small cash gift from a contractor and is uncertain whether the gift is a favour to cover incidental expenses, or a bribe. This creates a dilemma. If the gift is simply a favour, then to return it would offend the donor. On the other hand, if it is a bribe, then to keep it would incur an obligation. Here, a creative compromise might be to give the cash to a charity with the knowledge of the donor (and preferably in the name of the donor); in this way, the engineer would avoid giving offence and also avoid incurring an obligation. Since other courses of action may exist, some ingenuity might well be needed.

4. EVALUATE ALTERNATIVE SOLUTIONS (ANALYSIS)

When two or more courses of action are possible and they conflict with one another, all courses must be analyzed and compared so as to foresee their consequences. Obviously, any solution (or decision) must satisfy the law and the Code of Ethics. The solution must also satisfy the ethical theories. These tests can be summarized in the form of questions:

- **Legality.** Does the solution satisfy the law and the Association's Code of Ethics?
- **Utilitarian ethics.** What benefits accrue as a result of this solution? For whom? What hardships are involved? Are the benefits and hardships fairly distributed?
- **Duty-based ethics.** Can this solution be applied to everyone equally? Can the solution be published and withstand the scrutiny of your colleagues and the public?
- **Rights-based ethics.** Does this solution respect the rights of all participants? Does anyone suffer harm?
- **Virtue-based ethics.** Does the solution develop or support moral virtues? Is the solution a golden mean between unacceptable extremes? In particular, does the solution maintain the ideals of the profession?

5. DECISION-MAKING AND OPTIMIZATION

If all of the above questions are satisfied, then you likely have the optimum solution. However, if the various ethical theories do not agree, or if no solution seems acceptable, you may have to go back to step 2 and verify that the problem has been properly defined.

In some situations you may find that the arguments for conflicting alternatives are so equally balanced that no solution is clearly superior. When this happens, you should pose the following questions:

- Was the problem defined properly and clearly?
- Has all the necessary information been gathered?
- Have I sought advice from the people concerned?

- Has an alternative or compromise solution been overlooked?
- Have all the consequences of each alternative choice been fully evaluated?
- Is a personal benefit or conflict of interest affecting my judgment?

If the above questions can be answered satisfactorily and there is still no optimum course of action, you should probably select the course of action that does not yield a benefit to you as the decision maker. If the choices are equally balanced and the possibility of personal benefit exists, this choice will ensure that the decision is seen by others as ethically correct.

If no personal benefit is involved, then you must select and follow the ethical theory that is considered most appropriate. Although this involves a personal value judgment, you as the decision maker will have a clear conscience.

6. IMPLEMENT THE SOLUTION

Implementing the decision is the final step. Although the appropriate action will vary from case to case, it is usually advisable to act quickly and decisively when ethical decisions are required, especially if health, safety, or someone's reputation is at stake.

SOME FINAL COMMENTS ON THIS STRATEGY

This strategy may seem rather formal, but it is important to apply it a few times. Having applied it rigorously to a few practical problems, readers will begin intuitively identifying the unique aspects of a given ethical problem and selecting the appropriate solution. The assignments provided later in this text are useful exercises in applying this technique. After a few tries, readers will likely find that their problem-solving is much faster and more intuitive.

This strategy is sometimes criticized for being weighted too heavily toward utilitarianism. The strategy tries to satisfy all of the four ethical theories (or at least most of them); in that sense, this search for relative balance is in itself a form of utilitarianism. However, this should be an advantage, because engineers and geoscientists are accustomed to utilitarianism; after all, it is at the root of most engineering applications.

Many other formal methods have been proposed for ethical problem-solving, as a search of the philosophy literature under the key words "applied ethics" (or the older term, "moral casuistry") would show. Casuistry began in the seventeenth century and was a "sincere effort to apply rigorous standards of critical argument to the questions of moral conduct."[3]

However, the term "casuistry" is now used in a very negative way to denigrate arguments as poorly formed or hypocritical. This older method is very different from the problem-solving technique discussed here.

Other formal methods for resolving ethical problems exist, but none has been universally recognized by philosophers. The above strategy is therefore proposed as merely one more tool—a tool, however, that engineers and geoscientists may find especially useful when addressing ethical problems.

Engineers and geoscientists, because of their methodical nature, practical experience, and interest in problem solving and experimentation, should be well prepared to solve ethical problems.

DISCUSSION TOPICS AND ASSIGNMENTS

(Additional assignments can be found in Appendix CD-E on the CD-ROM included with this text.)

1. Examine the Code of Ethics for your province or territory (see Appendix CD-B or obtain the code on the Internet by connecting with your provincial Association). Does the code include all seven of the basic duties described in this chapter? Do you agree that every clause is essential? Does it contain any unexpected clauses? Compare your provincial Code of Ethics with the codes for other jurisdictions (or technical societies); then write a brief report on how your code could be improved to make it clearer yet more comprehensive (or why your code is the clearest and the most comprehensive).

2. Some Codes of Ethics include duties that do not appear in other codes. Examine the Code of Ethics for your province or territory to see if it contains clauses that require engineers to:

   • advertise in a dignified, professional manner.
   • report infractions of the Code of Ethics to the registrar of the Association.
   • consider the difficulty of the work and the degree of responsibility when setting fees.
   • refuse to pay commissions or reduce fees to obtain engineering work.
   • obey the law of the land.

   Should all of the above duties be included in your Code of Ethics, or should they appear elsewhere in the Act, or are they not appropriate for inclusion in either the code or the Act? For each of the above clauses, explain and justify your answer.

3. Mill's utilitarianism is sometimes classified as an adaptation of another ethical theory, called "hedonism," which is related to "epicureanism" and the opposite of "stoicism." Look up the key principle(s) of these three ancient schools of philosophy. Then explain why utilitarianism is a widely accepted guide to current ethical conduct, especially in engineering and politics, whereas these other theories are not widely followed today. (Note: The terms hedonist, epicurean, and stoic as used today are related to these ancient schools but do not convey their principles precisely.)

4. Critics of Mill's utilitarianism say that this ethical theory is *relativistic*, since it depends on case-specific comparisons and could yield different results under slightly different conditions. Even worse, utilitarianism can

encourage narrowly practical and/or selfish attitudes because acts are seen as ethical if they increase pleasure or reduce pain. Conversely, Kant's duty-based ethics is based on *absolute* rules that encourage honesty, unselfishness, ethical character, and self-sacrifice, and Kant expects everyone to follow the same rules, thus creating an ethical consistency. Debate the benefits and disadvantages of these two theories of ethics. Compare and contrast the relative nature of utilitarianism with the absolute nature of duty-based ethics by providing at least one example where each would differ and where each would agree.

5. All Codes of Ethics require engineers to treat one another with courtesy and good faith; most of the codes also forbid statements that maliciously injure the reputation or business of another engineer. Such clauses obviously have laudable aims; however, how does one determine when a violation of courtesy or good faith has taken place? Are these clauses enforceable? Is the prohibition against malicious statements a reasonable infringement on freedom of speech? How does one distinguish between malicious statements and statements that are critical of a fellow engineer but true, and perhaps necessary for technical reasons? For example, if a fellow engineer makes a serious calculation error, it would usually be beneficial to point it out. Write a brief summary comparing the benefits of clauses forbidding malicious statements with the potential for abuse of such clauses. Define the terms "malicious," "good faith," and "enforceability." Explain the difference (or the boundary) between useful debate and malicious lies.

NOTES

1. J.T. Stevenson, "Philosophy," *The Canadian Encyclopedia* 2003, Copyright, Historica Foundation of Canada, Toronto. <www.canadianencyclopedia.ca> (September 5, 2003).
2. Government of Canada, *Constitution Act, 1982, Part I, Canadian Charter of Rights and Freedoms*, Government of Canada document, Ottawa. Available from <www.canadianencyclopedia.ca> (September 5, 2003).
3. J. Haldane, "Applied Ethics," in *The Blackwell Companion to Philosophy*, N. Bunnin and E.P. Tsui-James (eds.), Blackwell Publishing, Oxford, 2003, pp. 494–98.

Chapter 7
Ethical Problems in Professional Employment

Most professional engineers and geoscientists in industry are employees and, from time to time during their careers, will face conflicts of interest or similar ethical dilemmas. In this chapter we discuss the limits of an employer's authority and of an engineer's obligations. We also examine some of the ethical problems that may confront professional employees in industry.

EMPLOYER AUTHORITY AND EMPLOYEE DUTIES

When an engineer accepts an employment offer, this creates a contract: the engineer, as an employee, agrees to use his or her ability to achieve the employer's legitimate goals; the employer, under the same contract, takes on a duty to treat the engineer in a professional manner but also clearly acquires the authority to direct the engineer. The need for authority is obvious, especially in large organizations, in which lack of direction could lead to chaos and bankruptcy. The employer has *management authority* to direct the company's resources, whereas the engineer has *technical authority* to exercise the special knowledge and skills that he or she has acquired through university education and practical engineering experience. In a well-run organization the distinction between management and technical authority will be well defined. The individuals involved will show mutual respect and will cooperate to achieve the employer's goals.

In most corporations, engineers are responsible for evaluating the technical feasibility of various courses of action; management has the authority to decide which course will be followed. Usually, this process works well and benefits the engineer, the managers, the shareholders, and society in general. However, a professional engineer may at some point be directed to do something that he or she considers morally wrong. For example, the engineer may be asked to alter calculations to show that a gear train has a slightly greater factor of safety against overload, so that the gear train meets the client's specifications. Or an engineer (or geoscientist) may be asked to approve the improper disposal of industrial waste water that is known or suspected to contain toxic chemicals. In many well-publicized cases, the pressure on management to generate profits was converted into pressure on an engineer to act unethically. Problems like these are not as common as they once were, but they still arise. The result is always an ethical dilemma: At what point does the engineer's duty to follow his or her conscience override obligations to the employer?

To resolve an ethical dilemma, you should follow the procedure described in the previous chapter. There are, however, different categories or degrees of moral conflict. These are outlined below.

Illegal actions. An engineer may be asked to perform an act that is clearly contrary to the law. The law may be a criminal law, a civil or business law, or a regulation made under the authority of an act (such as an environmental regulation). Engineers are asked, perhaps most frequently, to infringe trademark, copyright, or industrial design legislation. In situations like this, the necessary action is clear: the engineer must advise the employer that the action is illegal and must resist any direction to break the law. No employer has the authority to direct any employee to break the law.

Actions contrary to the Code of Ethics. An engineer may be asked to perform an action that, while not clearly illegal, is a breach of the Code of Ethics of the provincial Association. The employer may be unaware of the code's legal significance and may simply need to be informed of it. Here, the engineer should advise the employer of the appropriate section of the code and should decline to take any action on the employer's request. An employer who is an engineer or geoscientist is equally bound to follow the code. (Many other professions also obey Codes of Ethics.)

The employee has solid legal reasons for insisting on ethical behaviour: an employer cannot direct a professional engineer to take action that clearly violates the Code of Ethics. This extends to situations where an employer directs an engineer to disobey established design codes or standards. However, there is a grey area where no established code or standard exists. Almost every Code of Ethics anticipates this situation and advises engineers to "present clearly to the employer the consequences to be expected if the professional engineering judgment of the engineer is overruled by non-technical authority." In these situations, the engineer should make the objections in writing, for possible future reference.

Actions contrary to the conscience of the engineer. An engineer may be asked to perform an act that, while not illegal and not clearly a violation of the Code of Ethics or established codes and standards, nevertheless contravenes the engineer's conscience. These are, of course, the most difficult situations. Here is where the decision-making procedure described in the previous chapter should be most useful. The engineer must consider all relevant information and define the ethical problem as clearly as possible. Precisely how does the required action offend the engineer's conscience? The engineer must try to see the problem from the employer's perspective. Alternative courses of action must be generated and examined in light of the basic ethical theories discussed earlier, and the optimum course of action should be selected.

Of course, ethical problems may have personal consequences. Refusal to follow an employer's directive may result in disciplinary action or dismissal. The possibility of dismissal, the consequences of unemployment, and the remedies for wrongful dismissal should always be considered.

PROFESSIONAL EMPLOYEE GUIDELINES

The Codes of Ethics discussed in the previous chapter (and reproduced in Appendix CD-B) are useful guides to conduct, but they do not, of course, cover all problems in the professional workplace. In particular, Codes of Ethics do not address the problems that typically arise when hiring or terminating professional employees, or when establishing adequate salaries, benefits, hours of work, and so on. These issues are beyond the responsibility and authority of the provincial Associations. Some of these issues are mentioned in provincial labour laws. However, labour laws usually set minimum values, which rarely apply to professional employees.

Although advocacy groups such as the Canadian Society of Professional Engineers (CSPE) have been established to assist Canadian engineers, they do not yet provide extensive guidance of the sort offered by the corresponding American organization, the National Society of Professional Engineers (NSPE). (CSPE is discussed in more detail elsewhere in this text.)

In 1973, NSPE published *Guidelines to Professional Employment for Engineers and Scientists*; these guidelines still serve as a model for professional employment.[1] The crisis that stimulated their development was the U.S. government's cutbacks on aerospace expenditures in the late 1960s, including cancellation of the proposed supersonic transport aircraft (SST). Many engineers and scientists were suddenly and unexpectedly unemployed, and suffered severe financial hardship in the following years.

The NSPE guidelines, perhaps refreshingly, recognize that most Codes of Ethics have been developed for engineers in private practice and thus sometimes do not apply to engineers who are employees. Moreover, the guidelines recognize that employers play a major role in creating workplaces that can attract (and keep) professional employees and enable them to reach their full potential. The guidelines are voluntary and do not have any legally binding authority in either the United States or Canada; nevertheless, they are supported by at least twenty-six American or international technical societies. (The guidelines are provided in Appendix CD-D on the CD-ROM included with this text.)

The guidelines contain more than sixty detailed clauses, which are divided into four sections: "Recruitment," "Employment," "Professional Development," and "Termination." Nine of the clauses closely resemble the Codes of Ethics established by the provincial Associations. The other clauses discuss very specific, practical employment problems and are directed at both employers and employees.

The purpose of the guidelines is to establish a workplace climate that is based on "ethical practices, co-operation, mutual respect, and fair treatment" and that is productive for both employers and employees. Additional objectives are to safeguard the public, to encourage professionalism and professional growth, and to combat discrimination based on age, race, religion, political affiliation, or sex. These objectives are promoted through specific rules, organized as follows:

Recruitment. The procedures for hiring professional employees are addressed by four basic rules for applicants, which relate to the presentation of qualifications, attendance at interviews, confidentiality obligations, and acceptance of travel reimbursement and job offers. Eight rules are set for employers; these relate to the payment of applicants' expenses, clear descriptions of the job and the employment conditions, fairness in hiring, and preparation of written offers of employment.

Employment. This section suggests terms of employment for professionals. The guidelines, of course, require employees to obey all applicable laws and accepted ethical and professional practices, but they also go further in defining specific rules. Nine of these rules address the employee's responsibilities and contain many clauses—such as having "due regard for the health, safety, and welfare of the public"—that are very similar to the clauses found in most provincial Codes of Ethics. (In fact, these nine clauses could easily pass for a provincial Code of Ethics.)

This section also has twenty-one clauses relating to the duties of employers. These clauses consider important basic topics such as salary policy, benefits, levels of responsibility, performance evaluation, promotions, titles, merit, patents, pensions, overtime, transfers, and insurance. For example, one clause specifically recommends that dual ladders be established for promoting and compensating engineering and managerial employees. Another clause encourages employers to defend professional employees in any lawsuits that might arise as a result of their authorized activities. Still another discourages company policies that either require or forbid professional employees from joining a labour union.

Professional development. This section assigns responsibility for maintaining professional competence to both employees and employers. The employee is expected to set personal goals and show initiative; the employer is expected to promote this by providing challenging job assignments. The four rules for employees include the expectation that professional employees will seek professional registration (licensing) as soon as they are eligible and will join technical societies and participate in their activities. The four rules for employers ask them to provide support for employees who are trying to maintain their competence. This support could involve leaves of absence for courses or study, and permission to publish papers or present results at technical meetings.

Termination. Of course, even professional jobs terminate eventually. This section provides two rules that instruct employees to provide adequate notice and to maintain the confidentiality of the employer's proprietary materials and trade secrets. The five rules for employers call for the fair disclosure of reasons for termination, adequate notice (or payment in lieu of notice), assistance for terminated employees to find new employment, and avoidance of coerced retirements.

Although NSPE's guidelines do not have any legal authority, they are a useful aid for professional employees and give more specific advice than the provincial Codes of Ethics or the present activities of Canadian advocacy groups.

PROFESSIONAL EMPLOYEES AND LABOUR UNIONS

Professional employees have the right to negotiate pay scales, hours of work, and other conditions of employment. Sometimes an employer refuses to negotiate these basic conditions. The engineer is then faced with a difficult choice: resign, or take part in collective action against the employer. This creates a moral dilemma: the engineer has an obligation to the employer, but also has an obligation to himself or herself, and to the engineering profession as a whole, not to accept unprofessional working conditions or inadequate pay.

In both Canada and the United States, it was established long ago that engineers have the right to take collective action and even to form or join unions. Every province has a labour board and a labour relations act that can provide guidance and assist engineers who are contemplating collective action. Engineers who are part of company management are not permitted to organize collectively, and it would be illogical for them to do so. However, engineering employees face no such prohibition.

In every province there is labour legislation that establishes the right of employees to form unions and that obliges employers to negotiate in good faith. To this extent, unionization can be effective in situations where every other route is closed. Once employees formally ask the provincial labour board to recognize their union, the employer has no choice but to negotiate with the employees. However, the negotiations will usually take time. Furthermore, starting a union almost always results in confrontation; it also generates a bureaucracy and a plethora of formal procedures. So in most cases, unionization should be a last resort. Engineers should try to resolve problems with employers through negotiated contracts that do not involve formal unionization, for the simple reason that it usually will be less work.

Engineering employees who have decided that collective action is necessary may be tempted to join an existing staff or labour union within their company or industry. However, they are usually better advised to form a collective group composed entirely of engineers, if possible. This guarantees that the group's goals will always be consistent with the wishes of the engineers. Otherwise, there is a risk that the engineers will be a minority in the union and perhaps obliged to support labour action that is not in their best interests.

When engineers discuss unionization, a heated debate often results. Engineers are professionals and deserve to be treated as such. But in the modern world, most engineers are also employees, so when policies for negotiating pay or other terms and conditions of employment are absent, they are entitled to take the same collective action as any other employees. When engineers resort to collective action, they do so very reluctantly. Such actions do not generally indicate a failure on their part to act ethically; it is more likely to indicate that the employer has failed to establish fair policies and negotiating procedures.

UNETHICAL MANAGERS AND *WHISTLEBLOWING*

In rare instances, a company's management may appear to be acting unethically. An engineer who finds evidence of dishonesty, fraud, misrepresentation, destructive environmental practices, or similar unethical acts should immediately inform management of the problem and suggest remedial action. A dilemma arises for the engineer if management fails to respond to arguments based on ethics. The engineer's duty to the employer is thereby placed in direct conflict with the duty to the public welfare, which should be considered paramount, as all Codes of Ethics clearly state.

The engineer will then have to assess the situation carefully and balance the gravity of the risk to public welfare against the likelihood of overcoming management's resistance to remedial action. The engineer should act quickly on the problem, since management may interpret any delay as tantamount to agreeing with or condoning the unethical action. The engineer is generally faced with three possible courses of action: correct the problem, blow the whistle, or resign in protest.

Correct the problem. The engineer could continue to work for the company while trying to correct the problem and/or change company policy. This is the obvious first course of action. In fact, the engineer has an obligation under the Code of Ethics to inform the employer of the potential consequences if the engineer's judgment is disregarded or overruled. Therefore, the engineer must ensure that this obligation is satisfied before taking any other steps. Moreover, working to correct the problem is usually the most effective course of action, especially if the dishonest actions are minor and/or management is open to improvement and change.

Blow the whistle. The engineer could continue to work for the company while alerting external regulatory agencies that the company is acting dishonestly. This is commonly called *whistleblowing*, and it is an unpleasant and usually unfriendly act. However, in those rare situations where the engineer has full knowledge (preferably documented) of a clear and serious hazard to the public, where supervisors and management have refused to take action to correct the problem, and where attempts to correct the situation have failed, whistleblowing may be the last resort. This course of action is not recommended until all other routes have been tried. Whistleblowing and the problems associated with it are discussed in a later chapter.

Resign in protest. The engineer could resign in protest. This course of action may be necessary in serious cases where complicity may be suspected if the engineer remains with the company. The engineer should always consult a lawyer before resigning. There may be grounds for considering such a forced resignation as equivalent to wrongful dismissal.

INTRODUCTION TO CASE STUDIES

Chapters 7 to 13 provide many case studies for practice in ethical problem-solving. Each case presents the reader with a moral dilemma and asks for a decision that is substantiated by a Code of Ethics or by basic ethical concepts. Try to determine your own answer to each case before you read the author's suggested solution. A few words of caution are appropriate:

- These case studies involve both ethics and professional practice, and the suggested answers are based on the problem-solving technique described in Chapter 6. This technique aims to seek unanimity (or at least majority agreement) among four well-accepted ethical theories. However, since the technique seeks a relative balance, it leans heavily toward utilitarian ethics.

- Case studies are somewhat artificial, because the reader is given only a brief summary and cannot interview the participants to gather more information. For this reason, the opportunity to propose really creative alternatives is limited. In real-world cases you would seek and get all of the necessary facts. In spite of this obvious limitation, case studies are useful for developing skills in ethical decision-making.

- All of the case studies are based on real cases or reports of real cases, but all names and many details have been changed to provide complete anonymity. Any similarity to real people in comparable situations is entirely coincidental.

CASE STUDY 7.1—WOULD YOU APPLY FOR THIS JOB?

STATEMENT OF THE PROBLEM

Ralph X, a computer engineering student, is in his final year and is approaching graduation. He has applied for several jobs but has received no job offers. The placement office tells him that a good engineering job opportunity exists at a nearby tobacco company. Ralph reads the job description and reviews the company's profile through the Internet. The company manufactures cigarettes using recently built high-speed automated machinery, and the advertised job involves digital control. In fact, the first assignment would be to help optimize the control software to obtain maximum productivity from the new manufacturing machinery.

Ralph is delighted at the job description. It is precisely the type of job that he wants; furthermore, he would acquire good engineering experience and he could contribute a great deal of knowledge and enthusiasm. The job placement office says that no qualified graduates have applied for the job, so he could get an immediate interview if he is interested. As a bonus, the salary is very attractive, and Ralph desperately needs a job to pay his large student debt and to repay loans from his mother.

However, the thought of working for a tobacco company bothers Ralph's conscience. He has always hated cigarettes. He is a nonsmoker, and not just because cigarettes are expensive and addictive. His father, a heavy smoker, died of lung cancer about ten years earlier, when Ralph was twelve. This was a tragic episode in Ralph's life, and his father's death was attributed, at least in part, to smoking. Ralph really needs a job, but he would feel uneasy about manufacturing cigarettes. Also, how could he face his mother? She suffered financially and emotionally after his father's death, and he knows that she would be very sad to see Ralph working for a tobacco company.

QUESTION

Should Ralph apply for the engineering job at the tobacco company's cigarette factory?

AUTHOR'S SUGGESTED SOLUTION

Let us apply the problem-solving steps discussed in Chapter 6.

Recognize the problem and gather information. Ralph has already recognized that working for a cigarette manufacturer may bother his conscience and that he faces an ethical dilemma. To address Ralph's dilemma, we should first examine legality, as suggested earlier in this chapter. Activities that contravene the law or the Code of Ethics are clearly unethical. However, it is certainly legal to manufacture cigarettes. Therefore, Ralph would not be breaking the law or the Code of Ethics by manufacturing a legal product. The ethical dilemma must therefore be resolved by considering Ralph's personal attitudes and ethical standards.

Ralph has always hated cigarettes, and his father's early death has made him antagonistic to the tobacco industry. He needs a job, but he could continue his job search if he had to. Although he has student loans and owes money to his mother, no urgency exists. The banks are competing to offer him a line of credit as soon as he obtains his engineering degree, so he could survive financially for another six months or so.

Also, we should consider how society rates the tobacco industry, ethically. Cigarette manufacturers—like distilleries, breweries, wineries, and casinos—make products that can be addictive. These companies are extremely profitable, but a utilitarian analysis would show their ethical value to be marginal. Their products give a modicum of pleasure to many people but impose misery on those who become addicted. For example, many people (the author included) enjoy a glass of wine with a meal, and the taxes on these luxuries help support provincial governments. Do these benefits equal or exceed the misery of the small number of homeless winos who spend their lives in poverty and despair? Past experiments with alcohol and gambling prohibitions in the 1900s were unsuccessful, so in Canada (as in most Western countries), we have chosen to balance the ethical equation by controlling these industries more stringently and by alleviating the misery of addiction through welfare and social programs.

Define the problem. These facts give us a better picture. Ralph's dilemma can be summarized as follows: He needs the job and would benefit professionally from it. But if he takes the job, he may face emotional distress because of the product he is manufacturing. The product is legal (although ethically marginal). If he rejects the job, he will have to continue his job search and he will have to borrow money, but he could likely get a credit line from a bank.

Generate alternative solutions. Ralph should not make a final decision without considering alternatives. At least three acceptable courses of action are possible:

- Ralph could reject the job outright and keep looking for a better job. If necessary, he could borrow money.
- Ralph could arrange an interview to get more information, and delay the decision until after the interview. The interviewer might be able to place him in a job (such as a research job) that is unrelated to manufacturing or promoting cigarettes. But if Ralph is certain he will not take the job, it would be unethical for him to waste the interviewer's time.
- Ralph could apply for the job and accept it if it is offered, then grit his teeth and try to ignore any negative ethical issues.

A fourth alternative also exists: Ralph could apply and accept the job, and then work to undermine the company from within. This alternative is not discussed further, because it is clearly unethical. Besides the lies Ralph would have to tell to get the job, such a scheme is ethically equivalent to plotting to steal the company's assets or trade secrets.

Evaluate the alternatives. The above alternatives could be evaluated by applying the ethical theories discussed in the previous chapter:

- Mill's utilitarianism would direct Ralph to balance the benefits of the job against the personal discomfort that he would feel working for an organization that helped hasten his father's death.
- Kant would remind the reader that human beings should always be treated as an end or as a goal, and never as a means of achieving some other goal. Ralph should not sacrifice his career or self-respect to obtain a job. A job that is held in disrespect by friends or colleagues, or that would require him to deny his true self, could be demoralizing and would not be conducive to a productive career.
- Aristotle's concepts of developing virtue and seeking a golden mean between the extremes of excess and deficiency are not very helpful in Ralph's situation. There is little virtue in manufacturing cigarettes. However, Ralph might see a research job in the tobacco company as the golden mean between unemployment and an unacceptable job manufacturing cigarettes.
- Locke's rights-based ethics would likely contribute little to resolving Ralph's dilemma. Cigarettes themselves are not illegal, so the company has the right to manufacture them. Moreover, people have the right to smoke cigarettes, even if smoking is harmful, provided that they do not interfere with the rights of other people to clean air.

Decision making. Given the above analysis, what conclusion would you reach? You may weigh the utilitarian benefits differently, or you may be able to suggest more alternatives. It may be difficult to select a single *right* answer to this question; however, a decision made in haste or strictly for personal gain would almost certainly be a *wrong* one. The author concludes that the analysis favours rejecting the job opportunity. Some readers may disagree. The author would respect any solution based on gathering information and weighing the ethical nature of the alternatives.

CASE STUDY 7.2—ACCEPTING A JOB OFFER

STATEMENT OF THE PROBLEM

At a time of economic recession, an electrical engineering student, Joan Furlong, is nearing graduation. She is seeking a permanent position with an electronics company in digital circuit design and analysis. She is interviewed by several electronics and power companies. Her résumé clearly states her qualifications, job objective, and interests, which are mainly in digital circuit design. With her graduation day approaching, she receives an offer from the Algonquin Power Company for a job working on scheduling substation maintenance. The salary is good, so she writes back immediately and accepts. About two weeks later she receives a letter from Ace Microelectronics offering her a position on a new project in digital circuit design. The salary is roughly equal to the Algonquin offer, although the employment may end when the project does. Furlong is uncertain what to do. She sincerely wants to work in circuit design and not in scheduling maintenance. She identifies three possible courses of action.

Honour the Algonquin agreement. She could honour the agreement from Algonquin and decline the offer from Ace. She would thank Ace but tell them she has already accepted an offer, although she might be able to join them on a later project in a few years' time, once she has satisfied her obligation to Algonquin.

Revoke the Algonquin agreement. She could write to Algonquin, tell them her plans have changed, revoke her earlier agreement, and apologize for the inconvenience. She is aware of the Code of Ethics of her provincial Association, but she is not yet a member of the Association and does not feel bound to follow the code. Although she is a student member of the Institute of Electrical and Electronics Engineers (IEEE), the IEEE code does not seem to have any clause that pertains to this particular situation.

Revoke the Algonquin agreement but offer reimbursement. She could write to Algonquin as above, but offer to reimburse them for the recruitment expenses they have paid on her behalf.

QUESTION

From the ethical perspective, which of the above three options is best?

AUTHOR'S RECOMMENDED SOLUTION

This problem is not specifically addressed in provincial Codes of Ethics, although every code states or implies that an engineer has an obligation to act in "good faith" or with "good will" toward clients, employees, and employers. However, NSPE's guidelines discuss this problem specifically. The third clause of "Part 1—Recruitment" advises professional employees as follows: "Having accepted an employment offer, applicants are ethically obligated to honour the commitment unless they are formally released after giving adequate notice of intent."[2] When we apply the ethical theories to this case, all of them clearly support the NSPE directive. Let us examine the options in the order in which they were presented.

Honour the Algonquin agreement. Clearly, this decision is the most ethical one. All of the ethical theories discussed in the previous chapter would support this conclusion. Honouring agreements is a virtue, and utilitarians would argue that honouring agreements—even when they are not ideal— permits society to operate smoothly, benefiting everyone. Kant's philosophy would specifically tell Furlong that promises must be honoured: she made a promise to Algonquin and has a duty to fulfil it. Moreover, she may grow to enjoy the Algonquin job and end up feeling pleased that she followed her conscience. Furlong's personal rights are not an issue, because she freely accepted the offer. However, Algonquin has a legal right to enforce her acceptance.

Revoke the Algonquin agreement. Revoking or ignoring the obligation to Algonquin is clearly unethical. As explained earlier, Furlong has an obligation to Algonquin that cannot be erased with a simple apology. The company has probably sent rejection letters to the other applicants for the position and may stand to lose more than just its recruitment costs if its maintenance program is delayed. (The grave consequences of failing to recruit personnel are rarely understood by people outside the company.) The argument that Furlong is not yet bound by the relevant Code of Ethics is spurious, legalistic, and unacceptable as a justification for her actions.

Revoke the Algonquin agreement, but offer reimbursement. Sometimes in life we make serious mistakes, and the only way to remedy these mistakes is to admit them, apologize, and offer restitution. Even marriage—a vow at least as sacred as accepting a job offer—has provisions for divorce. If on reflection Furlong genuinely believes that accepting the Algonquin offer was a very serious error in judgment, she should admit her mistake and offer restitution. Algonquin will likely request the return of expenses paid during the recruitment (at least), and of course Furlong will never receive an offer from them again. So Furlong will pay a price for realigning her career path. This is not an ideal course of action; however, restitution acknowledges her ethical duty and ensures that the person who benefits most from this course of action (Furlong) compensates Algonquin for at least part of its losses.

In summary, the author leans strongly in favour of the first alternative—an offer of employment (like any legal agreement) should never be accepted unless it can be honoured. Furlong has a duty to fufil it. Moreover, she may grow to enjoy the job at Algonquin Power, and may be pleased that she followed her conscience. This is clearly the most ethical decision.

CASE STUDY 7.3— PART-TIME EMPLOYMENT (MOONLIGHTING)

STATEMENT OF THE PROBLEM

Philip Forte is a licensed professional engineer who has worked for Federal Structural Design for ten years. Unfortunately, for reasons that are not clear to either Forte or his employer, the company has not had many large contracts, and Forte's salary is very low. For the past ten years of his employment, his pay raises have rarely exceeded cost-of-living increases. As a result, he has been forced to take on extra employment in his spare time. He secretly brings the work to his office in the evening, where he uses the CAD system on his computer workstation. He is careful to pay for any office supplies or photocopying out of his own pocket, and he argues that the computer would be sitting idle in the evening anyway, so his employer is suffering no loss. In fact, Forte contends that his evening work is benefiting his employer, since it enables him to keep working for Federal Structural Design in spite of his low salary.

QUESTION

Is it ethical for Forte to carry on his part-time employment in this manner?

AUTHOR'S RECOMMENDED SOLUTION

Many engineers "moonlight" (that is, accept part-time employment)—so many, in fact, that guidelines for it have evolved over the years. In general, it is not unethical for an employee to work for more than one employer, although late-night work may require determination and stamina. However, every Code of Ethics requires the engineering employee to show his or her employer fairness and loyalty. This means that

- the employee should not compete for the employer's contracts,
- the time and effort spent on moonlighting should not reduce the employee's efficiency during the usual workday, and
- the employer must be fully informed of the moonlighting, in order to verify the situation.

It is clear that Forte is not acting ethically in this situation, since he has not informed his employer of the part-time employment. Thus, the employer is unable to judge whether the part-time work is competing or conflicting with Forte's full-time job. It is curious that Forte has remained with his employer for ten years with no promotion or significant increase in salary, for this implies that other, unstated factors are influencing the case. Perhaps Forte is exploiting the employer's facilities and contacts to generate a large part-time income, or perhaps the employer is exploiting Forte by forcing him to carry two jobs to survive financially. In any event, the engineer's secrecy about his part-time employment is clearly unethical. Furthermore, Forte is ignoring his obligation to the profession to insist on adequate pay for professional work (a clause in almost every Code of Ethics).

Moreover, if Forte is offering services to the general public when he moonlights, then in some provinces (such as Ontario) he must obtain a Certificate of Authorization, and the question of liability insurance must be addressed. Forte may be placing his career at more risk than he realizes.

CASE STUDY 7.4—ENGINEERS AS MEMBERS OF LABOUR UNIONS

STATEMENT OF THE PROBLEM

Jeanne Giroux is a licensed professional engineer working for Acme Automotive Manufacturing, a medium-size company that makes parts for cars and trucks. She works in the design engineering office (four engineers, six designers), where she supervises modifications to parts. This work involves stress analysis, detail drawing, and prototype testing—a wide range of duties. In her job, she has frequent contact with the machinists and other tradespeople in the manufacturing plant. She has noticed that the shop union is highly effective in negotiating terms and conditions of employment.

The union steward informs her that the employer is required by law to bargain in good faith with the union and that there are procedures for mediation and arbitration in case of stalemates in negotiations. In contrast, the design and engineering staff have enjoyed minimal pay raises in recent years, are limited to short vacations, and are required to work overtime (and occasional Saturdays) without additional pay during times of crisis. Although all four of the company's engineers are licensed, the provincial Association cannot assist them in negotiating with the employer. The Association has, however, provided them with a survey of engineering salaries. Giroux notes that the mean salary for the four staff engineers is almost 20 percent below the median salary for the appropriate group in the survey. The company administration manual does not mention procedures for staff pay raises.

The company's sales staff (none of whom are engineers) are paid partly on a commission basis but always receive much higher pay than the engineers. The union steward has told Giroux that the design and engineering staff can be included in the bargaining group as soon as at least six of the ten employees sign application forms. Giroux believes that all ten would probably join, if asked.

QUESTION

Is it ethical for engineers to join the shop labour union?

AUTHOR'S RECOMMENDED SOLUTION

Yes, it is ethical and legal for engineers to join labour unions, provided the engineers do not exert managerial control in the company. From the details of this case, it would appear that the design engineering staff are not considered part of management. However, although it may be ethical and legal to join the union, it may not be appropriate; the labour relations acts in some provinces specifically permit professional engineers to form bargaining units composed entirely of professional engineers. Moreover, it is not essential to have the support of an established union in order to form a bargaining unit. In most provinces this can be done directly through the provincial government's labour board. A majority of members of the group can request certification as a bargaining unit; the elected representatives then negotiate contracts for the entire group.

Establishing a union (a process referred to as "certification") requires a great deal of time, effort, organization, and paperwork. Small groups might find it more appropriate to make management aware of their dissatisfaction and to suggest three possible solutions:

- Negotiate individual contracts (a good choice for professional employees).
- Negotiate a collective contract (a good alternative to certification).
- Negotiate a company salary policy for the design engineering staff (an alternative that the employer would likely find more palatable, because it gives the company more control).

If the employer is unwilling to discuss any of these alternatives, the engineers could consider unionization as a last resort.

CASE STUDY 7.5—FALSE OR MISLEADING ENGINEERING DATA IN ADVERTISING

STATEMENT OF THE PROBLEM

Audrey Adams is a licensed mechanical engineer with marine experience working for a manufacturer of fibreglass pleasure boats. She has conducted buoyancy tests on all the boats manufactured by the company and has rated the hull capacity of each according to the procedure specified by Transport Canada. She notices that in the company's sales literature, a boat hull rated for a maximum of five people is consistently shown in photographs with six people on board. The sales literature seems to be otherwise correct. Adams knows that with six people on board, the boat would be safe in still water but could be flooded and sink in rough water. She believes that the sales literature is misleading and possibly hazardous.

QUESTION

What action should Adams take?

AUTHOR'S RECOMMENDED SOLUTION

The Code of Ethics for every provincial Association states that the welfare of the general public must be considered most important. In this situation, there is a potential hazard to the public and Adams has an ethical duty to take immediate action to reduce or eliminate that hazard.

Adams's first step should be to inform the engineering manager about the problem. Most likely, she would do this by writing an internal memorandum describing the errors in the sales literature. If these are simple errors or oversights by the sales personnel, they will be easy to rectify. Most companies are honest and would take immediate action to correct the sales literature.

In rare instances the company's management may be dishonest. For example, if Adams, while investigating the problem, were to discover that test results had been altered or that the manufactured boats were being sold with incorrect capacities stamped on the serial nameplates, the problem would be much more serious. In this situation, the company's management would be guilty of misrepresentation, which is possibly a criminal act. If a serious accident were to occur (such as an overloaded boat sinking, resulting in loss of life), the erroneous literature and incorrect capacities would become public knowledge, and the engineer would probably be investigated for possible unethical acts, incompetence, or collusion with management in the misrepresentation.

An engineer who discovers that the employer is dishonest must quickly dissociate himself or herself from any unethical activity and try to rectify its effects. In rare cases, that engineer must blow the whistle on the employer or resign in protest. Otherwise, the engineer risks being labelled as a participant in the unethical activity. Fortunately, extreme action is rarely necessary.

CASE STUDY 7.6—DISCLOSING PROPRIETARY INFORMATION

STATEMENT OF THE PROBLEM

An aeronautical research engineer from Company A conducts tests of a new aircraft tail assembly configuration in his company's wind tunnel and finds that in certain circumstances, devastating vibrations could develop in the configuration, leading to destruction of the aircraft. Later, at a professional meeting, Company A's engineer hears an engineer from Company B, a competitor, describe a tail assembly configuration for one of Company B's new aircraft that runs the risk of producing the same destructive vibrations that Company A's engineer has discovered in his own tests. Presumably there is an obligation—as a matter of both ethics and law—to maintain company confidentiality regarding Company A's proprietary knowledge. But at the same time, engineers have a duty to safeguard public safety and welfare. If the engineer from Company A remains silent, Company B may not discover the destructive vibrations until a dreadful crash occurs that perhaps kills many people. As an added complication, assume that the simple statement by Company A's engineer that the tail design "will not work" might not be believed by Company B's engineers.

QUESTION

What would you do if you were the engineer from Company A?

AUTHOR'S RECOMMENDED SOLUTION

On the one hand, under the Code of Ethics you have an obligation to your employer to maintain the confidentiality of proprietary information. Your company has paid a lot of money to test the tail assembly, and it would not be fair to your employer to turn this information over to Company B. Moreover, if Company B is as diligent in its testing and analysis as Company A, it will discover the vibration problem in due course, and you need take no action.

On the other hand, you also have an obligation under the Code of Ethics to consider the welfare of society as paramount and to report a condition that endangers public safety. If Company B should fail to discover the design flaw and a prototype aircraft later crashes as a result of this flaw, your lack of adherence to the Code of Ethics will be painfully clear.

You are therefore confronted with an ethical dilemma that involves two undesirable courses of action. In this situation, the duty to society must prevail over the loss of advantage, and Company B must be informed of the potential problem.

You should, of course, notify your employer before contacting Company B. You should also keep in mind the objective of this exchange of information. It would be unethical to convey the information in a way that harmed Company B's reputation. However, it is not your duty to release detailed data or to save money for your competitor. Your goal is merely to ensure that public safety is not endangered by Company B's defective design. Therefore, in consultation with your employer, you should determine what minimum information would achieve this purpose, and convey it in a direct and unambiguous way.

CASE STUDY 7.7— ALTERED PLANS AND INADEQUATE SUPERVISION

STATEMENT OF THE PROBLEM

Assume that you are a licensed civil (structural) engineer employed by a large retail company. You are responsible for designing and supervising the construction of new stores, as well as renovations or additions to existing stores.

Your boss, an architect, designs the floor plans to maximize their attractiveness and sales potential. The construction is either tendered out for contract or—for smaller jobs—completed by another branch of the company, which is directed by a competent manager. This manager was trained as a frame carpenter and was later certified as a technologist, but he is neither an architect nor an engineer.

After several years with the company, you have successfully participated in the design of two new, large stores and administered their construction by contractors. You have also developed plans for about twenty small projects, which have been carried out by the manager. At a certain point you discover that this manager has sometimes been deviating from your plans on the smaller jobs. In past technical discussions with him, you were always left with the impression that he had agreed to comply with your instructions. Now you realize that he has consistently been altering your plans—for example, by changing column spacings slightly, substituting different structural steel shapes, and using salvaged structural steel instead of new sections. You are genuinely uncertain of the extent of the changes; you are also uncertain about the factors of safety in the "as built" structures.

You confront the manager, somewhat angrily. He responds that he was simply "using cost-saving measures" and that "the changes didn't affect strength." Moreover, he insists that he has "twenty-five years of experience in project management" and that he has had "very few problems in the past." You report these facts to your boss, the architect, who is "simply too busy with project deadlines to worry about personnel problems" and she instructs you to "sort it out yourself." You realize that you are partly to blame for this problem because you were responsible for ensuring that your plans were being followed. The manager is an employee of your company, so you trusted him more than you would have trusted an external contractor, and you failed to inspect these smaller projects and prepare "as built" drawings.

QUESTION

What action should you take?

AUTHOR'S RECOMMENDED SOLUTION

Under every engineering Code of Ethics, the engineer's paramount responsibility is to ensure the safety of the public. The manager's assertion that "the changes did not affect strength" is not adequate, since he is not qualified to make that judgment. Clearly, if you are uncertain about the factors of safety in the final design, you must take some action on this problem. Regardless of the pressure of deadlines or the costs involved, you must convince your boss, the architect, that this problem must be remedied. Under the Code of Ethics for architects, she has an obligation to respect your expertise in determining structural strength.

You must determine what action to take, and the urgency. In order to guarantee safety, you clearly must review all of the projects, as quickly as possible, to determine the actual factors of safety in the as-built structures. Strengths will have to be recalculated where necessary and will have to be filed with the original design calculations to document the as-built strength. Wherever the as-built strength is marginal or inadequate, immediate structural repairs will have to be taken to increase strength. In the future, you will have to monitor the smaller projects more closely. Two incidents involving American structural projects are relevant to this case study—

Hyatt Regency Hotel. In 1981 in Kansas City, Missouri, two concrete walkways spanning the lobby of the Hyatt Regency hotel suffered a collapse that killed 114 people and injured more than 200 others. This was the deadliest structural failure in North America since the collapse of the Quebec Bridge in 1907 (discussed in Chapter 1). The cause of the Hyatt Regency collapse was eventually traced to a minor change in the design of the fittings supporting the upper walkway. Both walkways were to be hung from steel rods roughly 60 feet (18 m) long. However, such rod lengths are difficult to transport, so the fabricator suggested that each long steel rod be replaced by two shorter steel rods, both connected to the fitting that supported the upper walkway. This change seemed minor; in fact, however, it doubled the load on the fitting—a point that would have been obvious had anyone drawn a free-body diagram of the fitting. The design engineer did not recall seeing the change request, but the engineer's seal appeared on the revised drawings. The collapse was a tragedy for the victims. It was also costly for the insurance companies and for the engineers who lost their licences and were forced into bankruptcy.[3]

LeMessurier and the Citicorp Tower. William LeMessurier was hired as a structural consultant to the Citibank Tower in New York City, a fifty-nine-storey building with a structural steel frame. After the building was completed, LeMessurier had occasion to recheck his strength calculations, at which point he realized that when strong winds blew from a certain direction, the forces on the bolted joints would be significantly greater than he had earlier calculated. He chose to face his error directly. Wind tunnel tests confirmed his fears, and he revealed his concerns to the building designers and to the client, Citicorp. Citicorp agreed to immediate action, and repairs were made in record time before the building risked being demolished by a hurricane. Surprisingly, even though the repairs cost millions of dollars, LeMessurier was shielded from most of the financial loss and was highly praised for his prompt, ethical actions.

These two cases illustrate that minor changes can sometimes be critical. When faced with the possibility of structural flaws, the ethical response is to address the problem directly, determine the strength accurately, and take all necessary actions to ensure safety.

CASE HISTORY 7.1

THE *CHALLENGER* AND *COLUMBIA* SPACE SHUTTLE DISASTERS

The *Challenger* space shuttle explosion in 1986 is probably the most famous engineering tragedy of all time. Millions of people were watching the televised launch when *Challenger* exploded, resulting in the loss of seven lives, immense costs, and severe problems for the American space program. A second disaster happened in 2002 when the *Columbia* space shuttle burned up during re-entry, killing another seven astronauts. Investigations of these tragic events disclosed that even in high-profile, professional environments, engineers often face important ethical decisions.

Introduction

On January 28, 1986, the U.S. National Aeronautics and Space Administration (NASA) launched the space shuttle *Challenger* at Cape Canaveral, Florida. The launch had been delayed by bad weather, and since part of the cargo was a satellite that required a specific launch window, there was some concern

The Challenger space shuttle mission, unofficially called the "Teacher in Space Project," was launched on January 28, 1986. At 73 seconds after launch, a series of structural failures caused a fuel tank to explode. The shuttle and its crew of seven were lost, in the most heavily televised engineering failure in history. The subsequent inquiry revealed that engineers who tried to delay the launch for safety reasons were overruled by managers.

Source: CP Photo/AP Photo/Bruce Weaver.

about the delay. The weather overnight had been exceptionally cold (for Florida), and icicles had formed on the shuttle. Since the ice might break off during the launch and damage the tiled heat shield, the launch was delayed even further for inspections.

At 11:38 a.m., the rockets were finally ignited. At first, the shuttle rose according to the flight plan; however, at 59 seconds into the flight a plume of flame was evident near the booster rockets. By 64 seconds the flame had burned a hole in the booster, and at 72 seconds the booster's strut detached from the external tank. At 73 seconds into the flight, the loosened booster struck *Challenger's* right wing and then struck the fuel tank. The tank exploded. The shuttle was at an altitude of 14,600 m and travelling at about Mach 2 when the explosion occurred. Perhaps the explosion killed the crew members; however, later evidence showed that the crew module separated from the rocket during the explosion and was in free fall for 2 minutes and 45 seconds. It hit the ocean at a speed of about 320 km/h, killing any surviving crew members. Fragments of the shuttle continued to rain down on the rescue team for about an hour after the explosion.[4]

The *Challenger* explosion represented the first deaths of American astronauts during a mission (although there had been three American deaths in a ground test for the first Apollo mission, and three Soviet deaths when parachutes failed to deploy at the end of the first Soyuz mission). The *Challenger* launch, however, had an immediate, personal impact on millions of people around the world watching the launch on live television. The disaster was a serious setback for the American space program.

Investigation

American President Ronald Reagan convened a commission to investigate the *Challenger* explosion. The investigation involved over 6,000 people, and the resulting 256-page report was issued in June 1986.[5] After extensive deliberation, the commission concluded that the *Challenger* explosion was caused by the failure of a rubber O-ring seal between sections of a rocket booster. Hot gas from the rocket motor escaped past the O-ring (and past a secondary O-ring intended to double the factor of safety). This generated a lateral thrust that eventually broke a supporting strut. The strut's failure permitted the booster rocket to swivel, puncturing the central hydrogen fuel tank, which led, in turn, to an explosion of the shuttle's hydrogen fuel.

The investigators also learned that on the eve of the launch, engineers tried to have the launch delayed because they were uncertain how the O-ring seals would perform in such cold weather. The focus of the investigation turned toward NASA's management style, and the commission concluded that the decision to launch the *Challenger* had been flawed.

The O-ring Problem

When the rockets fire, they create enormous stress in the rocket casing, in all three dimensions. The joint between the rocket sections distorts, and the gap between the sections widens under this stress. The O-rings keep the joint sealed, preventing the hot gas from escaping. The O-rings, which are compressed in a groove, must be resilient enough to "spring back" to fill the gap and keep the joint sealed, as it widens due to joint stress and distortion. The ambient temperature is important because lower temperatures increase the hardness of the O-rings and decrease their resilience.

Roger Boisjoly was the Morton-Thiokol engineer most familiar with the O-ring design. He had conducted temperature tests on O-rings, and as early as July 31, 1985, he had recommended in writing that the problem of O-ring erosion (burning by the hot gases) be studied. Furthermore, he had warned that failure to "solve the problem with the field joint" could result in loss of a shuttle, probably on the launch pad.[6] He was authorized to set up an O-ring team, and in October 1985 he sought advice from 130 vendors and other seal experts. However, no help was forthcoming from these sources.[7]

The Evening Teleconference

The evening before the scheduled launch, it was predicted that the temperature at the launch site would be dropping as low as 20°F (–6°C). NASA engineering managers, concerned about the possible effects of the unusually low temperatures, initiated a discussion about the effect of the low temperature on the performance of the rocket boosters. A late-night teleconference followed, involving thirty-four people. Engineers and managers from Morton-Thiokol, the manufacturers of the rocket boosters, presented their concerns and recommendations to the NASA managers at the launch site. This critical evening conference is described in detail in a definitive book, The *Challenger Launch Decision*, written more than a decade after the tragedy.[8]

The teleconference focused on the performance of the O-ring seals at the circumferential joint in the rocket. Boisjoly stated that no previous shuttle had been launched at temperatures below 53°F (12°C), and that the rocket boosters recovered from that flight showed extensive damage to the primary O-ring, indicating that the O-ring had failed to seal properly. Fortunately, the secondary O-ring had contained the hot gas. The engineering managers at Morton-Thiokol advised NASA's launch staff that the low temperature could cause failure of the O-rings; they then recommended that the launch be delayed until the ambient temperature reached at least 53°F (12°C). Boisjoly's data also included results from a launch where the primary O-ring had failed when the launch temperature was 73°F (23°C).

At NASA, Lawrence Mulloy, an engineering manager at the next level (of a four-level launch approval protocol), questioned the recommendation to delay the launch. Mulloy pointed out the discrepancy in the data presented by Boisjoly from the previous boosters. One O-ring had failed during a fairly low-temperature launch, but one had failed during a fairly high-temperature launch. This perhaps indicated that temperature might not be the key factor in the joint failure. The Morton-Thiokol group asked for a brief delay so that they could discuss the question among themselves.

In the closed discussion with Morton-Thiokol engineers and managers, Boisjoly and the other engineers remained convinced that notwithstanding the apparent discrepancy in the results from previous launches, the cold temperature would seriously affect the O-ring performance. However, their inability to explain the discrepancy revealed that their knowledge and data on O-ring performance at low temperatures were inadequate. At this point in the discussion, Morton-Thiokol's vice-president, Joe Kilminster, intervened to prepare a formal recommendation to NASA. After some prodding, the four Morton-Thiokol engineering managers agreed to reverse the initial recommendation and approve the launch, effectively overruling Boisjoly. The teleconference with NASA resumed, and Kilminster announced the change in opinion and recommended that the shuttle launch go ahead. The shuttle was launched the next morning at 11:38 a.m. and exploded 73 seconds later.

Discussion of the Ethics

In the aftermath of the explosion, there was plenty of blame to go around. Boisjoly was appointed to the investigation team and was initially involved in redesigning the seal. He provided information freely to the president's commission, which led to severe friction with colleagues and superiors. Eventually, he began to feel isolated. He drifted out of contact with his colleagues—especially the NASA management—and finally resigned. In 1987 he filed lawsuits against Morton-Thiokol and NASA for personal damages.

Boisjoly was seen as an ethical whistleblower. One unanswered question is whether he could have done more to obtain cold-temperature O-ring data and present a more convincing case for launch delay at the crucial late-evening teleconference. Boisjoly insists that he made the proper ethical choices during his engineering career, often at the risk of his job. In 1988 he received the American Association for the Advancement of Science (AAAS) award for Scientific Freedom and Responsibility for his efforts to act ethically in the events leading to the *Challenger* shuttle disaster.[9]

The key decision to override the recommendations of the engineers, made by Kilminster on the eve of the launch, was clearly an ethical and management error. The best that can be said for the decision is that it was made under duress. The engineering managers were under intense pressure to meet schedules driven by political and financial priorities, with a space shuttle that was still experimental and not a tested, proven vehicle. The pressure had changed the management philosophy from "launch only when engineers can prove it safe to do so," to "launch unless engineers can prove it unsafe to do so." It took several years for NASA to redesign and recertify the rocket boosters and to get the space shuttle flying again.

The *Columbia* Disaster

On February 1, 2003, a second shuttle disaster happened: the Columbia broke up on re-entry from orbit. This disaster took place at a very high altitude, and observers on the ground saw the debris as several bright meteors streaking across the sky. An accident investigation board was convened with 13 members and a staff of 120. After intensive investigations, the board issued its final report on August 26, 2003.[10]

In simple terms, the accident happened because a piece of insulating foam broke away from the fuel tanks and struck the wing of the shuttle. The damage went unnoticed during flight, but the heat of re-entry was able to penetrate the left wing, weakening the internal structure. The shuttle disintegrated, and the pieces that did not burn at high altitude fell to the ground in a swath 1000 km long. The disaster was a pointed reminder that the space shuttle is not a tested airliner, but a vehicle still under development.

DISCUSSION TOPICS AND ASSIGNMENTS

(Additional assignments can be found in Appendix CD-E on the CD-ROM included with this text.)

1. Consider the circumstances of Case Study 7.6, in which it was recommended that Company A inform Company B that its aircraft tail assembly design had a serious vibration problem. If the problem posed no danger to passenger safety but merely increased fuel consumption and created noise in the cabin, would Company A still be ethically bound to inform Company B? Explain.

2. An engineer who accepts an offer of employment is creating a contract. That engineer is agreeing to use his or her ability to help the employer achieve legitimate goals. Say you have been hired to design electrical or mechanical components for a machinery manufacturer. During a recession, the employer decides to diversify into new areas to attract more business. What would your position be, ethically, if the employer asked you to participate in the design of:

   • bottling equipment for the beer and liquor industry?
   • medical equipment to make abortions safer and more convenient?
   • pill-making machines for the birth control or pharmaceutical industries?
   • security locks for the prison system?
   • equipment for nuclear power plants?

- slot machines for casinos?
- rifles or handguns for the Canadian armed forces?
- rifles for hunters?
- printing equipment for lottery tickets?

3. As an employee of a large Canadian manufacturing corporation, you have been assigned to assist the chief engineer on a six-person team that is to establish a branch plant in a developing country. Your task is to supervise the installation and commissioning of the manufacturing equipment. The local people who will be running the equipment are rural people with little or no education. As soon as you arrive on the site and familiarize yourself with the plan, which is well under way, you realize that the manufacturing line to be installed was removed from service in Canada because it created toxic waste. The waste must be disposed of by special incineration equipment that does not exist in this foreign country. Although the manufacturing line would not be permitted in Canada, the underdeveloped country does not have environmental laws that would prevent its installation and operation. You have some concerns about this project and discuss them with the chief engineer. He is sympathetic but points out that the manufacturing line ran in Canada for more than ten years before pollution laws stopped it, and no deaths were attributed to it. Moreover, the local people will be much better off when the line is running and there is useful employment for all concerned. What guidance does your provincial Code of Ethics provide for this problem? (See Appendix CD-B or obtain the code through the Internet.) Does the code apply to activities conducted in a foreign country? What alternative courses of action are open to you? Which course is best from the ethical standpoint?

4. You are working as a professional engineer in a small consulting company that gives its engineering employees considerable latitude in scheduling tasks, meeting deadlines, and reporting expenses. You are approached by the company president, who states that your professional attitude and attention to high standards have been recognized by senior management. The president also expresses concern about the lax attitudes of your colleagues, who seem to be abusing the freedom extended to them. Would it be ethical for the president to offer, and for you to accept:

- a secret assignment to monitor the behaviour of your colleagues and your immediate superior and report back to the president?
- a promotion to head engineer to replace your immediate superior, on the basis that the head engineer is not competent as a manager and should be replaced?

    Discuss, explain, and justify your answers on an ethical basis, using the techniques discussed in this text.

5. Earlier in this chapter, the actions of Roger Boisjoly in trying to alert Morton-Thiokol's management to the dangers of launching the *Challenger* space shuttle were discussed. The problem focused on a seal (an O-ring) that was unsuitable for low temperatures. In a totally unrelated case, William LeMessurier discovered an inadequacy in the design of the 59-storey Citicorp tower that created a risk of collapse in the event of strong winds from a specified direction. Compare and contrast these two incidents. Consider the degree of risk involved, the extent of the possible damage, the actions of the main participants, and the eventual outcomes. (The LeMessurier case study is discussed in Chapter 9, and both incidents are widely discussed on the Internet.)

NOTES

1. National Society of Professional Engineers (NSPE), *Guidelines to Professional Employment for Engineers and Scientists*, 3rd ed., NSPE, Alexandria, VA, October 31, 1989.
2. Ibid., Part 1 — Recruitment (employee clause 3).
3. G. Voland, *Engineering by Design*, Addison Wesley, Reading, MA, 1999, p. 433.
4. L.C. Bruno, "Challenger Explosion," from N. Schlager, ed., *When Technology Fails: Significant Technological Disasters, Accidents, and Failures of the Twentieth Century*, Gale Research, Detroit, 1994, p. 613.
5. *Report of the Presidential Commission on the Space Shuttle Challenger Accident* (in compliance with Executive Order 12546 of February 3, 1986), Washington, DC, 1986 <science.ksc.nasa.gov/shuttle/missions/51-l/docs/rogers-commission/ table-of-contents.html> (September 25, 2003).
6. R.M. Boisjoly, Interoffice Memo to R.K. Lund, Vice-President, Engineering, Wasatch Division, Morton Thiokol, Inc., July 31, 1985, in D. Vaughan, *The Challenger Launch Decision*, University of Chicago Press, Chicago, 1996, Appendix B, p.447. Also accessible through the Online Ethics Center for Engineering & Science, Case Western Reserve University <www.onlineethics.org/moral/boisjoly/MTImemo1.html> (September 25, 2003).
7. R.M. Boisjoly, *Ethical Decisions—Morton Thiokol and the Space Shuttle Challenger Disaster*, Online Ethics Center for Engineering & Science <www.onlineethics.org/essays/shuttle/pre-dis.html> (September 25, 2003).
8. D. Vaughan, *The Challenger Launch Decision*, University of Chicago Press, Chicago, 1996.
9. R.M. Boisjoly, J.S. Patel, and P. Sarin, *Roger Boisjoly on the Challenger Disaster*, Online Ethics Center for Engineering & Science <www.onlineethics.org/moral/boisjoly/RB1-7.html> (September 25, 2003).
10. H.W. Gehman, Jr., et al., *Columbia Accident Investigation Board Report*, Volume 1, August 2003, distributed by National Aeronautics and Space Administration and the Government Printing Office, Washington, DC, p. 25.

Chapter 8
Ethical Problems in Engineering Management

Managers hire, fire, delegate, and direct other professional employees. Most managers also control company resources, negotiate agreements with other businesses, and are responsible for ensuring that the company protects its assets and follows the law. Managers must also be alert to potential problems such as conflicts of interest, and must avoid becoming involved (or even appearing to be involved) in unethical practices. This chapter discusses some of these management problems and illustrates them with case studies and a historical example. Although the term "engineer" is used throughout for simplicity, the following discussions also apply to professional geoscientists.

ADHERENCE TO THE PROVINCIAL ACT

The most obvious responsibility of a professional engineering manager is to ensure that the professional engineering Act is being obeyed. The two most common infringements of the Act involve the use of unlicensed personnel to carry out the work of professionals, and the misuse of titles. It is unethical and unprofessional to allow such practices, and furthermore, it is illegal. These practices are contrary to the Act in every province and territory.

Unlicensed Personnel

In every province and territory, the professional engineering Act requires professional engineering (as defined in the Act) to be performed by, or supervised by, a professional engineer. Geoscience is similarly regulated in most provinces and territories. Employing unlicensed personnel to practise engineering or geoscience is illegal and dangerous. This problem is common in smaller industries, usually because of ignorance of the law.

If unlicensed people are working as engineers or geoscientists in your company, the manager must correct this practice even if it has endured for years, and even if hard feelings and antagonism are likely to result. If an unlicensed employee is clearly eligible for a licence, the manager should insist on proper registration. When this is not possible, the employee must be put under the supervision of a licensed professional to ensure that the work is properly regulated.

Misuse of Engineering and Geoscience Titles

The misuse of engineering and geoscience titles is also an offence under every Act. Many companies have positions with the word "engineer" or "geoscientist" in the title, such as "project engineer" or "chief geoscientist." These imply that the person holding the position is a licensed professional. Two situations may arise:

- If an unlicensed person is using a misleading title and practising engineering or geoscience, the problem is extremely serious—this is an offence punishable under the Act. Such situations must be rectified immediately.
- If an unlicensed person is using a misleading title but is performing tasks that do not require a licence, the title must be changed to eliminate the ambiguity. Tact and diplomacy are appropriate if the practice goes back many years, but a new job title is needed. The title may be elegant, but it must not contravene the Act.

Using a misleading title contravenes the Act, so the manager's failure to change the title might be interpreted as "contributing to the illegal practice of the profession."

HIRING AND DISMISSAL

The engineering manager usually hires and dismisses engineering and geoscience staff as required. The manager should therefore be familiar with the provincial regulations for hiring and dismissal. The following paragraphs review only a few of these.

Employment Contracts and Policies

The best method for employing professional engineers is through clear-cut employment contracts that specify duties, contract duration (either fixed length or indefinite), remuneration, pay raises, vacation entitlement and statutory holidays, and so forth. Some contracts even include definitions of just cause for termination, along with terms and amounts of severance pay. When a company has so many professional employees that personal contracts cannot be negotiated with all engineers, the company must institute clear policies that deal with these issues. NSPE's *Guidelines to Professional Employment for Engineers and Scientists* (in Appendix CD-D) discusses many topics that should be included in professional employment policies.

Terminating Employment for Just Cause

A manager must take responsibility for terminating or discharging employees when their services are no longer required. These terminations must be in accordance with the employment contract or published company policies. In addition, employees may be discharged for just cause, which is defined below.

> Those matters which would allow an employer to terminate an employee, without notice or severance pay, are as follows:
>
> 1. serious misconduct;
> 2. habitual neglect of duty;
> 3. serious incompetence, not just management dissatisfaction with performance;
> 4. conduct incompatible with his or her duties or prejudicial to the company's business;
> 5. wilful disobedience to a lawful and reasonable order of a superior in a matter of substance;
> 6. theft, fraud or dishonesty;
> 7. continual insolence and insubordination;
> 8. excessive absenteeism despite corrective counselling;
> 9. permanent illness; and
> 10. inadequate job performance over an extended period as a result of drug or alcohol abuse and failure to accept or respond to the company's attempt to rehabilitate.
>
> If one of these elements of misconduct exists, and is ascertained even after the employee has been discharged, the company can rely on that misconduct and not pay the employee any severance allowance.[1]

Wrongful Dismissal

When an employee without an employment contract is dismissed and the reason does not constitute just cause (as defined above), there is a risk of wrongful dismissal. These cases have the potential to end up in the courts, so legal advice is extremely useful for both the employee and the manager.

In a comprehensive article on wrongful dismissal, lawyer Howard Levitt described six situations that could also be considered wrongful dismissal even when technically, the employee had not been dismissed: forced resignation, demotion, a downward change in reporting function, a unilateral change in responsibilities, a forced transfer, and serious misconduct of the employer toward the employee.[2]

In summary, it is important for a manager to be alert to the challenging complexities associated with supervising the work of others. A manager needs leadership ability, sensitivity, and a professional attitude. A knowledge of the law (or access to legal advice) is also useful—advice should always be sought before the hard decisions are made, not after the fact.

REVIEWING WORK AND EVALUATING COMPETENCE

The law requires engineers, whether they are employees or managers, to practise only within their limits of competence. Engineers should not undertake— and managers should not assign—work that is not within the competence of the engineer.

Reviewing Work for Accuracy

Most engineers expect their calculations to be routinely reviewed by a second engineer for accuracy. This is especially true in the aerospace and nuclear power industries, where errors could be extremely costly and could have serious liability implications. Important calculations, and the assumptions on which they are based, are always double-checked for errors, and key decisions are never made on the basis of a single engineer's unchecked calculations.

This routine check greatly increases the accuracy of the calculations, improves confidence and safety, and lowers the risk of failure and liability. The person who performed the original calculations is always informed prior to the review, given the opportunity to clarify any doubtful points, and shown the results.

Reviewing Work to Assess Competence

It is also common practice to evaluate the performance of all employees on a regular basis. Typically it is the engineering manager who performs this evaluation. To evaluate competence, it may be necessary to review an engineer's work at other times as well.

However, a manager should never ask a professional engineer to review the work of another engineer without the knowledge of the engineer who prepared the work. This precept, which is included in most Codes of Ethics, is simple common courtesy and should apply to any professional employee. A secret review is like a trial in absentia, and this is generally contrary to our system of natural justice. The work of a professional, by definition, requires specialized knowledge, so additional information or explanations may be required when the work is checked. A professional reputation is a valuable asset founded on many years of study and experience. The review of any professional's work must not be conducted in a careless or cavalier manner that could inadvertently damage someone's professional reputation.

In summary, reviews of an engineer's work for accuracy or competence are routine and do not require the engineer's permission; however, these reviews must not be done without informing the engineer. Engineering managers should be especially sensitive to the need for this common courtesy.

Maintaining Competence

Most provincial Codes of Ethics specifically instruct engineers to maintain their competence. (This is discussed in more detail in Chapter 15.)

Furthermore, employers have an obligation to encourage engineers to maintain professional competence. Note that the NSPE guidelines (in Appendix CD-D) suggest that the engineer and the employer are mutually responsible: the employee must show initiative, and the employer should provide a supportive environment and appropriate job assignments.

For example, whenever a company takes on a new project or a new computer system, its engineers will usually have to upgrade their skills and/or knowledge. It is appropriate for the employer to arrange and pay for a review course or a workshop and for the engineer to make full use of that opportunity. This is a win/win arrangement for the engineer and the employer.

When an engineer has simply failed to keep skills current and has drifted into incompetence, but refuses to admit the fact or to exert the effort to become more effective, the manager faces a more difficult decision. An incompetent engineer should not be practising engineering; in fact, every provincial Act cites incompetence as a basis for taking away an engineer's licence. The task of the manager, as in any problem-solving exercise, is to gather information, generate alternatives, examine those alternatives, and seek the optimum course of action. The outcome will depend on the facts of the individual case, but the manager must deal with the professional employee fairly and ethically.

DISCRIMINATION IN ENGINEERING EMPLOYMENT

The engineering manager plays a key role in hiring and dismissing engineers and in evaluating their performance. This means that he or she is in the front line of the battle against discrimination. Discrimination should not be a problem in Canada, since the Charter of Rights and Freedoms prohibits discrimination on almost any basis unrelated to competence. Although much progress has been made in overcoming discrimination in recent years, certain groups in engineering, such as women, Aboriginal peoples, and people with disabilities, are still underrepresented in engineering. The problem is especially obvious where women are concerned—they are a majority in the general population, yet they are definitely a minority in the engineering profession.

Women and minority groups have a legal right to be treated fairly. Although they would not expect preferential treatment, artificial obstacles must not be created for them. This topic is discussed in more detail in Chapter 13, "Fairness and Equity in Engineering."

MANAGING PATENTS, TRADEMARKS, DESIGNS, AND COPYRIGHT

All engineers should have a basic knowledge of the law of intellectual property, which encompasses patents, trademarks, industrial designs, copyright, and—most recently—integrated circuit topography (more commonly known as circuit design).

In particular, engineering managers should know how to search databases for patents, trademarks, and designs. The Canadian Intellectual Property Office (CIPO), a branch of Industry Canada, maintains this country's database, which can be searched through the Internet. In the United States, the Patent and Trademark Office (PTO) also maintains a patent database searchable by Internet. These databases are a valuable source of useful design information.

Engineering managers must also protect the intellectual property created by engineers under their direction. Infringement seems more common now, perhaps because illegal copying is now so very fast and easy. The ethical aspects of infringement of intellectual property are discussed below in general terms. For more detailed information or for answers to legal questions, consult CIPO or Industry Canada or see your lawyer. Table 8.1 summarizes the various forms of intellectual property and how long they may be protected.

Patents

A patent is awarded only if the invention is new, useful, and innovative. A patent protects the way that a device operates, unlike an industrial design registration, which protects the appearance. Improvements to inventions can also be patented; however, obtaining a patent on an improvement does not confer the right to use the original invention. Patent rights last twenty years, but a small maintenance fee is required to keep the patent in force; otherwise it enters the public domain.

At the start of a design project, the CIPO database should be searched, both as a source of useful ideas and inspiration, and to avoid infringing on existing rights. Once a new design or invention has been developed, the manager should consider registering it to protect the inventor's (and the employer's) rights. For a nominal fee, patent agents and other private companies will assist in patent searches and applications. For more information on patent procedures, consult the CIPO publication, *A Guide to Patents*.[3]

Designs

Industrial designs can be protected through a process similar to patent or copyright protection. In this case, however, only the aesthetic appearance is protected. Registration of an industrial design protects the shape, configuration, pattern, or ornament applied to a finished article, which is typically made in quantities by machine. For example, the pattern of decoration on the knives, forks, and spoons of a silverware set would commonly be registered as an industrial design. After ten years, design rights expire and the design enters the public domain.

It is neither unethical nor illegal to use designs that have entered the public domain. In fact, engineers are encouraged to use them, since this is often a fast and profitable way to develop attractive new products. However,

TABLE 8.1 — Summary of Protection for Intellectual Property

Intellectual property	What is protected	Duration of protection
Patents	New, useful, and innovative devices, machines, processes, or compositions of matter (or improvements to existing inventions). A patent protects the way something operates or is made.	20 years
Industrial designs	The shape, configuration, pattern, or ornamentation applied to a finished article, made in quantities by hand, tool, or machine. An industrial design protects the appearance or ornamentation. (Some industrial designs may also be protected as trademarks.)	10 years
Integrated circuit topographies	The patterns or configurations of electronic circuits used in integrated circuits (including three-dimensional configurations).	10 years
Copyright	Written literature and artistic, dramatic, and musical works, including computer programs. Works reproduced mechanically or electronically (films, photos, recordings, communication signals, etc.) are given protection with a shorter duration. (Exceptions exist.)	The life of the author/creator, plus 50 years. Mechanically and electronically copied works are usually limited to 50 years.
Trademarks	Logos, slogans, names, symbols, or designs (or any combination of these) used to identify a company's goods or services in the marketplace.	15 years, renewable indefinitely.
Trade Secrets	No protection exists for trade secrets. If someone independently discovers the same secret, the other person may patent it and prevent you from using it.	Uncertain.

if you have developed original designs, you should protect your rights by registering them. For more information, CIPO publishes *A Guide to Industrial Designs*.[4]

Circuit patterns are a special form of design and are the newest form of intellectual property. They are called, in general terms, "integrated circuit topographies," and they have their own separate registration process. Integrated circuit topographies are defined as the configurations that form integrated circuits, including the three-dimensional aspect of such designs, and they can be extremely valuable. Registration provides legal protection for ten years. For more information, CIPO publishes *A Guide to Integrated Circuit Topographies*.[5]

Copyright

Copyright protects written works, including computer programs and artistic, dramatic, and musical works. These creative works are protected for the life of the author/creator plus fifty years beyond death. However, works that are reproduced mechanically or electronically, such as motion-picture films, photographs, recordings, communication signals, and so on, are given fewer years of protection—typically a maximum of fifty years from the date of creation. (Some exceptions to these general rules exist.) Copyright differs from other forms of intellectual property in that it is obtained by any creator (author, artist, performer, photographer, etc.) immediately upon creation of the work. Registering the copyright is optional; doing so simply gives slightly more certainty in enforcing the rights.

Illegal copying is, of course, the most common and troublesome form of infringement. Photocopy machines, audio recorders, and video recorders are standard features in many homes and workplaces. The Internet is another major tool of copyright infringement. New digital formats have been developed in recent decades, including CD, DVD, and MP3, along with the tools for quickly making copies of massive amounts of data (for example, by "burning" data on CDs). These developments have made it a simple matter to infringe the copyright of written, visual, and audio material. Technologies for protecting Web content are being developed but are lagging far behind the technologies for stealing it. Software piracy and video and audio copying are costing large corporations many billions of dollars. For example, music companies in the United States have been unable to find any technical means to stop infringement and have recently begun suing to recover lost revenues. In 2003, people who downloaded music from the Internet (some as young as twelve) were shocked to receive subpoenas to appear in court. The prevailing attitude among computer users is astonishment that this sort of copying could be illegal.

Although a casual attitude to copyright infringement may be tolerated in elementary schools to encourage computer use, such attitudes are not acceptable in university, in employment, or in professional practice. The first Internet generation—the graduates who are now entering our profession—may have been rewarded by their high school teachers for submitting reports that were openly copied from the Internet. However, university professors and other professionals insist on much higher standards. When a professional person prepares a report—or indeed any written, graphic, or otherwise copyright material—the person must identify and cite the sources of any (and all) material included that is not original. Failure to do so can have serious penalties.

Claiming credit for work that is not your own is called "plagiarism" and is usually both an infringement of copyright and a serious academic offence. Plagiarism has several levels of severity, from flagrant copying of an entire report and passing it off as one's own work, to failing to cite sources thoroughly. In universities, the penalty for plagiarism is expulsion. In the business world, the legal penalty for infringement of copyright depends on how much the copyright owner lost financially.

Engineering managers must be alert to the potentially serious consequences of illegal copying, software pirating, and plagiarism. These are all copyright infringement and may lead to legal problems. Engineering reports must cite sources, just as academic works do. Illegal use of the work of others, or claiming credit for the work of others, is not just unethical; it may lead to legal problems. Pirated software must not be used in the engineering office. (This topic is discussed in more detail in Chapter 12.) For more information, the CIPO publishes a *Guide to Copyrights*.[6]

Trademarks

Trademarks are the commonly used logos, slogans, names, symbols, or designs that identify a company's goods or services in the marketplace. Trademark registration gives protection for fifteen years. Unlike other forms of intellectual property, trademarks can be renewed indefinitely—more specifically, for as long as they still serve the purpose of identifying a company's goods or services in the marketplace. Trademark infringement is fairly rare in engineering, although high-priced consumer goods (such as prestige watches and fashion accessories) are often duplicated by "knock-off" companies.

Should the need arise, you can conduct trademark searches on the Internet at no cost by connecting to the CIPO website. A simple search may be worthwhile to check that a new product's suggested name is not already registered as a trademark. For more information, CIPO publishes *A Guide to Trade-marks*.[7]

The following example illustrates the importance of searching the trademark database. The University of Waterloo recently introduced a Mechatronics Engineering program. The university contacted the Canadian Council of Professional Engineers (CCPE) to ensure that the program would satisfy the accreditation process and that the program had the legal right to use the term "engineering." (CCPE is the registered trademark holder of the terms "engineer" and "engineering.") However, the term "mechatronics"—which indicates that the new program is a combination of mechanical, electronics, and robotics subjects—was not searched until after the program name had been widely advertised. University officials were surprised to learn that, about a decade earlier, the term had been registered in Canada as a trademark by a German company. (Negotiations over the use of the term are underway as this text goes to press.)

Trade Secrets

Of course, intellectual property can also be protected simply by keeping it secret. Trade secrets may be effective for products, processes, and material compositions; obviously, they are irrelevant for trademarks. In general, secrecy is maintained by requiring employees to sign employment contracts with confidentiality clauses. Trade secrets have no legal status in patent law, so breaches of confidentiality must be enforced under contract law or tort law.

CASE STUDY 8.1— MISREPRESENTATION AS A LICENSED ENGINEER

STATEMENT OF THE PROBLEM

You are the manager of the engineering design department for a fairly large consulting engineering firm. As part of your job, you hire and dismiss department staff members, including engineers, designers, CAD operators, and clerical workers. Six months ago you hired Jorge Xavier, who had recently moved to your area from another province. During the employment interview, you emphasized that it was essential that he be licensed. The letter of appointment sent to him stipulated that he was being hired as a professional engineer. After Xavier started work, you had a sign placed on his door and had business cards printed, both of which had the P.Eng. designation after his name.

You are startled to receive a complaint from a client who claims that Xavier is not a licensed professional engineer. The client is furious that you and your company would send unqualified people to work on her project. You contact the provincial Association of professional engineers, which confirms that Xavier does not hold a licence. Now you are furious.

QUESTION

Who is responsible for this problem? Can you fire Xavier for just cause? Would it make any difference if

- *Xavier is licensed in another province but has neglected to apply for a licence in your province?*
- *Xavier has applied to obtain a licence in your province, but that licence is still being processed by the provincial Association?*
- *Xavier has never been licensed in any province?*

AUTHOR'S RECOMMENDED SOLUTION

This case involves a breach of the professional engineering Act and its Code of Ethics. A professional licence is valid only for the province in which it was issued. A licence cannot be transferred from one jurisdiction to another. When a person moves to a different province or territory, a licence application must be submitted to the Association in that jurisdiction. The process is routine, and additional licensing conditions (such as writing the Professional Practice Exam) are rarely required. A new licence will generally be awarded to the applicant, with a minimum of inconvenience.

There can be little doubt that Xavier is guilty of practising professional engineering without a licence. He has used the business cards that clearly say

P.Eng. without protest or correction, and he is not licensed in the province where he is working. Consequently, he is committing an infraction of the Act, although the fact that you, as manager, had the business cards prepared could be considered a mitigating factor. You will be guilty of a breach of the Code of Ethics if you permit Xavier to continue to practise engineering.

It is essential to determine what work Xavier has done for the client. If he has been in a junior or training position during his first six months with the firm, and if his work has been supervised by another engineer, as would usually be done (at least initially), then the risk to the client or to the public may be minimal. Damages may be limited to the embarrassment and the possible overbilling of fees.

However, if Xavier has been making independent decisions on engineering projects, then the engineering firm would undoubtedly be liable for any problems that might arise from those decisions. You must review the engineering decisions that Xavier has made, and discuss this liability problem with the company lawyer. In any discussion of liability, it would be made clear that you, as manager, are responsible for verifying the qualifications of those who work for you.

As to whether Xavier should be dismissed, and the grounds for that dismissal, it depends on which of the three situations applies.

- If Xavier has failed to apply for a licence after six months of employment but has a valid licence from another province, then it could be argued that this constitutes either professional misconduct or habitual neglect of duty, both of which are recognized as just cause for dismissal.
- If Xavier has applied for a licence and that licence has for some reason been slow in coming, and if he has a valid licence from another province, then he has probably complied with your requirements and dismissal would probably be unjust.
- If Xavier has never been licensed in another province, and if he has been unable or unwilling to obtain a licence in your province, then he has been dishonest in his employment interview with you, and such fundamental dishonesty would be just cause for dismissal.

Xavier clearly contravened the Act when he used the P.Eng. designation without having been licensed. Therefore, he may also be subject to a charge under the Act, even if he does have a valid licence from another province. This charge would be prosecuted in the provincial courts; however, the Association would initiate the charge under the authority of the Act.

However, you as manager must bear much of the responsibility for any embarrassment or liability the firm suffers. You stated the requirement for a licence clearly, but you did not follow up to verify that in fact Xavier had a valid licence and that he had the legal right to use the P.Eng. designation. A company that offers engineering services to the public has a duty to verify that its engineers keep their licences up to date.

CASE STUDY 8.2—CONCEALING A CONFLICT OF INTEREST

STATEMENT OF THE PROBLEM

You are an engineering manager in a fairly large company, and you have been asked to sit on a ten-member standards committee that sets performance and safety specifications for the automotive equipment your company manufactures. The committee comprises three industry representatives (including you), three government representatives, and three engineering professors, and it is chaired by a representative from an engineering society associated with the automotive industry.

One of the other industry representatives has proposed a revision to the specification for a component you manufacture. The change would make a fairly modest improvement in quality; it would also require specialized manufacturing expertise and equipment. During the meeting on the specification, you realize that if the revision is approved, your company will benefit greatly since it has the necessary expertise; it will also create hardships for some of your competitors.

You believe that the person proposing the revision would benefit in a similar way. You are uncertain whether you should mention all of this to the committee. You did not propose this revision, but it would improve the quality of the product, and any benefit your company receives would arise strictly by chance.

QUESTION

Do you have an ethical obligation to inform the committee that your company stands to benefit from this revision? Do you have an obligation to point out that the person proposing the revision may also benefit?

AUTHOR'S RECOMMENDED SOLUTION

You have a clear conflict of interest, which you must disclose to the committee. The Code of Ethics states that an engineer must place the welfare of society above narrow personal interest. The main function of a standards committee is to serve the public welfare, not the financial interests of its members. After you have disclosed your conflict of interest, it might be acceptable for you to answer questions and/or express your opinion of the revision. However, you definitely should not participate in the formal vote on the revision.

You are under no obligation to speak about the member who is proposing the revision unless you believe that a deliberate fraud is being perpetrated. This does not seem to be the case. More likely, your disclosure of your own conflict will encourage the other industry representatives to declare their own conflicts.

Conflicts of interest are common on such committees for the simple reason that the best-informed people are those involved in the design and manufacture of the components concerned. This makes it especially important to be alert to unfair and unethical advantages that may result from such positions of trust.

CASE STUDY 8.3— DISCLOSING ERRORS IN PLANS AND SPECIFICATIONS

STATEMENT OF THE PROBLEM

You are the engineering manager for Acme Assembly, which designs, fabricates, and assembles machinery. You have received a contract to construct twenty gearboxes that have been designed by Delta Designs, a company that sometimes competes with Acme. However, Delta is extremely busy and does not have the capacity for this work at the present time.

One of your engineers notices that the sizes of shafts and gears on the drawings seem rather small for the torque and power ratings of the gearboxes; rough calculations seem to confirm that assessment. You call the chief engineer at Delta, who tells you he is too busy to double-check the drawings. He has full confidence in his designers and says you should get on with the job. He points out that you are employed in this contract as the fabricator, not as the designer, and should not be reviewing his work.

QUESTION

Do you have an ethical obligation to continue to pursue this apparent discrepancy? Would it make any difference if failure of the gearboxes could result in injury or death rather than just financial loss?

AUTHOR'S RECOMMENDED SOLUTION

Under the Code of Ethics, an engineer has an obligation to ensure that the client is fully aware of the consequences of failing to follow the engineer's advice. Here, a single telephone call probably would not satisfy this requirement, either ethically or legally. You should follow up the telephone call with a letter that describes your concerns and that also requests written instructions to proceed.

If the chief engineer at Delta instructs you, in writing, to proceed with the fabrication, you should do so unless you consider the flaws in the design to be obvious and serious. This might indicate a problem of incompetence, negligence, or fraud on the part of Delta's chief engineer. If these suspicions are supported by other evidence, it would be appropriate to ask your lawyer and/or your provincial Association for advice on how to proceed. (See the discussion elsewhere in this text on whistleblowing.)

The potential for injury or death in the case of failure is important, because failure to safeguard the safety of the public could be considered professional misconduct on your part. If serious injury or death is likely or even possible, the chief engineer's complaint about your reviewing his work is irrelevant. A review of the design is appropriate in situations like these. Through your diligence you have sought to protect the public—and to safeguard the chief engineer's reputation as well.

CASE STUDY 8.4— DISCLOSING PRELIMINARY MINING DATA

STATEMENT OF THE PROBLEM

You are a professional geologist responsible for all of the ore assays in a mine. You report directly to the mine's chief executive officer, who is an accountant by training. You have just finished evaluating initial ore assays for a newly opened part of the mine. These show much lower ore content than anticipated. The CEO is very disappointed at the news. You reassure him that the results are preliminary and that more thorough results will be available in a week or so. The CEO had hoped to present good news about the exploration to shareholders at a meeting to be held in the next few days.

The CEO tells you to keep the results confidential and not to report or discuss them until after the shareholders' meeting—not even with the company's own employees.

QUESTION

Is it ethical to hide this information from the shareholders, who are the owners of the company?

AUTHOR'S RECOMMENDED SOLUTION

This question is important, but it is not so much an ethical question as a question about the corporation's management structure, which apparently is being misunderstood here. The shareholders are indeed the ultimate owners of the corporation, but they do not run it. The shareholders elect directors, who form a board of directors. In turn, the board appoints the officers of the company—the president, CEO, treasurer, and so on—and it is these people who are responsible for the day-to-day operations of the company. Employees take direction from these company officers. So the simple answer to this question is that the CEO is responsible for directing the company's operations and that the geologist has no legal or ethical duty to report directly to the shareholders.

In the mining and oil and gas industries, geological data are extremely sensitive information and can be the basis for important financial decisions. Unauthorized disclosures can lead to abusive stock market tactics. Most boards authorize only the CEO to issue public statements, which means that a geologist has no authority to dislose data to the public without the CEO's approval.

In fact, all public disclosures from mineral companies must follow the strict guidelines set out by the Canadian Securities Administrators (CSA). The CSA guidelines are a fairly recent set of rules introduced after the Bre-X mining fraud (discussed in Chapter 2). Every geoscientist involved in preparing mineral studies should be familiar with the CSA guidelines, which regulate all public statements relating to mineral projects, be they oral or written (including news releases, prospectuses, and annual reports). The guidelines also require all disclosures to be based on a technical report prepared by a "qualified person" (as defined in the document and discussed in Chapter 2). Furthermore, they require this report to adhere to a particular format. So the answer to this case study is very clear—any disclosure of the ore assay results by the geologist would be unethical and probably illegal.[8]

CASE STUDY 8.5— PROFESSIONAL ACCOUNTABILITY IN MANAGEMENT[9]

STATEMENT OF THE PROBLEM

Ethel Eager, P.Eng., is a mechanical engineer at a well-known specialty chemicals company. The company makes consumer products in Canada for the North American market. It also has plants in the United States, which compete with Canadian plants for North American production mandates.

Eager started out five years ago in a junior production position, reporting to Cam Complacent, P.Eng., the production supervisor. When Eager started at the Canadian plant, it was highly successful. However, over the five years of her employment, the plant has become steadily less competitive relative to other firms and its sister plants in the United States. When Complacent retired recently, Eager was promoted to fill his job.

Having passed her Professional Practice Exam while working at the company, Eager is aware of the importance of professional ethics in engineering. Over the past five years she has noticed several unusual practices and events in the plant and in the office. For example, supplies often run out before forecast, inventory is invariably balanced by assuming losses, and there are frequent shortages in customer shipments. In the human resources area, she has noticed a tendency to "horseplay" on the graveyard shift, as well as what she would consider to be instances of racial and sexual harassment. Also, procedures for recording the hours that employees actually work are very casual, and overtime is high.

These discrepancies disturbed her, and Eager had approached her boss, Cam Complacent, about them several times. Each time, he played down her concerns and said being "easy" on these subjects helped keep morale and productivity up. Although Eager was personally convinced that some employees were cheating their employer by taking products home and misrepresenting their hours of work, as a junior employee she had decided to take her manager's advice to keep quiet.

However, shortly after she replaced Complacent as supervisor, Eager was informed, early one Monday morning, that there had been a major theft at the plant on the weekend. A truck had pulled up to the warehouse without being challenged, loaded up, and disappeared. Fortunately, the police soon caught the two thieves, who turned out to be employees, one of them a relative of a senior employee. Indeed, the police soon found that a network of employees was involved. They now want to interview Eager about further investigations.

Meanwhile, Eager has just received a fax from the company's vice-president for North American manufacturing, who wants to investigate why the Toronto plant's costs have been so high and why productivity has been so low relative to the company's other plants. The fax concludes: "Understand major theft has occurred. Will be in Toronto tomorrow to review your situation." The future of Eager and her plant looks grim.

QUESTION

Should Eager be held accountable for the employees' actions? What lessons, if any, can be learned from this case?

AUTHOR'S RECOMMENDED SOLUTION

As a middle manager and a professional engineer, Eager is accountable to her superiors, possibly to the police, and to her profession, because she knowingly allowed a dishonest environment to flourish. All of the stakeholders involved—Eager, her superiors, her peers, her employees, and even her suppliers—have suffered or will suffer. Because she is a professional engineer, Eager has a duty under the provincial Code of Ethics to all of these stakeholders to act at all times with devotion to high ideals of personal honour and professional integrity. She also has a duty to expose, before the proper tribunals, unprofessional or unethical conduct by another engineer.

Although there are mitigating circumstances in this case (e.g., Eager's relative inexperience and her employer's lack of an ethics program), Eager has learned two valuable lessons:

- the meaning of accountability
- that there are no small ethics problems

In hindsight, Eager now knows that turning a blind eye to the problems at her plant was wrong. She must also realize that there would have been benefits to dealing with her concerns as they arose and that now there are consequences to having ignored them. She should have explained to Complacent that as a professional engineer, she was duty bound to act on her concerns. She should have suggested to him that together they discuss the subject with senior management. Had Complacent been unwilling to consider this approach, as a last resort, Eager could have considered going alone to senior management or obtaining advice from the provincial Association.

CASE STUDY 8.6—STUDENT PLAGIARISM

STATEMENT OF THE PROBLEM

Oliver T. is an engineering student in the last week of his co-op summer work-term at a large manufacturing company. Oliver has enjoyed the job, worked hard, learned much, and been well paid. He is on very good personal terms with his boss and is fairly certain that his boss will rate his work as excellent. However, when he returns to university next week, he must submit a written work report describing a project he undertook during his work term. This university requirement has serious implications. The report is marked by a professor, university standards are applied, the grade is recorded, and Oliver's graduation could be delayed if the report is unsatisfactory. Oliver had several interesting work assignments that were suitable as report topics, but so far he has written nothing, and he is beginning to worry. He has no reason or excuse for failing to prepare a report.

In the last week of his work-term, Oliver mentions his problem to his boss. The boss has a file cabinet containing copies of work reports written by former work-term students. Some of the reports are stored on floppy disks. Surprisingly, the boss suggests that Oliver could save a lot of time if he simply revised and submitted an older report as his own.

QUESTION

Should Oliver submit a work report written by a former student as his own? If he does so, and is caught, what penalties does he face? Does it reduce his culpability that the boss suggested the idea? Is the boss guilty of any unethical action?

AUTHOR'S RECOMMENDED SOLUTION

A person who submits a report written by someone else is committing plagiarism. Directly or indirectly, every Code of Ethics or ethical theory condemns plagiarism as unethical. (In common parlance, this is an ethical "no-brainer.") Obviously, Oliver should not yield to the temptation to submit a plagiarized work report.

We all have a duty to prevent plagiarism, yet plagiarism continues to occur, with profoundly negative results. In fact, in some businesses and institutions (especially universities) plagiarism has been increasing recently. This increase is probably caused by the easy availability of written material in digital form on the Internet. Fortunately, the Internet also provides tools for detecting plagiarism, as discussed below. Most universities now have severe penalties in place that make plagiarism highly unattractive.

Plagiarism undermines both the educational process and the cooperative program. Co-op employment is not just a job. Co-op integrates work experience with academic study, and work reports test and illustrate this integration. If Oliver is caught plagiarizing a work report, the consequences will typically be as follows.

Academic penalties. Plagiarism is a serious academic offence that usually results in suspension for a first offence and expulsion for a second offence. A suspension or expulsion is usually shown on grade transcripts. Either will delay graduation far more than a late report.

Future recommendations. Obviously, professors and co-op employers will hesitate to recommend students for job openings or graduate programs if the students have been suspended or expelled for plagiarism. Good recommendations are also needed to obtain an engineering licence.

The fact that the boss was willing to help Oliver commit plagiarism would not lessen the penalties for Oliver; in fact, it might extend the disciplinary action to include the boss (assuming that the boss is a licensed engineer). The boss has breached the Code of Ethics and could be reported to the provincial Association. Also, the university would almost certainly bring such collusion to the attention of the boss's employer, and might refuse to allow the employer to participate in the co-op program in the future.

How to avoid plagiarism. If your work includes any material (including sentences, photos, drawings, or figures) from any other source, cite the complete source—it is easy to do. Failure to cite sources is plagiarism. In particular, any material cut and pasted from websites must be fully identified with a proper reference that cites the URL and the date. Authors who submit reports containing Web material that is not fully cited are guilty of plagiarism.

How to detect plagiarism. Written material plagiarized from the Internet is easy to detect. A key word search using words from the plagiarized report and almost any search engine will generally turn up the source very quickly. In addition, a Web service for detecting plagiarism has been developed and is available to professors for a nominal fee.[10] Many universities are also developing in-house solutions for plagiarism that involve scanning parts of submitted reports to create a database for searching. Even one passage in a report could be proof of plagiarism, if the source has not been cited.

Where to learn more. Guides to avoiding plagiarism can be found by a simple Web search using "plagiarism" as the search argument. Several excellent sites exist.

CASE HISTORY 8.1

THE VANCOUVER SECOND NARROWS BRIDGE COLLAPSE

This case history reminds engineers that serious failures can occur during construction; in fact, the risk of failure may be higher for temporary supports, forms, and scaffolds, because they are rarely analyzed as thoroughly as the main, permanent structure. Engineering managers in particular should be reminded to double-check the calculations for critical components, especially when the original work was carried out by less-experienced engineers.

On June 17, 1958, two spans of the Vancouver Second Narrows Bridge collapsed during construction. Eighteen workers were killed. This tragic accident was caused by the lateral buckling of beam-webs in a temporary tower that was supporting the partially completed bridge. A fairly simple calculation would have shown that the beam-webs were unsafe. The following description of the tragedy is reprinted, with permission, from W.N. Marianos, Jr., "Vancouver Second Narrows Bridge Collapse."

Background

The Second Narrows Bridge connects Vancouver, British Columbia, with its northern suburbs across Burrard Inlet, the city's harbor. The structure was built for the British Columbia Toll Highways and Bridges Authority. The main bridge, a steel cantilever truss structure, has a total length of over two thousand feet (610 m). Unlike older, simpler, and shorter bridges whose spans or sections rest independently on their piers or abutments, those of a cantilever bridge run continuously over or extend beyond the piers. The main bridge has three spans: a 1100-foot-long (335 m) centre span balanced by two side spans, one 465 feet (142 m) and one 466 feet (142 m) long. Four steel truss spans, each 276 feet (84 m) long, make up the northern approach to the main bridge. The structure was designed by Swan, Wooster and Partners, a Vancouver engineering firm. Dominion Bridge Company was the contractor for the construction of the steel spans. The foundations and bridge piers were constructed by Peter Kiewit Sons and Raymond International.

Details of the Collapse

By mid-June 1958, the approach spans were in place and erection of the northern side span of the main bridge was in progress. The length of the span required the use of two temporary supports for construction, since the side span would not be self-supporting until its full length was in place. Each temporary support, called a falsework bent, consisted of two columns, one under each side of the span. The columns were built on temporary piers in the harbor. These piers were supported by a group of foundation piles. The load from each column was distributed to the foundation piles by a grillage—a two-layer grid of steel beams. The lower set of beams sat on top of the foundation piles. The upper layer, a set of four beams set side by side, supported the column bases.

Eighteen people were killed when failure of temporary construction supports caused the Vancouver Second Narrows Bridge spans to collapse on June 17, 1958.

Source: © Bettmann/CORBIS/Magma.

On June 17, the first side span was supported on a permanent concrete pier at one end, and was overhanging the first falsework bent, designated "bent N4," at the other. At 3:40 P.M. that afternoon, bent N4 collapsed, plunging the partially completed span into Burrard Inlet. The falling metalwork pulled the permanent pier it was resting on out of line, which caused the adjacent approach truss to collapse as well.

Immediately after the accident, the government of British Columbia appointed a royal commissioner, Sherman Lett, chief justice of the provincial supreme court, to determine the cause of the collapse. The commissioner selected five leading engineers to investigate and report on the matter: F.M. Masters and J.R. Giese of the United States; J.R.H. Otter and Ralph Freeman of Britain; and A.B. Sanderson of Canada. Materials testing and special investigations were conducted at the University of British Columbia and testing laboratories in Vancouver.

The commissioner's report concluded that the collapse was caused by failure of the four upper grillage beams. The webs (the vertical portion) of the beams buckled laterally, causing the collapse of the falsework bent columns.

Faulty design of the falsework or temporary columns led to the grillage failure. The commission discovered two major errors in the Dominion Bridge Company's grillage design calculations. The first mistake was in checking the grillage beam shear strength (the capacity of a beam to carry a load in its vertical plane; shear stress tends to tear a beam vertically, usually at supports or at points of concentrated load). The cross-sectional area of the entire beam was used in the calculation rather than just the areas carrying the load. This mistake would lead the grillage designer to believe the beam strength was about twice as much as it actually was.

A second calculation, which checked the need for web stiffeners, was also incorrect. Stiffeners are metal plates welded to beam webs to give them additional stiffness and resistance to buckling. The contractor's engineer had used the one-inch (2.5 cm) thickness of the beam flanges (the horizontal elements) rather than the actual 0.65-inch (1.6 cm) web thickness. This led to the erroneous conclusion that no stiffeners were needed.

A separate investigation by Dominion Bridge Company came to the same conclusion—that incorrect calculations led to a fatally inadequate grillage design. One of the errors in calculation was even discovered before the accident, but no corrective action was pursued. The two engineers responsible for the calculations were both killed in the collapse.

Wood blocks and plywood pads had been included in the grillage to provide some bracing of the beams. Laboratory tests indicated that these blocks were only marginally effective at best. Most of the wooden blocks were not even located at the most effective bracing points. The investigation performed at the University of British Columbia also indicated that the ability of the beam webs to resist buckling was not adequately predicted by the usual design formulas for column buckling.

In his report, the royal commissioner laid the blame for the collapse on the Dominion Bridge Company. The commission found the contractor negligent for "(a) failing properly to design and substantially construct false bent N4 for the loads which would come upon it . . . ; (b) failing to submit to the engineers plans showing the falsework the contractor proposed to use in the erection . . . ; and (c) leaving the design of the upper grillage of false bent N4 to a comparatively inexperienced engineer, and failing to provide for adequate or effective checking of the design and the calculations made in connection with the design."

The commissioner also found that a failure in the construction process had contributed to the accident. His report pointed out that the bridge design engineers, Swan, Wooster and Partners, had a responsibility to make sure the contractor submitted the falsework plans and calculations for their approval, as required by the project contract. The engineers certainly knew that the bridge was under construction, and they had prepared the section of the project specifications that required engineer's approval of the temporary falsework structures. Commissioner Lett concluded that "there was a lack of care on the part of the engineers in not requiring the contractor to submit plans of the falsework." Ironically, the satisfactory performance of Dominion Bridge on earlier projects may have contributed to the design engineer's laxness in pursuing the falsework plans and calculations for review.

The commissioner recommended that on future large bridge projects the consulting engineers recommend allowable stresses for temporary construction support structures, and that the contractor be required to submit all construction plans and calculations for approval prior to construction. The contractor, however, would always remain legally responsible for the adequacy of construction methods and temporary structures.

Impact

After the inquiry, construction resumed on the bridge. Two concrete bridge piers damaged in the collapse had to be rebuilt. This required the careful removal of two thousand cubic yards (1529 m^3) of reinforced concrete. The collapsed superstructure spans were salvaged, and some undamaged members were reused. Erection of the bridge continued according to the original plan, with the notable addition of careful checking and review of all construction calculations and plans. The additional time and materials required to reconstruct the damaged portions of the bridge added four million dollars to the original contract price of sixteen million dollars.

The editors of *Civil Engineering* magazine noted that the collapse "illustrates the ever-present risks that are inherent in construction, due to human error. The failure emphasizes the need for utilization of all possible checks on construction procedures."

Today, the leading bridge design firms continue to carefully review and check the contractor's construction plans and calculations. The collapse of the Vancouver Second Narrows Bridge was neither the first nor the last incident of mistaken temporary construction calculations leading to disastrous consequences. The accident vividly highlights the importance of independent checking of critical aspects of the construction process.[11]

DISCUSSION TOPICS AND ASSIGNMENTS

(Additional assignments can be found in Appendix CD-E on the CD-ROM included with this text.)

1. Assume that you are the manager of a new project, and one of your first responsibilities is to make time and cost estimates for the project. The project is fairly complex, so your calculations result in very high estimates—so high that you fear the project may be discontinued. Some older engineers on the project say that many earlier projects would have been cancelled if the true extent of their final costs had been known early in the game. Moreover, they argue that no one can ever be really sure of what something is going to cost; after all, these are only estimates. In the earlier projects, a very optimistic face was put on the cost estimates, and even though the final costs exceeded the estimates, the projects were successful.

The older engineers urge you to reduce your estimates so that the project will not be cancelled. But you have put a lot of careful work into your estimates and believe your figures are as correct as any estimate of the future can ever be. Therefore, if you reduce the estimates, you know you will be lying. Furthermore, you know your own reputation in the company will be flawed if it becomes apparent that you shaved your estimates. However, you fear that some of the people in your project team may be laid off if your project is cancelled. You are caught in a dilemma, and as a manager, you must decide one way or the other. Explain how you would try to solve this ethical dilemma. Write a brief summary of your decision and your reasons for it.

2. Renée Langlois is a professional engineer who has recently been appointed president of a large dredging company. She is approached by senior executives of three competing dredging companies and asked to cooperate in bidding on federal government dredging contracts. If she submits high bids on the next three contracts, the other companies will submit high bids on the fourth contract and she will be assured of getting it. This proposal sounds good to Langlois, since she will be able to plan more effectively if she is assured of receiving the fourth contract. Is it ethical for Langlois to agree to this suggestion? If not, what action should be taken? If she agrees to this suggestion, does she run any greater risk than the other executives, assuming that only Langlois is a professional engineer?

3. Assume that you are a City Engineer—the manager of engineering for a Canadian city. The city has instructed you to develop plans and cost estimates for a very large sports complex. You discuss the plans with the city's chief administrative officer (CAO) and the city treasurer, who are arranging a loan to build the sports complex. They inform you that the sports complex is to be a jewel of pride for the city, and that it is a good time to build it, because interest rates are low. Over the next year, you notice that they are often wined and dined by representatives of the financial institutions who are competing to lend the funds to the city. Since you work fairly closely with the CAO and the treasurer on city business, you notice that they often receive small gifts, especially from the XYZ financial company with whom they are negotiating, and on several occasions they have attended NHL hockey games and played rounds of golf as guests of the XYZ executives. On at least one occasion, the CAO was flown to Florida for a golf holiday with XYZ executives. Eventually, the CAO and treasurer arrange a loan with the XYZ company and ask city council to approve the loan. The treasurer informs the mayor and city council that the interest rate will be 4.2 percent and that payments will total about $100 million over the thirty-year life of the loan. City council approves the loan, and construction starts on the sports complex.

About six months after construction starts, reporters for the local newspaper start to ask questions about the loan payments. The city's loan agreement with XYZ is very complicated, and a full schedule of payments is not shown. The newspaper hires a financial analyst to examine the agreement. The analyst's report, published on page 1 of the newspaper, states that the actual loan rate is about 8 percent and that payments will total about $225 million over the thirty-year life of the loan. Eventually, the CAO and treasurer apologize for the mistake, and say that the XYZ company assured them verbally of the lower rate but did not submit the final contract to them for signing until about an hour before city council met to approve it. As a result, they had no time to double-check the figures and, since they had such a good working relationship with XYZ, they believed the lower rate still applied.

Using the engineering Code of Ethics for your province or territory as a guide, what if any rules of ethical conduct have been broken by the CAO and treasurer? What action, if any, should the city take with respect to their employment? If, after the details became public, the CAO admitted that the golf holiday in Florida might appear to be a conflict of interest, and sent a cheque to the XYZ company to repay most of the travel costs, would it cancel the conflict of interest? As the City Engineer, you were not responsible for checking the loan agreement, but is your professional reputation affected? If you are also a city taxpayer, how do you feel about paying much higher taxes for the next thirty years because of this engineering project?

4. Using the Internet, search and find the Canadian patent for any well-known invention. The Canadian patent database is available through the Canadian Intellectual Property Office (CIPO) at <www.cipo.gc.ca> (as of September 20, 2003). Find the link to the American database. Continue your Internet search to see if there is a U.S. patent for the same invention. Are the patents identical? Compare the two databases. The American database is larger, but by how much? Which database is easier and faster to search?

5. Engineers are creative people and are usually employed to create intellectual property (patents, copyright, trademarks, integrated circuit plans, etc.). Most engineering employees are asked to sign a waiver of intellectual property rights, so that if an engineer invents a new device that is subsequently patented, the engineer is named as the inventor but the rights to the invention are assigned to the employer. Discuss the ethical aspects of this requirement. Is this an unfair infringement on the engineer's rights? What should happen if the engineer invents a new device in the evenings or on weekends when not on duty? Does it make any difference if the new device is in the same area as the engineer's employment or in a totally unrelated area? What action could a new employee take to avoid this conflict? What do other sources, such as NSPE's Guidelines for Professional Employment, say about this issue?

NOTES

1. Howard A. Levitt, *The Law of Dismissal in Canada*, as quoted in *CSPEAKER*, Canadian Society of Professional Engineers (CSPE), September 1981, pp. 1–4. Reprinted with permission of Howard A. Levitt.
2. Ibid.
3. Canadian Intellectual Property Office (CIPO), *A Guide to Patents*, Industry Canada, Ottawa.
4. CIPO, *A Guide to Industrial Designs*, Industry Canada, Ottawa.
5. CIPO, *A Guide to Integrated Circuit Topographies*, Industry Canada, Ottawa.
6. CIPO, *A Guide to Copyrights*, Industry Canada, Ottawa.
7. CIPO, *A Guide to Trade-marks*, Industry Canada, Ottawa.
    NOTES: All of the above guides [references 3 to 7] are available from the CIPO website:
    <strategis.ic.gc.ca/engdoc/main.html> (September 13, 2003). U.S. patents are available through the U.S. Patent and Trademark Office (PTO) at:
    <www.uspto.gov/patft/index.html> (September 13, 2003).
8. Canadian Securities Administrators (CSA), *Standards of Disclosure for Mineral Projects, National Instrument 43-101*. Document NI-43-101 can be found on the CSA websites in British Columbia (www.bcsc.bc.ca), Ontario (www.osc.gov.on.ca), Quebec (www.cvmq.com), and Alberta (www.albertasecurities.com)(September 15, 2003).
9. Case Study 8.5 is adapted from James G. Ridler, P.Eng., "Accountability: At the Core of Professional Engineering," *Engineering Dimensions*, vol. 18, no. 1, January–February 1997, pp. 40–41. Used with permission of James G. Ridler.
10. Turnitin (plagiarism detection service) <www.turnitin.com/static/home.html> (September 15, 2003).
11. W.N. Marianos, Jr., "Vancouver Second Narrows Bridge Collapse," from N. Schlager, ed., *When Technology Fails: Significant Technological Disasters, Accidents, and Failures of the Twentieth Century*, Gale Group, Detroit, 1994, pp. 191–95. Reprinted by permission of The Gale Group.

Chapter 9
Ethical Problems in Private Practice and Consulting

In this chapter we discuss the role of the professional engineer or geoscientist in private practice as well as several important aspects of such practice: the consultant's relationship with clients, the ethical aspects of advertising, the sealing of drawings, the competition for contracts, and the review of others' work. This chapter provides several case studies that illustrate some ethical pitfalls of private practice, as well as a case history that shows how a professional engineer solved a serious ethical dilemma.

THE CLIENT–CONSULTANT RELATIONSHIP

The consulting engineer or geoscientist needs to develop a good working relationship with the client. Usually the client engages the consultant to monitor an engineering project—for example, the design or construction of a building. This creates a three-way relationship between the client (owner), the contractor (designer or builder), and the consultant (engineer). Typically, the client requires the consultant's advice to ensure that the contractor's work is adequately performed and of good quality.

The relationship between the client and the consultant will, of course, depend on the personalities of the individuals, the type of project, and the problems encountered. These factors combine to create a wide spectrum of client–consultant relationships. D.G. Johnson describes three points along this relationship spectrum:

The **"independent" model.** The client explains the problem and then turns over decision-making power to the consultant, who takes charge of the problem and makes decisions for the client. The consultant does not provide technical knowledge to the client, but acts in place of the client, keeping the clients' interests in mind, in a paternalistic but independent way. This is one end of the client–consultant spectrum, and it is generally unacceptable, since it robs the client of the ability to make any choices.

The **"balanced" model.** The consultant interacts with the client, by providing engineering advice and evaluating the risks and benefits of various alternatives, but the client makes the choice of the action to follow. This relationship is similar to the ideal patient–physician relationship, where the professional may have the knowledge and expertise to solve the client's problem, but the client must be informed of the possible choices, and their benefits and risks, before making a decision to proceed with treatment. In a balanced relationship, the client and consultant must treat each other as equals. The consultant has a responsibility to provide engineering expertise to the client, but the client retains power to make the key decisions. The balanced relationship is the approximate mid-point of the spectrum, and is generally the optimum client–consultant relationship.

The "agent" model. The consultant is simply an agent or "order-taker" for the client, and contacts the client for instructions before acting. This is the other end of the client–consultant spectrum, and it is also generally unacceptable, since the client does not make full use of the engineer's knowledge. This relationship may also be seen as demeaning by the consultant.[1]

In any project, the client–consultant relationship will be situated somewhere along the spectrum described above. The precise point will depend on the personalities and relative knowledge of the participants and on the types of problems encountered. The goal is to keep information flowing between client and consultant to ensure that both are fully aware of the crucial areas of the work.

ADVERTISING FOR ENGINEERING WORK

An engineer in private practice may need to advertise. This brings up a thorny issue that plagues all the professions in North America. Every province places some restrictions on how engineers can advertise their services; usually these are found in the Code of Ethics or in professional practice guidelines. These restrictions are intended to ensure fairness and honesty when clients evaluate professional qualifications and experience.

Advertising fills our newspapers, magazines, radio programs, and television screens, and even "pops up" on our webpages. Engineers would obviously consider it demeaning and unprofessional to promote their services as if they were soap powder, soft drinks, or chewing gum. However, advertising that communicates facts about the availability, experience, and areas of expertise of an engineer in private practice is fair and unobjectionable. "Calling card" or "business card" advertising of the type seen in the back pages of most engineering publications is certainly an acceptable, professional form of advertising. In recent years, clearer guidelines have been established for advertising, and other forms are acceptable provided they maintain the professional image that the Associations have worked for decades to achieve.

Most Codes of Ethics simply state that the professional engineer or geoscientist should maintain the dignity, honour, and integrity of the profession; specific rules for advertising are defined in professional practice guidelines. For example, in the *Guideline for Ethical Practice*, APEGGA (Alberta) requires all advertisements, proposals, presentations, and solicitations for professional engagement to be "factual, clear and dignified."[2] In Quebec, Regulation 10 under the Act gives very precise rules concerning the information that may be conveyed on business cards and stationery, in newspapers, magazines, and directories, and on signs on work premises, offices, and vehicles.

In Ontario, regulations made under the authority of the provincial Act permit advertising provided that it is done in a professional and dignified manner; that it is factual and does not exaggerate; and that it does not directly or indirectly criticize another licensed engineer or the employer of another licensed engineer. The regulation also expressly forbids the use of the engineer's seal or the Association's seal in any form of advertising; this means that these seals cannot be used on business cards or letterhead.[3] The seal has a legal significance (explained later in this chapter) that is totally incompatible with advertising. The Association's name and logo may be used on business cards and letterhead, but only to signify membership in the Association. In its *Guideline to Professional Practice*, PEO (Ontario) publishes several advertising rules, which are reproduced with permission.

Advertising may be considered inappropriate if it:

i.   claims a greater degree or extent of responsibility for a specified project or projects than is the fact;

ii. fails to give appropriate indications of cooperation by associated firms or individuals involved in specified projects;

iii. implies, by word or picture, engineering responsibility for proprietary product or equipment design;

iv. denigrates or belittles another professional's projects, firms or individuals;

v.   exaggerates claims as to the performance of the project; or

vi. illustrates portions of the project for which the advertiser has no responsibility, without appropriate disclaimer, thus implying greater responsibility than is factual.[4]

In general, advertising is acceptable if it is factual and truthful, and communicates accurate information about qualifications, experience, location, or availability, in a dignified manner.

ENGINEERING COMPETENCE

Engineering competence gained through education or experience is a valuable asset. The client is paying for that competence when hiring the engineer. An engineer who accepts an assignment that is beyond his or her level of competence could be guilty of either unprofessional conduct or incompetence. Either of these could lead to disciplinary action.

This does not mean that an engineer must be an expert in every phase of a proposed project before accepting it. However, the engineer must know whether he or she can become competent, through study or research, in a reasonable period of time. Alternatively, the engineer must know whether a colleague or consultant with the needed expertise can be hired without delaying the project. The essential point is that the client's project must not be placed at risk (or become needlessly expensive) because of the engineer's lack of competence.

As an engineer you are expected to know your level of competence; you are also expected to expand your knowledge and experience and maintain your competence (as discussed in Chapter 15). You must be realistic about your abilities—a difficult task at the best of times. However, no one knows the limits of your knowledge better than you do yourself.

USE OF THE ENGINEER'S SEAL

Each provincial Act provides for engineers to obtain and use a seal on approved documents. This seal is usually an inked rubber stamp which indicates that the person named on the stamp is licensed in that province or territory. The terms "seal" and "stamp" are interchangeable. The Act (or a regulation) typically requires that all final drawings, specifications, plans, reports, and other documents involving the practice of professional engineering, when issued, be dated and bear the signature and seal of a professional engineer. The use of the seal is not optional; it is a standard requirement under the provincial Act (or regulations). For example, Ontario's regulations require engineers to "sign, date and affix the . . . seal to every final drawing, specification, plan, report or other drawing prepared or checked" by the engineer.[5]

An APEGGA (Alberta) guideline explains the significance of the seal:

> A professional stamp or seal affixed to a document is intended to indicate that the document has been produced under the supervision and control of a fully qualified professional member of APEGGA, or that it has been thoroughly reviewed by a professional member of APEGGA who accepts responsibility for it. Professional stamps and seals shall be affixed, signed and dated only after the responsible member is satisfied that the document or component, for which he or she is professionally responsible, is complete and correct.[6]

An engineer or geoscientist who signs or seals documents that are not based on thorough knowledge may be guilty of professional misconduct. In a case cited recently, the British Columbia Supreme Court ruled that an engineer was liable in a dispute over an improperly designed residence foundation. The court stated: "By affixing his seal to the drawings and by his letter to the defendant municipality . . . the defendant [engineer] . . . certified that the foundation drawings conformed to all the structural requirements of the 1980 National Building Code."[7] Clearly, the court considered the seal on the drawings a guarantee of their accuracy and conformance with codes. But in a somewhat different type of case, the Supreme Court of Canada ruled that "[t]he seal attests that a qualified engineer prepared the drawing. It is not a guarantee of accuracy. The affixation of the seal, without more, is insufficient to found liability for negligent misrepresentation."[8]

These contradictory rulings suggest that there is still some debate over fine legal points. Accordingly, the best strategy for the professional engineer is to avoid the courts entirely. Do not affix your seal to a document unless you are willing to accept full responsibility for it, based on detailed knowledge of the document and of the project to which it applies, and unless you are completely satisfied with the document's accuracy.

Relying on the Work of Others

During your career, you will frequently have to depend on the work of others, be they employees, fellow workers, suppliers, or consultants. Their work will typically be presented to you in the form of a document (drawing, report, logbook, etc.). When you are responsible for a major project and must use others' work, how should you verify that work? If a project fails because you relied on inaccurate information provided by others, who is responsible?

APEGGA (Alberta) has recently issued a comprehensive guideline on this topic, which states that work done by others must never be accepted blindly; due diligence must always be applied to it. The degree of diligence will depend on "the type of information, how it is applied, the member's professional experience, and the impact of poor quality work on the final product." If any doubts remain concerning the validity, reliability, applicability, or thoroughness of work provided by others, that work should be set aside and due diligence or verification should continue.

This particular guideline is too long to present here. However, many of its key points can be condensed into fairly simple questions, which constitute a cursory summary of a very thorough process. When you must depend on the work of others, ask yourself the following:

- **Necessity.** Is it necessary for you to rely on the work?
- **Applicability.** Do you understand the purpose, methods used, and limitations of the work?
- **Credibility.** Are the qualifications, experience, and reputation of the authors, and of the sponsoring institution or company, acceptable?
- **Quality of documentation.** Was an established methodology used? Are the results supported by data? Are sources available? Are data, facts, interpretations, assumptions, opinions, and anecdotes clearly separated and identified as such?
- **Corroboration.** Are results corroborated by duplicate tests? Have computer results been compared with actual tests or other software output and/or hand calculations?
- **Limitations.** Are limitations or caveats clearly stated and understood? In particular, for statistical results, are extrapolations or interpolations required? Have statistics been interpreted properly?
- **Age.** Have current methods, codes, and industry standards been used? Have new, untested or speculative methods been used?
- **Integrity.** Is the document original (or unaltered)?[9]

If you assert the due diligence suggested in the list of questions above, then you can reasonably rely on the information—that is, the information can reasonably be expected to be safe, and decisions based on the information are unlikely to cause harm or damage.

Such care or diligence is essential, for a basic reason: When you use information from others in making a key decision, or include that information in a document and then sign, seal, and date it, you assume professional responsibility for it. That is, by your use, you confirm that you have exerted reasonable care, under the circumstances, to have confidence in the information.

Conversely, if you ignore this due diligence, and you erroneously use unreliable information or include it in a document, then you cannot later blame others (such as the sources of the information) for what may result.

Checking Engineering Documents

It is important to define what "checking" means. If it means scanning a document for obvious errors and then passing it back to a colleague, who signs, seals, and assumes responsibility for it, this is simple courtesy and you have no liability.

However, if checking means examining, signing, and sealing a document, or otherwise assuming responsibility for it, then it really means *approving* the document, and approval should never be given unless you have analyzed the work carefully enough to be confident that it is correct.

How would you respond to these two hypothetical scenarios?

- **Request from a friend**. You are approached by a friend who is not licensed. She asks you, as a favour, to check, sign, and seal an engineering or geoscience drawing to satisfy municipal bylaws. She assures you that the drawing is completely correct and that your signature and seal are mere formalities.

- **Pressure of work.** You are the designated professional on the Certificate of Authorization (or permit to practise) for an engineering or geoscience firm. The firm has many projects underway, and you are unable to monitor all of them adequately. Your employer asks you to sign and seal documents for a project, of which you were previously unaware.

In each of the above situations, you would refuse to sign or seal the document until you have performed an adequate analysis of the work and are willing to accept full responsibility for it. Do not be led into a trap by external pressure. For some documents, a proper check would require complete duplication of the analysis. (Obviously, if you completely redo the work, it is appropriate for you to assume responsibility for it.)

As a professional you must always refuse to sign and seal documents prepared by unlicensed practitioners, so that they can practise without a licence, or so that they can avoid the scrutiny or cost of a full engineering analysis. Also, regardless of the urgency of requests to sign and seal documents, do not assume responsibility for work that is beyond your area of expertise, or for work that you have not reviewed thoroughly for accuracy.

Preparation and Approval

If one engineer has prepared a document or drawing and another engineer has approved it, then both seals should be affixed. If this is not possible, or not expected, then only the approving engineer should seal it. This seal indicates that he or she takes the responsibility for the document or drawing. Where final drawings cover more than one engineering discipline, it is typically recommended that the drawings be sealed by the approving engineer (typically a chief engineer or project leader) and by the design engineer for each discipline. The seals of the design engineers should be "qualified" by an explanatory note that indicates clearly each engineer's area of responsibility.

Sealing of Preliminary Documents

Preliminary documents, drawings, or specifications are usually not sealed; instead, they are clearly marked "preliminary" or "not for construction." Only the final drawings are sealed. Similarly, an engineer should not seal a document that has no engineering content. Sometimes, to satisfy the requirements of a regulatory agency, a preliminary document may need to be sealed. In this situation the comment "preliminary" or "not for construction" should appear prominently.

Sealing of Reports and Detail Drawings

The engineer generally has responsibility for a project as a whole, and his or her seal must appear on the major reports, specifications, and drawings that describe the project. Individual pages of a report (or drawings included in a report) need not be sealed, provided the report as a whole has been signed, sealed, and dated.

Usually the engineer is not expected to seal every detail drawing. However, the drawings must be prepared under the engineer's control and supervision, and he or she assumes responsibility for them whether they are sealed or not. For example, the special case of structural steel is described in Ontario's *Guideline to Professional Practice*:

> In the case of structural steel, the steel supplier provides shop drawings for review by the structural engineer. The steel supplier has selected standard connections from the handbook published by the Canadian Institute of Steel Construction according to the moments and forces given by the engineer for each connection. The selection of "standard connections" is not considered to be part of professional engineering practice and such shop drawings need not be sealed. Shop drawings depicting special connections do require sealing.
>
> Some design engineers require a seal on all shop drawings and erection diagrams. This is their prerogative. Alternatively, a letter signed by a professional engineer stating that the shop drawings have been prepared under his or her supervision may be acceptable.[10]

Sealing of Masters and Prints

The master drawings must, of course, be complete and unambiguous, since they are usually the major reference for describing the concepts and the details of the structure, machine, process, or whatever is being designed. An engineer in private practice must have an effective procedure in place for controlling the issuing of preliminary and final drawings so that the client's security and confidentiality are protected and no confusion can arise. The appropriate time to seal a drawing is when it is approved and released for fabrication or construction. Modifications to final drawings must be rigidly controlled and documented. This control is aided by sealing only the prints and not the master. In this way, the prints can be checked for modifications when they are sealed.

Sealing of Computer-Generated Drawings

Computer-aided design has simplified the production of drawings but has also created security problems. Faulty data and unauthorized modifications are usually difficult to detect. Electronic seals must be controlled with passwords so that unauthorized or preliminary prints cannot be confused with master copies. Professionals in private practice must maintain some form of control within the firm in order to prevent unauthorized copying or modification of computer files containing design drawings, reports, specifications, and so on. It is recommended that until better standards and procedures have been developed, final drawings in computer file form be protected by password or by storing them on digital media, in secure locations. In addition, seals should be applied—by hand—only to prints made from these files. The prints or hard copies of the files would then constitute the master document. Copies of the engineer's seal generally should not be reproduced in any form, whether by computer or by stick-on labels, since this is equivalent to losing secure control of the seal.

CONFIDENTIALITY

Under the Code of Ethics, professionals are obligated to keep the client's affairs confidential. A client will sometimes ask the engineer to sign a confidentiality agreement. Since engineers intend to maintain confidentiality anyway, they are usually willing to sign these agreements. The requirement for confidentiality can create ethical problems, however. Consider the following two cases:

- If an engineer in private practice is hired by a new client who is a competitor of a former client, the engineer may have a conflict of interest. The engineer must not accept a contract that requires disclosure of a previous client's affairs, be they technical, business, or personal. This applies especially to proprietary information and to trade secrets that could result in financial loss to the former client, if disclosed. Even if the competitor does not expect the professional to disclose the information, the appearance of a conflict of interest may remain. In this situation, any confidentiality agreement signed with the former client should be reviewed. Undertaking work for the competitor is obviously risky.

• Another problem with confidentiality agreements arises in environmental projects. Where danger to the public is involved, the Code of Ethics (or environmental regulations) may require the engineer to reveal information. Consider a case where an engineer advises a client to remedy an environmental hazard. What should happen if the client refuses to do so? If the engineer has signed a confidentiality agreement and later blows the whistle to the authorities, this could be interpreted as a breach of contract. Clearly, the engineer is facing a serious ethical dilemma: breach the contract, or obey the law. Ontario's *Guideline to Professional Practice* suggests a compromise: include a clause in the confidentiality agreement stating that if the client should fail to act on certain hazards within a specified period of time, the engineer is entitled to fulfil any reporting requirements that are specified in law, after first notifying the client.[11] An engineer practising in the environmental area should get legal advice on the proper wording of such agreements.

CONFLICT OF INTEREST

As noted earlier, a conflict of interest arises in a professional relationship when the professional has an interest that interferes with the service owed to the client. For example, an engineer who recommends that a client purchase goods or services from a company in which the engineer has partial ownership (of which the client is not aware) has created a serious conflict of interest that is contrary to the Code of Ethics.

Conflicts can be much simpler than this. An engineer may be tempted to suggest that the client adopt a course of action where the main benefit is to reduce the engineer's workload. Unless there is a similar reduction in fee, the engineer has a conflict of interest that must be disclosed fully to the client. In every instance of conflict (or potential conflict) of interest, the engineer must make a full disclosure to the client of the engineer's personal interest, whatever that may be. If the client agrees that the conflict is insignificant, the work can proceed, but now the client is making a fully informed choice.

A client who later learns that an engineer benefited personally and secretly from a decision that was ostensibly based on technical factors would be justified in contacting the provincial Association to lodge a complaint of professional misconduct.

REVIEWING THE WORK OF ANOTHER ENGINEER

The issue of reviewing another engineer's work is especially sensitive when engineers are in private practice. As a general rule, an engineer must be informed when his or her work is to be reviewed, but it is not necessary to obtain the engineer's permission for that review. In all instances, the welfare of the client or the general public must come before the engineer's personal wishes. Ontario's *Guideline to Professional Practice* summarizes the situation as follows:

> The Code of Ethics permits engineers to be engaged to review the work of another professional engineer when the connection of that engineer with the project has been terminated. Before undertaking the review, reviewers should know how the information will be used. Even when satisfied that the connection between the parties has been terminated, reviewers should, with the agreement of the client, inform the other engineer that a review is contemplated. They should recognize that the client has the right to withhold approval to inform the engineer, but [should] satisfy themselves that the reasons for the owner's decision are valid before proceeding with the review.

> If a client asks an engineer to review the work of another engineer who is still engaged on a project, either through an employment contract or an agreement to provide professional services, the reviewer should undertake the assignment only with the knowledge of the other engineer. Failure to notify the engineer under this circumstance constitutes a breach of the Code of Ethics. On the other hand, should a second engineer be engaged by another person (say, a building department) to provide professional engineering services on the same project, he or she would have no obligation to advise the original client of the commission.

> Senior engineers are often asked to review a design prepared by another engineer. (Most engineers are expected to have their work routinely reviewed as part of an ongoing quality control and professional development process.) If reviewers find that design changes are necessary, they should inform the design engineer of these findings and the reasons for the recommended changes. During the design stage, reviewers (who are acting as the client's agent in this case) and engineers may agree on changes to the engineers' proposal. However, design engineers must not agree to any change or alternative suggested by reviewers that could result in an unworkable installation, be in conflict with the relevant codes, or create a risk of damage or injury.

> Reviewers must administer the design contract and evaluate engineers' work at arm's-length, so that the engineer of record maintains full responsibility for the design.

> ...

> It is emphasized that the acceptance of mutually agreed-upon changes does not relieve the original design engineer of responsibility for the design or work under review.

> Once the review has been completed, there is no obligation or right for the reviewers to disclose their findings to the other engineer. In fact, in most cases, disclosure of the findings would not be permitted by the client. Reviewers' contractual obligations are to the client. However, reviewers should seek the client's approval to inform the engineer of the general nature of the findings, and if appropriate, should try to resolve any technical differences.[12]

DESIGN CALCULATIONS

A client may request that an engineer submit calculations that were done to support a recommendation. This amounts to a review of the engineer's work, but obviously it is done with the full knowledge and cooperation of the engineer. The client has an ethical right to review these calculations and to make a copy for a permanent record. However, the time necessary to prepare the calculations in a format understandable to the client should be included as part of the contracted service.

Occasionally the computation techniques or the data on which the computation is made may be proprietary and the engineer may not wish to divulge them. In this situation, the conditions for reviewing the calculations should be negotiated beforehand and the extent of disclosure should be understood in advance. The usual procedure is to provide the proprietary data to the client with the clear understanding that they will be kept confidential.

COMPETITIVE BIDDING FOR SERVICES

A detailed procedure for selecting an engineer in private practice was described in Chapter 5. The procedure involves three stages and separates the process of selecting the best-qualified engineer (or firm) from the process of negotiating the fees. This prevents many problems that commonly arise when engineers are selected on a competitive basis by lowest bid.

However, it should be emphasized that seeking professional services by lowest bid is neither illegal nor unethical, and no one should be dissuaded from the procedure by the misguided belief that competition is harmful. Quite the opposite: ingenuity thrives on healthy competition. However, there is a danger in competitive bidding, as explained in Ontario's *Guideline to Professional Practice*:

> With professional services there are ultimately only two elements which a client is retaining, i.e. the engineer's knowledge and time. Shortchanging on a professional engineering fee will result in the substitution of less skilled engineers or less time put into the assignment, thus potentially shortchanging the project.[13]

However, some competitive activities in obtaining contracts are considered unfair and unethical. For example, any agreement to pay a kickback, gift, commission, or consideration, either openly or secretly, would be considered an unfair and unethical (and likely illegal) method of obtaining contracts. Many codes also describe supplanting a colleague as unethical, where supplanting is defined as intervening in the client–engineer relationship of a colleague and, through inducements or persuasion, convincing the client to fire the engineer and hire the intruding engineer.

NEGLIGENCE AND CIVIL LIABILITY

The engineer in private practice is usually careful to avoid the two main sources of liability: breach of contract and negligence. These are usually inadvertent, and both must be distinguished from professional misconduct and incompetence (which are discussed in Chapter 14). A breach of contract is a failure to complete the obligations specified in a contract, whereas negligence is a failure to exercise due care in the performance of engineering. It is possible to obtain protection for breach of contract by incorporating a practice, and protection against negligence by purchasing liability insurance; however, it is not possible to avoid disciplinary action for negligence, incompetence, or professional misconduct. Ontario's *Guideline to Professional Practice* summarizes the situation as follows:

> An individual engineer can protect personal assets against an action for damages for breach of contract by incorporating the practice. After incorporation, it is the company that is the contracting party and not the individual. As far as protection from liability for negligence there is nothing available to an engineer other than careful, thorough engineering and insurance.[14]

ACEC CODE OF CONSULTING ENGINEERING PRACTICE

The Association of Consulting Engineers of Canada (ACEC) has a code of practice that applies to member firms of ACEC and requires them to fulfil their duties with honesty, justice, and courtesy toward society, clients, other consulting engineers, and employees. The code is reproduced below:

ACEC Code of Consulting Engineering Practice

Members of the Association of Consulting Engineers of Canada shall fulfil their duties with honesty, justice and courtesy towards Society, Clients, other Consulting Engineers and Employees.

Society

- Members shall practice their profession with concern for the social and economic well-being of Society.

- Members shall conform with all applicable laws, by-laws and regulations.

- Members shall satisfy themselves that their designs and recommendations are safe and sound and, if their engineering judgement is overruled, shall report the possible consequences to clients, owners and, if necessary, the appropriate public authorities.

- Members expressing engineering opinions to the public shall do so in complete, objective, truthful and accurate manner.

- Members should participate in civic affairs and work for the benefit of their community and should encourage their employees to do likewise.

Clients

- Members shall discharge their professional responsibilities with integrity and complete loyalty to the terms of their assignments.

- Members shall accept only those assignments for which they are competent or for which they associate with other competent experts.

- Members shall disclose any conflicts of interest to their clients.

- Members shall respect the confidentiality of all information obtained from their clients.

- Members shall obtain remuneration for their professional services solely through fees commensurate with the services rendered.

Other Consulting Engineers

- Members shall relate to other consulting engineers with integrity, and in a manner that will enhance the professional stature of consulting engineering.

- Members shall respect the clientele of other consulting engineers and shall not attempt to supplant them when definite steps have been taken towards their employment.

- Members shall compete fairly with their fellow consulting engineers, offering professional services on the basis of qualifications and experience.

- Members engaged by a client to review the work of another consulting engineer, shall inform that engineer of their commission, and shall avoid statements which may maliciously impugn the reputation or business of the engineer.

Employees

- Members shall treat their employees with integrity, provide for their proper compensation and require that they conform to high ethical standards in their work.

- Members shall encourage their employees to enhance their professional qualifications and development.

- Members shall not request their employees to take responsibility for work for which they are not qualified.[15]

FIDIC CODE OF ETHICS

The International Federation of Consulting Engineers (FIDIC) has adopted a Code of Ethics that guides the conduct of individual consulting engineers and member firms of consulting engineers. In addition, many FIDIC policy documents on consulting engineering practice are available on their website. For our purposes the Code of Ethics is the most relevant FIDIC document; it appears to be in complete agreement with the ACEC Code of Ethics and other codes discussed in this text. It is reproduced below with permission.

FIDIC Code of Ethics

The International Federation of Consulting Engineers recognises that the work of the consulting engineering industry is critical to the achievement of sustainable development of society and the environment.

To be fully effective not only must engineers constantly improve their knowledge and skills, but also society must respect the integrity and trust the judgement of members of the profession and remunerate them fairly.

All member associations of FIDIC subscribe to and believe that the following principles are fundamental to the behaviour of their members if society is to have that necessary confidence in its advisors.

Responsibility to Society and the Consulting Industry

The consulting engineer shall:

- Accept the responsibility of the consulting industry to society.

- Seek solutions that are compatible with the principles of sustainable development.

- At all times uphold the dignity, standing and reputation of the consulting industry.

Competence

The consulting engineer shall:

- Maintain knowledge and skills at levels consistent with development in technology, legislation and management, and apply due skill, care and diligence in the services rendered to the client.

- Perform services only when competent to perform them.

Integrity

The consulting engineer shall:

- Act at all times in the legitimate interest of the client and provide all services with integrity and faithfulness.

Impartiality

The consulting engineer shall:

- Be impartial in the provision of professional advice, judgement or decision.

- Inform the client of any potential conflict of interest that might arise in the performance of services to the client.

- Not accept remuneration which prejudices independent judgement.

Fairness to Others

The consulting engineer shall:

- Promote the concept of "Quality-Based Selection" (QBS).

- Neither carelessly nor intentionally do anything to injure the reputation or business of others.

- Neither directly nor indirectly attempt to take the place of another consulting engineer, already appointed for a specific work.

- Not take over the work of another consulting engineer before notifying the consulting engineer in question, and without being advised in writing by the client of the termination of the prior appointment for that work.

- In the event of being asked to review the work of another, behave in accordance with appropriate conduct and courtesy.

Corruption

The consulting engineer shall:

- Neither offer nor accept remuneration of any kind which in perception or in effect either a) seeks to influence the process of selection or compensation of consulting engineers and/or their clients or b) seeks to affect the consulting engineer's impartial judgement.

- Co-operate fully with any legitimately constituted investigative body which makes inquiry into the administration of any contract for services or construction.[16]

CASE STUDY 9.1— BENEFITING FROM A CONFLICT OF INTEREST

STATEMENT OF THE PROBLEM

Edward Beck is a consulting engineer in a small town. He has been elected to sit on the town council as a councillor, a part-time job that he does mainly as a form of public service. Beck has also been hired by a developer to draw up plans for the street layout and water and sewage facilities for a new residential subdivision in the town. The developer's submission to town council includes Beck's drawings and specifications. Later, during a town council meeting, Beck votes to approve the subdivision. During the discussion, Beck does not publicly state his relationship with the developer, nor does he conceal it. His signature and seal are on some of the plans submitted to council. Everyone knows that he is the only engineer in town who does this type of work, and he is certain that they would prefer to see local people hired for this project.

QUESTION

In voting to approve this project, has Beck acted unethically?

AUTHOR'S RECOMMENDED SOLUTION

This situation sometimes occurs in small towns with few engineers, where a conflict of interest cannot be avoided. Engineers certainly should not be disqualified from projects because they are performing a public service as members of town councils. However, in this case it is not enough that "everyone knows" that Beck has a business relationship with the developer. Beck created a serious conflict of interest when he voted to approve plans that he himself prepared. He should have made a clear, unequivocal statement of his involvement in the project and his relationship with the developer, then withdrawn from the debate and abstained from the vote. By participating in a formal vote without declaring the conflict of interest, Beck has exposed himself to the possibility of a complaint to the provincial Association and to possible disciplinary action.

CASE STUDY 9.2—ADVERTISING COMMERCIAL PRODUCTS

STATEMENT OF THE PROBLEM

Alonso Firenze is a consulting engineer to the Acme ATV Company, which manufactures all-terrain recreational vehicles. While preparing a television campaign to increase sales of the vehicle, the company's ad man suggests that Firenze appear on camera to endorse the vehicle's safety features as a professional engineer and safety expert. The ad man points out that Firenze has conducted extensive tests, studies, and surveys on the vehicle and can speak with authority. In addition, Firenze is a handsome guy and would enjoy the exposure to the general public.

QUESTION

Would it be unethical for Firenze to appear in the television commercial and make a statement endorsing the recreational vehicle?

AUTHOR'S RECOMMENDED SOLUTION

This case study illustrates the potential risk, both to the profession and to Firenze's reputation, when professional standards are subordinated to commercial pressures. The proposal raises ethical issues in several ways:

Accuracy of the information. Television commercials are not documentary programs. Most documentary programs attempt to evaluate an issue in a balanced way, and seek out the opinions of experts. In contrast, the main goal of commercial advertising is to increase sales, and the participants are typically actors who read scripts prepared by others. This explains why commercial television advertising has a rather sordid history of half-truths. Advertising appeals to one's emotions rather than to logic. Firenze is being asked to participate in an activity that is not intended to convey a balanced, objective view of the product. He also runs the risk of having his comments taken out of context by the advertiser. The result may be a commercial message that has the appearance of credibility but not the substance of an impartial evaluation.

Conflict of interest. Perhaps Firenze would be appearing in the commercial to please the employer, or perhaps to advance his personal career. Whichever the case, a conflict of interest may result. The advertising is not intended to help him carry out his main function as a professional engineer and safety expert.

Demeaning the profession. Will viewers of the commercial conclude that Firenze is appearing because of his personal convictions, or because the employer has instructed him to perform? Will they believe that Firenze is giving his honest opinions, or will they believe he is reading a script?

Summary. Almost all Codes of Ethics specifically direct engineers to enhance the public regard for the profession, and require engineers to make public statements only when they are founded on adequate knowledge and firm conviction. Some provincial Associations publish very specific guidelines to ensure that advertising does not demean the profession.

Participation in a documentary program on the vehicle's safety would be a suitable professional activity (and perhaps more engineers should participate in such programs). But if Firenze agreed to serve as a company "pitch man," viewers would be more likely to conclude that he was reading a script written by others. This would probably reduce public esteem for the engineering profession and therefore would breach the Code of Ethics.

CASE STUDY 9.3—CONTINGENCY FEE ARRANGEMENTS

STATEMENT OF THE PROBLEM

As an engineer in private practice, you are considering whether to offer your services on a contingency basis, an arrangement whereby you would be paid a percentage of some outcome. Two clients wish to retain you:

- Client A wants to retain you to act as an expert witness in a lawsuit against a third party. The lawsuit, if successful, should result in the award of a very large sum as a settlement.
- Client B has shown a tentative interest in retaining you to recommend changes to the energy usage in a manufacturing process. You believe Client B would be more responsive if fees were contingent on the savings. After an initial study of the problem, you believe that the energy savings could be immense.

QUESTION

Would it be ethical to offer your services on a contingency basis to either of these clients, with the understanding that you would be paid a percentage of the legal settlement (Client A) or a percentage of the value of the energy savings (Client B)?

AUTHOR'S RECOMMENDED SOLUTION

These two cases seem similar but are distinctly different.

An expert witness is permitted to express opinions, whereas a nonexpert witness must confine his or her testimony to known facts. Therefore, an engineer testifying as an expert witness must have an impartial attitude toward the outcome of the case. However, as a recipient of a percentage of the potential settlement from Client A, you would have a conflict of interest and your testimony would be suspect. Therefore, it would be unethical to accept this case on a contingency basis. You should bill Client A for your time and expenses so that the reimbursement is independent of the outcome of the case.

The case of Client B is somewhat different, since there is no need for impartiality. In fact, your bias toward reducing energy consumption could be very beneficial to the client. Also, you have a duty to yourself and to your colleagues to charge an adequate fee. From your study, you evidently believe that this fee will be adequate. Therefore, the proposal to base the fee on a contingency is not unethical. However, a word of warning: there might be a perception of unethical behaviour unless the results can be measured accurately and impartially and can be achieved without degrading the client's product or facilities. Therefore, although this method of setting a fee is not unethical, some risks are associated with it. You would be well advised to use one of the more common billing methods (as described in Chapter 4) unless the client expresses a preference for the contingency method and the savings can be clearly and unequivocally measured.

CASE STUDY 9.4—ADHERENCE TO PLANS

STATEMENT OF THE PROBLEM

A professional engineer in private practice is engaged by a building contractor to prepare drawings for the forms and scaffolding needed to construct a reinforced concrete bridge. The forms and scaffolding must sustain the weight of about 1400 tonnes of concrete until the concrete is cured. The engineer prepares the drawings and signs and seals the originals, which he gives to the contractor. The contractor later engages the engineer to inspect the completed structure. The engineer finds that the contractor has made several major deviations from the plans. He is not sure whether the structure is safe or unsafe. The contractor has stated that time is of the essence, and concrete is to be poured in the next forty-eight hours. The engineer feels an obligation to the contractor because of their previous professional relationship and hopes that it will continue.

QUESTION

What should the engineer do?

AUTHOR'S RECOMMENDED SOLUTION

Two issues are at stake here. Once the engineer passed the sealed original drawings to the contractor, control was lost. It is possible that changes were made to the originals that, if unsafe, could create serious problems for the engineer. As a general rule, only prints should be signed and sealed so that modifications will be evident. In this case, apparently no changes were made. However, the contractor did not construct the forms and scaffolds according to the plans, and the engineer is now faced with the unpleasant task of informing the contractor that the deviations from the plans must be evaluated to ensure that they are safe. This will undoubtedly require some calculations and perhaps a second inspection. The engineer should notify the contractor in writing that concrete must not be poured until the review and reinspection is complete and that the structure could constitute a hazard to workers and the general public. The strength analysis should be carried out as quickly as possible. If the forty-eight-hour deadline cannot be met, the project must not proceed until all safety concerns have been satisfied. It is perhaps useful to point out that the contractor could have consulted the engineer about the changes earlier in the construction so that the delay could have been avoided. Failures occur more often in temporary structures, because they do not receive sufficient analysis. The collapse of the temporary supports for the Vancouver Second Narrows Bridge illustrates this point. (Review the case history in Chapter 8.)

CASE STUDY 9.5—FEE REDUCTION FOR SIMILAR WORK

STATEMENT OF THE PROBLEM

Susan Johnson is a professional engineer in private practice. She is hired by Client A to design a small, explosion-proof building for storing flammable paints, chemicals, and explosives. The work is carried out in her design office, and copies of the plans are provided to her client. After construction is complete, she is approached by Client B, who has seen the building and has a similar requirement. Client B suggests that the fee should be substantially reduced since the design is already finished and only minor changes would be required.

QUESTION

Would it be ethical for Johnson to reduce her fees as suggested? Would it be good business practice?

AUTHOR'S RECOMMENDED SOLUTION

First, Johnson should clarify whether she or Client A owns the copyright for the drawings. (This should have been specified in the original agreement with Client A.)

Second, establishing fair and reasonable fees depends on five factors:

- level of knowledge and qualifications required;
- difficulty and scope of the assignment;
- responsibility that the engineer must assume;
- urgency with which the work must be accomplished (will overtime payments or extra personnel be required?); and
- time required (number of person-hours).

Of these five factors, only the last two (urgency and time required) are likely to be reduced because of the earlier project. If Client A were requesting a second building of the same design, then it might be appropriate to pass on some of the savings in time. However, Client B benefits from receiving a design that has been tested and is likely to be more dependable and easier to construct. The level of knowledge, the qualifications, and, most important, the responsibility the engineer must assume are unchanged. In summary: it would be unfair to Johnson and poor business practice to accept a substantial fee reduction for providing the drawings for this structure.

CASE STUDY 9.6—ALLEGED COLLUSION IN FEE SETTING

STATEMENT OF THE PROBLEM

A large corporation wants to expand its manufacturing facilities and interviews three consulting engineer firms to design and supervise the construction of the new plant. Each consulting engineering firm states in its proposal that fees would follow the schedule published by the provincial Association. The corporation decides that it could reduce the cost of the project by conducting some initial studies itself and by providing its own engineers to assist in supervising the construction of the plant. The corporation then asks each consulting firm to quote on how much its consulting fees would be reduced if the corporation provided this assistance.

The three consulting firms meet, discuss the corporation's request, and then submit the same amount as a fee reduction. The corporation complains to the provincial Association that the engineers are colluding in their bids and that this conduct is unethical if not illegal.

QUESTION

Is it unethical for the three consulting firms to agree on the fee reduction to be allowed for the assistance?

AUTHOR'S RECOMMENDED SOLUTION

The contracting procedure recommended by the Association of Consulting Engineers of Canada (ACEC) for engaging consulting engineers is fairly well established and was described earlier in this text. The procedure provides for

The 59-storey Citicorp Tower in New York City was designed by William LeMessurier. Because of lot restrictions, the building was supported by four massive pillars, as shown in the photo. In May 1978, a few months after the building's completion, LeMessurier learned that a design fault could cause the building to collapse under certain severe wind conditions. The gripping story of his successful efforts to strengthen the building before the start of the storm season is a tale of professional ethics at its best.

Source: © Alan Schein Photography/ CORBIS/Magma.

competition on the basis of qualifications, experience, scheduling, and service; however, it discourages competition based on price alone. Although competitive bidding is not illegal or unethical, corporations must make the basis for selection clear when issuing the request for proposal.

In this case, the corporation seems to have been following the ACEC procedure in the early stages, and the engineers responded appropriately. It seems that the corporation misunderstood the procedure. Price competition is not part of the ACEC contracting procedure, although fee negotiation is appropriate once a consulting firm has been selected. Here, the cooperative action by the firms is not unethical. At no time was it made clear that the corporation wanted competition on a fee basis, and the complaint after the fact reveals an apparent misunderstanding on the part of the corporation.

CASE HISTORY 9.1

LEMESSURIER AND THE CITICORP TOWER

This event shows an exceptional engineer resolving a serious ethical dilemma. This case is inspiring, and differs from most of the case histories in this text, in which negligent or incompetent decisions led to disaster.

It is instructive to note that the designer, William LeMessurier, applied a problem-solving process several times in this case. Early on, before the public became aware of the design flaw, LeMessurier faced an ethical dilemma. He pondered several alternatives, including concealing the flaw. He decided that the most ethical route was to face the problem squarely. He then went through the problem-solving process three more times (at least) to resolve technical, implementation, and financial problems.

Introduction

Structural engineer William LeMessurier was hired as a consultant to advise the designers of the fifty-nine-storey Citicorp Tower in New York City. The design was severely constrained by the building site. Specifically, one corner of the lot was occupied by a church. It was agreed that the Citicorp Tower could occupy the space above the church; however, it could not touch the ground on the church's corner of the site. The building was therefore supported by four massive pillars or stilts rising 114 feet (34.5 m) from the ground. These four pillars supported the building in the middle of each of the four walls, rather than at the corners, where supports are typically located. Above the stilts, the building rose fifty-nine storeys, with a crown structure that peaked at 914 feet (278 m).

LeMessurier created a simple, innovative design for supporting the corners: a steel frame with six sets of structural steel braces running diagonally upward from the columns at the middle of the walls to the corners, each brace welded to the floor-beams of nine stories. The diagonal braces support the corners and give the finished building a distinctive appearance. In addition, the building was the first major skyscraper to incorporate an active damper to reduce sway during high winds. This was a 400-ton concrete block on the top of the building. Such dampers are typically controlled by sensors that measure the building's lateral acceleration and use hydraulic pressure to drive the massive block. The reaction forces cancel some of the building sway (and reduce structural stress).

LeMessurier's analysis of the loads and design of the main supporting structure were essential to guarantee the strength of the building. His analysis was given to the design engineers, who determined the structural details. The design engineers were not required to inform LeMessurier of all of the subsequent design details, provided that the structural strength appropriately exceeded the predicted loads. The building was completed in 1977.

First Problem: Recognizing and Resolving the Ethical Dilemma

In May 1978, a few months after the building's completion, LeMessurier learned that the design engineers had actually used less expensive bolted joints in the construction, instead of the welded joints he had recommended. This was not a serious change, provided the bolted joints were as strong as the welded joints. But in June 1978, motivated by a question from a graduate student studying the Citicorp Tower design, LeMessurier reviewed the design to prepare a lecture explaining the analysis. He came to a shocking realization:

when strong winds blew from a quartering direction (that is, at an angle to the wall, hitting two sides simultaneously), the stresses in the structural members and the forces on the bolted joints would be significantly greater than he had earlier calculated. This quartering load condition had not been part of the New York building code when he calculated the stresses during the design phase, and the design engineers had not considered it either. Although the diagonal braces would withstand quartering loads, the bolted joints specified by the design engineers likely would not.

LeMessurier now suspected that the finished building was under-strength and might pose a serious hazard. He faced an ethical dilemma. By revealing the design flaw, he would be risking almost certain humiliation and financial ruin. Yet by concealing his knowledge, he would be placing tenants and neighbours of the building at risk of disaster. According to one source, LeMessurier "contemplated, ever so briefly, destroying his notes or even killing himself . . ."[17] LeMessurier evaluated the alternatives and decided that the ethical choice was clearly to investigate the design flaw directly, remedy it if necessary, and accept the consequences, however cruel they might be.

Second Problem: Determining the Hazard and the Best Technical Remedy

LeMessurier consulted Alan Davenport, an award-winning Canadian civil engineer. (Among his many achievements and commissions, Davenport carried out the wind-tunnel tests for the CN Tower.) Davenport had run earlier wind-tunnel tests for the Citicorp Tower; he ran the tests again using winds at an angle to the wall surfaces. Davenport's results reinforced LeMessurier's concerns. A severe storm could cause one of the mid-level joints to fail; if this happened, the entire structure would come cascading down. Statistically, such a storm could be expected to occur about every sixteen years. Moreover, although the active damper might help reduce the loads and stresses, the damper required electric power, which might easily be disrupted during a severe storm.

The best technical solution was fairly clear—the joints could be reinforced by heavy steel strapping, which would be welded across the joints from the inside. But who would pay for such repairs, and how could they be completed without panicking the tenants and neighbours of the huge building? Moreover, time was of the essence: The design flaw became clear to LeMessurier in July 1978; the hurricane season typically reaches New York in September.

Third Problem: Gaining Support and Implementing the Remedy without Panic

LeMessurier acted on the last day of July 1978. His initial contacts were through the lawyers for the company that had hired him as a consultant and that company's insurance company. Many meetings followed, and Leslie

Robertson, also a prominent structural engineer, was retained to review LeMessurier's calculations and conclusions. (Ironically, Robertson was the structural designer for the twin towers of the World Trade Center, which collapsed minutes after the terrorist attacks of September 11, 2001.)

Citicorp, the tenants of the building, were next to be informed, and their response was critical. Fortunately, Citicorp realized the value of LeMessurier's advice, and although questions of cost and inconvenience arose, the gravity of the problem took precedence for everyone concerned. Citicorp authorized repairs to start almost immediately. At the same time, weather experts were hired to monitor and predict wind conditions, and New York City officials were contacted to work out an evacuation plan for the area around the Citicorp Tower. A press release was issued. Worries were diminished somewhat when it was explained that the repairs were intended to cope with higher winds. This was partially true, since slightly higher winds were predicted for the autumn of 1978.

Robertson became a key figure in designing and supervising the repairs. Each bolted joint along the diagonal braces was concealed by a plywood shelter while the repairs were made. The welding was done at night to minimize the inconvenience and allay the concerns of the tenants. Stand-by generators were installed to ensure that the electrical supply to the active damper could not be interrupted by a storm.

Repairs to the Citicorp Tower had not been completed by September 1978, when the weather office observed Hurricane Ella moving toward New York. Fortunately, the hurricane moved back out to sea without incident. By October, the repairs were complete. The building's strength now significantly exceeds the original design objectives. It is now believed that the tower will be able to withstand a wind storm that, statistically, can be expected to occur only once every seven hundred years! It is now one of the safest structures in the world.[18]

Final Problem: Paying the Bill

Citicorp willingly paid for the repairs but also informed LeMessurier that it expected to be reimbursed. After negotiation, LeMessurier's liability insurance company agreed to pay $2 million (USD). Citicorp eventually agreed to accept that amount as the settlement, and also to exonerate LeMessurier. Citicorp did not reveal the total costs, but a rough estimate would be at least double the insurance coverage.

LeMessurier fully expected his liability insurance premiums to rise steeply; after all, his design error had resulted in a large claim and a great deal of anguish. However, the insurance company agreed that LeMessurier had acted promptly and ethically; it also agreed that if he hadn't done so and if the building had collapsed, the company would have been liable for a much greater amount in death and injury claims. In fact, perhaps not surprisingly, LeMessurier's liability insurance premiums were reduced.[19]

Conclusion

Obviously, LeMessurier could have ignored any responsibility for this problem. Initially, he had no firm knowledge that the structure was deficient, and with good luck (no serious wind storms) the building might have survived indefinitely. However, as an ethical engineer, his perspective was simple and clear: if you have a licence from the state to hold yourself out as a professional, you have a corresponding responsibility. If your structure poses a risk to the lives of others, you must do something about it.

DISCUSSION TOPICS AND ASSIGNMENTS

(Additional assignments can be found in Appendix CD-E on the CD-ROM included with this text.)

1. Assume that you are a consulting engineer in a partnership. Your partner suggests that your business cards and stationery need some discreet advertising and suggests that the following be printed on them:

   • a stylish logo
   • your engineering seal, reduced in size to fit
   • the slogan "The best in the business!"

   Which (if any) of these advertising components would conform to the constraints on advertising established in your provincial or territorial Code of Ethics (or provincial advertising regulation)? Explain your answer with reference to your Act or to your Code of Ethics (or professional practice guidelines).

2. You have been hired as a machine design consultant to a soap manufacturer to suggest methods of speeding up a liquid detergent production line. In the course of your work, you inadvertently gain access to confidential company documents and discover that the company is adding minute quantities of a known carcinogen to the detergent without listing it as an ingredient. You know this substance has been banned. This confidential information is totally irrelevant to the job you were hired to perform, and you have discovered it entirely by chance. Do you have an obligation to act on this information? If so, what action would you take? Explain.

3. A rural town in a resort area has been instructed by the provincial government to replace an old wooden bridge for safety reasons. The town council hires a consulting engineer, Ali, to design a concrete bridge to replace the unsafe wooden structure. Because of poor soil conditions, pilings are required. The resulting design will clearly be very expensive to construct. One of the town councillors discusses the matter with a neighbour, Baker, who is also a consulting engineer and who has a summer cottage in the township. Baker suggests that in view of the soil problems, a culvert might serve the same purpose and would be much cheaper than the bridge. Since Ali is a concrete specialist and is not capable of carrying out the redesign for the steel culvert, he is paid for his work on the concrete bridge design and

replaced by consulting engineer Gambon, who designs a large culvert structure. The culvert is subsequently constructed at a fraction of the predicted
cost of the concrete bridge. Ali is disappointed that construction of the concrete bridge did not proceed (and that fees for supervising the construction
therefore were not paid). He alleges that there was unethical conduct by
Baker or Gambon or both. In this example, were the actions of any of the
engineers unethical? Would the replacement of Ali by Gambon be considered "supplanting" (as defined in this chapter and in some Codes of Ethics)?

NOTES

1. D.G. Johnson, "Engineering Ethics," in I.J. Gabelman, ed., *The New Engineer's Guide to Career Growth and Professional Awareness*, IEEE, Piscataway, NJ, 1996, p. 173. © 1996 IEEE. Reprinted with permission of IEEE.
2. Association of Professional Engineers, Geologists and Geophysicists of Alberta (APEGGA), *Guideline for Ethical Practice*, APEGGA, Edmonton, March, 2003, p. 22. <www.apegga.org> (December 7, 2003).
3. Professional Engineers Act, Regulation 941, Section 75, RSO 1990, c. P.28.
4. Professional Engineers Ontario (PEO), "Advertising," *Guideline to Professional Practice*. Revised 1996, p. 21. Reprinted with permission of PEO. (PEO advises that this guideline may be revised in 2004.)
5. Professional Engineers Act, Regulation 941, Section 53, RSO 1990, c. P.28.
6. APEGGA, *Guideline for Ethical Practice*, p. 9.
7. Quoted in J.M. MacEwing, "Legal Significance of the Engineer's Seal," *Canadian Consulting Engineer* July–August 1996, p. 8.
8. Ibid.
9. Association of Professional Engineers, Geologists and Geophysicists of Alberta (APEGGA), *Guideline for Relying on Work Prepared by Others*, APEGGA, Edmonton, June 2003, p. 13 <www.apegga.org> (December 7, 2003).
10. Professional Engineers Ontario (PEO), "Use of the Seal," *Guideline to Professional Practice*. Revised 1996, p. 19. Reprinted with permission of PEO. (PEO advises that this guideline may be revised in 2004.)
11. Professional Engineers Ontario (PEO), "Recommended Confidentiality Agreement," *Guideline to Professional Practice*, p. 15.
12. Professional Engineers Ontario (PEO), "Rules of Practice," *Guideline to Professional Practice*. Revised 1996, p. 7. Reprinted with permission of PEO. (PEO advises that the guideline may be revised in 2004.)
13. Professional Engineers Ontario (PEO), "Selection of an Engineer," *Guideline to Professional Practice*. Revised 1996, p. 9. Reprinted with permission of PEO. (PEO advises that this guideline may be revised in 2004.)
14. Professional Engineers Ontario (PEO), "Contractual Liability," *Guideline to Professional Practice*. Revised 1996, p. 20. Reprinted with permission of PEO. (PEO advises that this guideline may be revised in 2004.)
15. Association of Consulting Engineers of Canada (ACEC), *1997 Directory of Member Firms*, ACEC, Ottawa, 1997, p. xxvii. Reprinted with permission of ACEC.
16. International Federation of Consulting Engineers (FIDIC) website, <www1.fidic.org/about/ethics.asp.> (October 3, 2003). Reproduced with permis-sion of FIDIC.
17. J.R. Chiles, *Inviting Disaster: Lessons from the Edge of Technology*, HarperCollins, New York, NY, 2002, p. 196.
18. G. Voland, *Engineering by Design*, Addison Wesley, Reading, MA, 1999, p. 398.
19. C. Whitbeck, *Ethics in Engineering Practice and Research*, Cambridge University Press, Cambridge, UK, 1998, p. 146.

Chapter 10
Environmental Threats and Hazards

Water, land, sea, and sky—these are the pride and wealth of Canada. Yet despite its beauty and utility, the Canadian environment is vulnerable to pollution and also to abuse by self-serving or negligent opportunists. This chapter gives an overview of the environmental hazards threatening Canada. The chapter should be read in conjunction with Chapter 11, which defines the role that engineers and geoscientists can play in combatting these hazards.

This chapter begins with an overview of pollution in North America, goes on to discuss some of the more notorious environmental hazards, and concludes with some case histories of environmental tragedies from Canada and around the world.

CANADA'S ENVIRONMENTAL HEALTH

Engineering, science, and technology have been of immense benefit to Canada and to humanity in general. It is said that medicine has given people health; that the humanities have given people pleasure; but that engineering has given people the time to enjoy them both. However, industrialization brings problems even while improving our lives. The lifestyle of North Americans involves high resource consumption and extremely high energy usage. The result is air pollution, water pollution, acid rain, the greenhouse effect (global warming), ozone depletion, and the growing problem of waste disposal. As ethical professionals, engineers and geoscientists must play a role in monitoring and reducing this overconsumption and in curtailing the emission of dangerous pollutants. This chapter tries to put these threats in perspective.

In spite of the well-known dangers of pollution, millions of tonnes of pollutants are discharged in North America every year. The release of pollutants into the environment is monitored by the Commission for Environmental Cooperation (CEC), an agency set up by the governments of Canada, Mexico, and the United States as part of the North American Free Trade Agreement (NAFTA). In April 2003 the CEC released its seventh annual report on industrial pollutants in North America. This 376-page report (based on data for the year 2000—the latest year for which such data could be compiled) found that overall, Canada and the United States had reduced industrial releases and transfers of chemicals by 5 percent in the six years from 1995 to 2000. In 2000, total emissions were down to 3.3 million tonnes of pollutants, generated by 22,036 industrial facilities—still an immense amount of pollution.[1]

The decrease was mainly the result of reductions in discharges by large facilities. Unfortunately, many smaller facilities are increasing their chemical releases at an alarming rate. "It's very discouraging," comments Victor Shantora, the CEC's acting executive director. "The 'small-p' polluter might not grab the same headlines as a large power plant or chemical manufacturer, but their effect is being felt throughout the North American environment."[2]

Figure 10.1 shows the top twenty-five polluting provinces and states in terms of total emissions. In this ranking, the top polluters are Texas and Ohio, with Ontario third and Quebec twenty-fifth. No other Canadian provinces

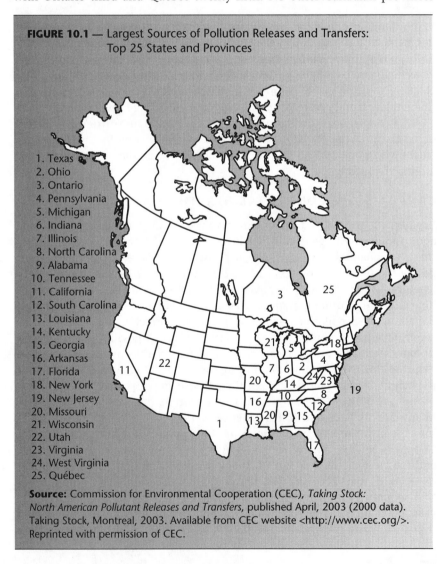

FIGURE 10.1 — Largest Sources of Pollution Releases and Transfers: Top 25 States and Provinces

1. Texas
2. Ohio
3. Ontario
4. Pennsylvania
5. Michigan
6. Indiana
7. Illinois
8. North Carolina
9. Alabama
10. Tennessee
11. California
12. South Carolina
13. Louisiana
14. Kentucky
15. Georgia
16. Arkansas
17. Florida
18. New York
19. New Jersey
20. Missouri
21. Wisconsin
22. Utah
23. Virginia
24. West Virginia
25. Québec

Source: Commission for Environmental Cooperation (CEC), *Taking Stock: North American Pollutant Releases and Transfers*, published April, 2003 (2000 data). Taking Stock, Montreal, 2003. Available from CEC website <http://www.cec.org/>. Reprinted with permission of CEC.

TABLE 10.1 — Ranking of Provinces in Order of Pollution Releases and Transfers

Rank	Province	Pollution* (kg)
1.	Ontario	200,507,310
2.	Quebec	49,807,808
3.	Alberta	25,478,251
4.	British Columbia	13,621,833
5.	New Brunswick	7,608,604
6.	Manitoba	7,040,181
7.	Nova Scotia	5,333,568
8.	Saskatchewan	1,834,876
9.	Newfoundland	536,970
10.	Prince Edward Island	354,328
	Total	312,123,729[3]

SOURCE: Commission for Environmental Cooperation (CEC), *Taking Stock: North American Pollutant Releases and Transfers*, published April 2003 (2000 data). Taking Stock, Montreal, 2003. Available from CEC website <http://www.cec.org/>. Reprinted with permission of CEC.

*Notes: (1) Pollution data are given in kilograms and are estimates of releases and transfers of chemicals as reported by facilities. They should not be interpreted as levels of human exposure or environmental impact.

(2) The rankings are not meant to imply that a facility, state, or province is not meeting its legal requirements. In combination with other information, the data can be used as a starting point in evaluating exposures that may result from releases and other management activities that involve these chemicals.

(3) The data are taken from the American *Toxics Release Inventory* (TRI) and the Canadian *National Pollutant Release Inventory* (NPRI). Data from Mexico are not available for 1995–2000.

appear in the top twenty-five. This is a significant change from the previous report (based on 1994 data), which placed Canadian provinces much higher relative to the American states.

Table 10.1 summarizes pollution emission data for the ten provinces. Clearly, pollution is a direct result of industrial activity, with Ontario, Quebec, Alberta, and British Columbia leading the list.

The CEC data, illustrated in Figure 10.1, indicate that pollution is contributing to environmental degradation at a significant rate. The following paragraphs describe these environmental threats. A more detailed (and more mathematical) analysis of these problems is provided in the comprehensive text *Environmental Science: The Natural Environment and Human Impact*.[3]

WASTE DISPOSAL

The most common degradation of the environment involves the indiscriminate disposal of wastes—be they solid, liquid, or gaseous—as by-products of manufacturing, processing, or construction. Controlling waste disposal is usually within the authority of an engineer or geoscientist, who should seek methods that are least harmful to the environment.

In the early part of the twentieth century, little was known about the insidious effects of heavy metals, dioxin, asbestos, pesticides, and other toxic substances. People believed that the environment was a vast sink that could accept any amount of waste without becoming contaminated. Waste was tipped into dumps as cheaply and quickly as possible, with little regard for the environment.

We now realize that in many parts of Canada, ill-considered methods of industrial and domestic waste disposal are creating a crisis. Large cities are transporting waste hundreds of kilometres for disposal, and small towns are beginning to realize that the local dump or "nuisance ground" is a source of disease, a fire hazard, and a danger to groundwater. Dumps are gradually being replaced by closely monitored landfills, which accept only low-hazard solid waste and cover it every day with a layer of soil to reduce odours and pest problems. Some landfills are lined with plastic to prevent waste fluids from entering the groundwater, or at least to reduce this problem. These "sanitary landfills" are a great improvement over town dumps, but they are not the ideal solution to waste disposal.

Incineration has one advantage: the volume of waste is reduced, so only the ash needs to be buried. However, the gases and particulate emissions released by the incinerators are still objectionable. Moreover, if chlorine-based organic compounds are burned, trace amounts of toxic chemicals may be dispersed.

The best way to solve the waste disposal problem is to reduce the volume of waste through more efficient use of resources, and to reuse or recycle waste materials whenever possible. For example, waste automobile tires are now being used as fuel to heat cement kilns; in this process, even the ash is consumed since it becomes part of the product.[4] Composting, pyrolysis, and density-based separation of organic and nonorganic materials have been successful in various applications. More research is needed in the recycling of waste materials.

Every province now has environmental protection legislation (see Chapter 11) and has begun recycling programs. All of this has reduced some solid waste. However, some liquid waste—especially toxic, flammable, explosive, radioactive, or hazardously reactive chemicals—is still being dumped illegally because of the shortage of proper incineration facilities and high-hazard disposal areas. Hazardous liquid waste can pose a serious threat to health if it leaks into the underground water table. This has happened many times in the past. The names "Love Canal" and "Minimata" are now synonymous with the unethical disposal of industrial waste and with the human tragedy that followed. (Case histories of both are provided near the end of this chapter.) In the United States alone, 32,000 hazardous waste disposal sites had been identified by 1991; thirteen years later, very few of these have been cleaned up.[5] Any solution to this problem will require both technical ability and political awareness.

AIR POLLUTION

Air pollution has many components, but the best known are sulphur oxides and nitrogen oxides. Sulphur oxides are created mainly by the burning of fossil fuels such as coal and petroleum, although they are also produced by many other industrial activities. SO_2 is a foul-smelling gas that reacts with oxygen in the atmosphere to form SO_3, which then combines immediately with water to yield sulphuric acid in the form of droplets. The highest SO_2 values have been reported in the northeastern region of North America and in Europe, where high-sulphur fossil fuels are burned in large quantities. In most large cities, SO_2 emissions have been reduced recently, as a result of the shift from high-sulphur coal to low-sulphur natural gas.

Sulphur oxides are detrimental to plant life. They also corrode metals, dis-colour fabrics, and degrade building materials. Severe damage to plant life can be observed many miles downwind from certain smelting operations. It seems that a combination of sulphur oxides and air particles is especially damaging to human health, partly because of the action of small particles in conveying sulphuric acid into the lungs. SO_2 is a serious lung irritant, and dramatic episodes such as the "London smogs" have been attributed to the combination of SO_2 and particulates. The London smog of 1952 established the link between atmospheric pollution in smog and increased mortality.[6] The worst smog in Canada occurred in southern Ontario in 1962 and lasted five days. This episode is believed to have been a London smog; however, it was named the "Grey Cup smog" because it caused the 1962 Grey Cup football game to be postponed because of poor visibility.[7]

Even in the absence of sulphur, the burning of fossil fuels causes serious air pollution in urban areas. Exhaust gases typically contain unburned hydro-carbons (HC), carbon monoxide (CO), nitrogen oxides (NO_x), and "normal" combustion products such as SO_2 and H_2O. In the atmosphere, many of these products react chemically to produce new contaminants. Because these processes are stimulated by sunlight, the resulting products are referred to as photochemical oxidants. Two of the principal photochemical oxidants are ozone (a lung irritant) and peroxyacetyl nitrate (PAN, a lung and eye irritant). Ozone is constantly being created in the atmosphere by natural processes, but not to a degree great enough to constitute a pollution hazard.

Nitrogen oxides are also a problem in air pollution. There are several known oxides of nitrogen, but the important ones from the standpoint of air pollution are nitric oxide (NO) and nitrogen dioxide (NO_2). The term NO_x is commonly used to refer to nitrogen oxides collectively. NO_x is a product of almost any combustion process that uses air, since nitrogen is the chief com-ponent of air. To a great degree, the formation of NO_x is the result of high combustion temperatures, principally from motor vehicles, which in indus-trialized urban areas account for 50 to 60 percent of atmospheric NO_x. The Los Angeles type of photochemical smog is caused mainly by NO emissions from cars; occurs on warm, sunny days when traffic is heavy; and reaches a peak in the early afternoon. It therefore differs from the London smog, which forms on cold winter nights as a result of SO_2 produced by coal combustion.[8]

The nitrogen oxides participate actively in photochemical reactions with hydrocarbons, thus helping produce photochemical smog. NO_2 plays a double role in air pollution: it is a component in the formation of photochemical smog, and it is toxic in its own right. NO is much less toxic than NO_2, but NO is readily converted into NO_2 in the atmosphere: by reacting with oxygen in the presence of water, it becomes nitric acid, which is extremely toxic to any growing organism. The adverse effects of air pollution on humans and animals include serious lung disorders, reduced oxygen in the blood, eye and skin irritation, and damage to internal organs.[9] Damage to painted surfaces, cars, and buildings is mainly the result of acid rain, which is the topic of the following section.

Most provinces and the federal government have Clean Air Acts that specify emission standards and ambient air-quality standards. Air pollution control is mainly a provincial responsibility, although the federal government regulates trains, ships, and gasoline.[10] Government regulations must, of course, be followed in any engineering practice, and reducing emissions of SO_x and NO_x should always be a primary objective, wherever possible.

ACID RAIN

The problem of acid rain captured Canadians' attention in the 1980s. Both sulphur oxides and nitrogen oxides are implicated, because they form sulphuric and nitric acids in the atmosphere and cause rainfall to become more acidic than it otherwise would be. Neutral water ideally has a pH of 7.0, but "normal" rainfall in remote areas that are unpolluted has a pH of about 5.6 because of the presence of small amounts of acid of natural origin.[11] Rain is typically called "acid rain" when the pH falls below 5.0. In many areas of Canada and the northeastern United States, the rain has a pH value as low as 4.0.

When acid rain falls, it harms mainly fish, trees, farms, buildings, and cars. Aquatic life begins to be affected when the pH falls below 5.0, and most fish are killed when a pH of 4.5 is reached. The result is hundreds of lakes in northeastern North America (and in Scandinavia) that are devoid of fish, and thousands of other lakes that are threatened.[12] Lesions on plants caused by simulated acid rain have been observed when the pH drops lower than 3.4, although subtle effects may be occurring at higher pH levels. Humans may also be harmed, because the acidity leaches magnesium, aluminum, and heavy metals out of the soil and concentrates them in drinking water. Fish are especially susceptible to dissolved aluminum, and this may be a risk to humans as well. Acid rain does severe damage to limestone buildings and monuments, since it dissolves the key chemicals in limestone.

The areas with the most acid rain lie downwind from the areas that produce most of the sulphur and nitrogen oxides, so the downwind people obviously feel that the upwind people should take corrective action, whatever the cost. Current annual emissions of SO_2 amount to about 20 million tonnes in the United States and about 2 million tonnes in Canada. Coal-fired thermal electric power plants produce about 70 percent of American emissions and about 20 percent of Canadian emissions.[13] Sulphur oxides produced from burning coal (especially high-sulphur coal) are the main source of acid rain.

These pollutants also result from the smelting of nickel, lead, and copper ores, which is done extensively in Ontario and several states in the American Northeast. A secondary cause of acid rain is NO_x from power stations and automotive emissions, mainly on the American west coast.

Acid rain is an international problem. The heavier flow of pollutants is believed to be from the United States into Canada, because of the greater industrial activity south of the border. However, the smelters around Sudbury, Ontario, are major sources of the sulphur dioxides that cause acid rain. The federal governments in Canada and the United States have agreed to control acid rain, but it is engineers—especially those in the power generation and smelting industries—who must monitor this problem and work to alleviate it. The most effective way to reduce acid rain is to reduce acid emissions, although low pH levels in lakes can also be reversed by adding lime to neutralize the acid. Reducing emissions is costly, but an early study shows that in addition to a cleaner environment, the economic benefits are about equal to the cost of the controls.[14]

WATER POLLUTION

Some rivers are less polluted now than they were in the nineteenth century, when there was a serious water pollution crisis. In the mid-nineteenth century, 20,000 people in London, England, died of cholera. As Donald Carr says, "in the Western world this was the greatest pollution disaster of history."[15] Typhoid and cholera epidemics stemming from water contaminated by sewage were widespread. Water pollution has at least six possible sources:

- Disease-causing bacteria.
- Organic waste decaying in the water, reducing the dissolved oxygen content.
- Fertilizers that stimulate plant growth and also depress oxygen levels.
- Toxic materials, such as heavy metals and chlorinated hydrocarbons (DDT, PCB).
- Acidification, as mentioned earlier.
- Waste heat, which can also reduce dissolved oxygen levels.[16]

Cholera and typhoid are practically unknown in Canada today as a result of sewage treatment and the use of chlorine to kill bacteria in drinking water. Nevertheless, we must be vigilant in monitoring water quality. The pollution caused by lawn fertilizer in summer and highway salt in winter is said to come from "non-point" sources (that is, from everywhere) instead of from "point" sources such as sewage treatment plants, power plants, and factories, which can be more easily monitored.

Agriculture also creates dangerous pollution when the runoff from dairy or pig operations is too close to the intake of municipal water systems. Such a problem occurred in Walkerton, Ontario, in May 2000 (as discussed in Chapter 2). Before the cause was traced, seven people died and more than 2,300 became ill.

Every summer, many beaches in Canada are closed to swimmers because of high bacteria and fecal content in the water. This is a scandal for a country that boasts the largest endowment of fresh water in the world. Laws to control pollution have been passed at all levels of government. In the twenty-first century a key principle is that "the polluter pays" to remedy the problems. As a professional engineer, you have a duty to society to ensure that environmental laws are followed. In most cases it is not enough to treat pollution at the "end of the pipe." Waste must be eliminated earlier in the process, by increasing efficiency and by reducing, reusing, or recycling raw materials.

GLOBAL WARMING AND OZONE DEPLETION

"Global warming" is a popular term used to indicate a change in the planet's climate caused directly or indirectly by human emissions of greenhouse gases into the atmosphere. There is evidence that the global mean surface air temperature rose between 0.3°C and 0.6°C in the hundred years between 1890 and 1990. This may sound like a small change, but it is estimated that the end of the last ice age was accompanied by a temperature increase of only about 2°C. Obviously, even a slight warming of the entire planet could cause severe climatic changes. Evidence points toward the "greenhouse effect" as the cause of global warming. This effect is caused by emissions of carbon dioxide (CO_2), methane (CH_4), nitrogen dioxide (NO_2), ozone (O_3), and the chlorofluorocarbons ($CFCl_3$ and CF_2Cl_2, called CFCs).

Chlorofluorocarbons were first manufactured in large quantities in the 1940s, when wartime factories converted to producing household appliances. Production of CFCs increased because they are superb refrigerant gases (hence the common term "freons"). Only recently have we learned that as greenhouse gases, freons are much more potent than CO_2. One molecule of CFC has the same greenhouse effect as 10,000 molecules of CO_2.

Since the end of the nineteenth century, the amount of CO_2 in the atmosphere has increased by about 23 percent, to 340 parts per million (ppm). This increase is attributed to two factors: the burning of fossil fuels, and deforestation in the tropics. Carbon dioxide is a vital factor in the earth's heat balance because it traps heat in the atmosphere. If there were substantially less CO_2 in the atmosphere, very little heat would be retained and the surface of the earth would be coated with ice. Conversely, if the amount of CO_2 were to double, the earth's average temperature might increase by 3 or 4°C, with serious consequences. How quickly the amount of atmospheric CO_2 could double depends on the rate at which we burn fossil fuel, but estimates have ranged from 88 to 220 years.[17] Based on observed trends and computer models, researchers have predicted several climatic changes:

- Global temperatures in the year 2050 will be higher than at any time in the past 150,000 years.
- The rate of change of global temperature will be larger than any rate seen in the past 10,000 years.

- There will be significant rises in sea levels, leading to flooding of low-lying areas.
- Increased evaporation may cause significant decreases in the levels of lakes and reservoirs.
- Global warming will not be uniform in either space or time. Warming will be more intense over land than over sea and will be greatest in the northern hemisphere in winter.[18]

The rise in sea level could be dramatic. The rise of 15 cm in this century has already caused some beaches on the Atlantic coast to erode at the rate of 0.9 to 1.5 m (3 to 5 feet) per year.[19] It has been predicted that over the next hundred years, if the average global temperature increases by 1.5 to 5.4°C, the sea level could rise another 8 to 25 cm because of the melting of glaciers. The huge glaciers in Antarctica, which make up 85 percent of the total ice on earth, are not involved in this calculation, because it is believed that they are currently subtracting water from the earth's oceans and thus might remain in balance.[20] The Arctic ice cap is also not involved at all; because most of it is floating, its melting would not substantially change the sea level.

The computerized climate models are more worrisome. Global warming would drastically shift existing climatic patterns. The hydrological cycle might be intensified, with higher temperatures and greater rainfall at northern latitudes; simultaneously, rainfall at mid-latitudes might decrease. Canada might prosper, but the wheatfields of the United States might turn into dust bowls. Those who work with the climate models hasten to remind everyone that these models are not yet very accurate: global warming seems to be predicted with reasonable confidence, but its accompanying regional effects are uncertain. Recent events seem to validate predictions of large-scale climate disruption. The floods in Quebec (1996) and in Manitoba (1997), the ice storm in Ontario and Quebec (1998), the massive, heat-induced electrical power failure in Ontario and the northern United States (2003)—all of these were certainly unusual. A recent severe drought in British Columbia (2003) led to huge forest fires in the interior, followed by torrential rains and flooding on the west coast.

Chlorofluorocarbons play a sinister role in environmental degradation. They are the primary cause of ozone layer depletion. Ozone in the stratosphere helps screen out damaging ultraviolet rays (ozone at ground level is a pollutant and irritant). CFCs combine with ozone to create gaps in the stratospheric ozone layer, thus permitting ultraviolet rays to reach the earth's surface, where they harm plants and increase rates of skin cancer. This was first observed in 1985 by scientists from the British Antarctic Survey. Measurements since then show that ozone depletion has been increasing in severity.[21]

International agreements have been negotiated—and in Canada, provincial laws have been passed—to phase out the use of CFCs. This will help protect stratospheric ozone and will also lessen the greenhouse effect.

Regrettably, a criminal black market in illicit CFCs has sprung up, mainly for servicing out-of-date refrigeration plants. This illegal trade is hindering moves toward safer refrigerants.[22, 23]

What should engineers do about all this? As a start, any process or operation that produces needless carbon dioxide or other greenhouse gases should be discouraged. More importantly, refrigeration lines containing CFCs—which have been used in refrigeration systems for decades—should never be voided into the atmosphere. And, of course, any criminal trade in CFCs should be reported to the police.

ENERGY CONSERVATION AND NUCLEAR POWER

During the 1970s, two oil shortages caused long line-ups at gasoline stations. By the 1980s, there was an oil glut. Gasoline became cheaper, and we lost interest in energy conservation. However, analysts warn that a new and different form of energy crisis looms in this century. It has been predicted that coal reserves will last another three or four centuries; however, one source predicts that oil and natural gas reserves will be 90 percent depleted by 2020.[24] A still more recent prediction suggests 2050.[25] Alternative energy sources (solar, wind, wave, geothermal) have added very little to the global energy supply. It seems that nuclear fission and nuclear fusion (should it ever be developed) will be essential if we are to maintain our present standard of living and extend it to the citizens of developing nations.[26] Between 1980 and 1989 the proportion of the world's energy needs satisfied by nuclear energy doubled from 2.5 to 5 percent.[27] The World Nuclear Association (WNA) reports that in 2003 more than 440 commercial nuclear reactors were operating in thirty-one countries and were supplying 16 percent of the world's electricity.[28] Although nuclear advocates sound aggressively self-serving in their predictions, it does appear that the world is going to need nuclear energy.[29]

Atomic Energy of Canada Ltd. (AECL), the designer of the CANDU nuclear reactor, contends that the CANDU is safer and more reliable than the Americans' light-water reactors, since the CANDU is fuelled by natural uranium and moderated by heavy water. In a recent article in the *Ottawa Citizen*, Geoffrey Wasteneys, a retired consultant on electric power production, made the following comments about the CANDU system:

> The CANDU reactor . . . is an outstanding Canadian development that enabled the production of electricity by the use of natural uranium, thus avoiding the requirement for a vastly expensive process of enrichment.

> This enabled many countries that possessed uranium resources to produce electric power by a nuclear process without having to deal with the United States or the Soviet Union, which together possessed a virtual monopoly on the production of enriched uranium, a byproduct of their production of nuclear bombs.

There is no serious problem with the spent CANDU fuel: It is stored for possible later utilization. (It includes plutonium, and currently, salvaged plutonium from nuclear bombs is being used as fuel.)

An aggressive U.S. marketing policy and propaganda by such firms as Westinghouse that derided the Canadian system as "a scientist's hobby" shut the CANDU out of all but marginal world markets such as India, Pakistan, Argentina, Korea, Romania and China. In consequence, heavy development costs were not fully supported by sales. The decline in the use of nuclear power stations in some countries relates to design faults in present systems and cheaper crude oil.

There has never been a serious problem with a CANDU reactor. The breakdowns have related to the (non-nuclear) steam cycle. Most of the required renovations at Ontario Hydro stations are nearing completion. There are no nuclear power stations in Quebec. The atrocious Darlington cost overruns were due to initial design faults, to some extent the outcome of political intervention.[30]

Nuclear energy has the great advantage of using fuel that is compact and plentiful. Even so, public concerns over operating safety and the disposal of radioactive waste are hindering plans to construct new nuclear generating plants. A meltdown at a nuclear plant would devastate surrounding cities and towns. The risk of it actually happening is extremely remote, but even a slight risk generates fear.

Another serious worry connected with nuclear power concerns the long-term disposal of highly radioactive waste. It will be necessary to keep the waste out of circulation for thousands of years because of the extremely long half-lives of some of the elements, such as plutonium. At present, the plan is to store such waste in stable geological underground layers from which water has been absent for millions of years. As with most matters related to nuclear energy, waste disposal is the subject of bitter debate. Supporters of nuclear energy admit that there can be no absolute guarantee that humans will not be exposed to the waste for thousands of years into the future. However, they emphasize that the risk of future exposure is very small. Opponents of nuclear energy have insisted on a guaranteed method for keeping nuclear waste permanently out of contact. They point out that no such guarantee can be made and declare that this is reason enough to phase nuclear energy out of existence.

Clearly, nuclear plants must be designed and managed carefully, professionally, and ethically—and this must be transparent to the general public. Nuclear plant designers and operators must pay close heed to the public's anxiety and must work to demonstrate (and improve even further) the safety of the CANDU reactor, since the predicted future depletion of oil and gas reserves seems destined to create a greater dependence on nuclear energy.

Coal-fired power does not seem to be a reasonable alternative to nuclear energy, because of the associated problems of air pollution. In the aftermath of the Three Mile Island and Chernobyl disasters, coal power has been compared with nuclear power to see which is more dangerous. One writer asserts that their dangers are similar, even if we accept the high estimate of 39,000 future cancer deaths from Chernobyl. He estimates that the death toll from the use of coal in the former USSR is between 5,000 and 50,000 per year. Many of these deaths result from the mining and transporting of coal, which requires 100 times as much material handling as uranium with an equivalent energy output. In the United States, 100 or more coal miners die every year, and nearly 600 of the 1,900 deaths in railway accidents each year are the result of transporting coal. However, the big killer is air pollution, although it is impossible to say with certainty whether any given death has been specifically caused by coal burning. Even so, it has been estimated that 50,000 people in the United States die every year as a result of air pollution, mostly resulting from the burning of coal. The estimate of 5,000 to 50,000 deaths from burning coal in the former USSR is derived by extrapolating these data, assuming similar populations and similar pollution conditions.[31]

The debate over the safety of our energy sources will not be resolved in this textbook; however, the issue illustrates the importance of professional ethics, conserving energy, increasing efficiency, and avoiding waste. In all of these, ethical actions are essential.

EXPONENTIAL POPULATION GROWTH

Population growth is generally viewed as an achievement, not a problem. However, the consumption of nonrenewable resources and the resulting environmental degradation are proportional to the size of the population.

The world population is at record levels and continues to grow quickly. Around 1800, there were roughly 1 billion people on this planet. After the Industrial Revolution, new machines and medicines and improvements in nutrition caused life expectancy to begin to rise and infant mortality to begin to fall. As a result, the world's population passed 2 billion in 1930, 3 billion in 1960, 4 billion in 1975, 5 billion in 1987, 5.6 billion in 1994,[32] and 6.35 billion in 2003.[33] A simple graph of these numbers indicates that the world's population is growing exponentially.

Engineers and geoscientists know that exponential growth cannot be sustained by finite resources—even when resources are as vast as this planet's. Moreover, the expectations of the citizens of developing countries are also increasing sharply. Satellite television is now showing these people the conspicuous consumption of the developed world, and it is only human nature for the world's poor to want a share in this bounty. The pressure to consume resources will reach disastrous proportions within the lifetime of most readers. Obviously, we must use resources more efficiently if we hope to avoid catastrophe, or at least delay it.

Wildlife and plant populations do not increase in size indefinitely. Sooner or later, environmental resistance limits the population.[34] Every species in a given habitat has an equilibrium point for its population. For example, food shortages (such as dwindling prey) limit the population of wild animals. Less food leads to increased mortality—especially infant mortality—until the birth rate, the death rate, and the food supply are in equilibrium. Human population is a little different, since we can devise ways to improve our habitat and in this way raise the equilibrium point. But even for us there are limits, since resources are finite.

Many population forecasts have been made, using different assumptions. It is expected that the world's population will peak between 2050 and 2100, at which time an equilibrium point, estimated between 10.6 billion[35] and 12 billion,[36] will have been reached. The following summarizes—bleakly—the problem of population growth:

> The question of how many people the Earth can support is now at the top of the international agenda. In the next 35 years, even if the current trend of declining fertility rates continues, the United Nations forecasts that the Earth's population of 5.7 billion will balloon to nearly nine billion. Ninety-five percent of that growth will take place in the developing world. The upshot, say many experts, is that shantytowns like Nairobi's Mathare Valley and refugee-producing conflicts like the recent slaughter in Rwanda will proliferate—creating a 21st century of growing anarchy, warfare and disease in which masses of Third World migrants will be scrambling to get inside the protected citadel of the industrialized West. If those forecasts prove accurate, Canadians and other citizens of the developed world may face a stark choice: whether to open their borders to millions of new refugees, or to slam the door shut and turn their backs on the spreading misery.
>
> ...
>
> While the steep rise in the world's population in the last half of the 20th century has brought calls for zero, or even negative, population growth, many conservative economists insist that there is no crisis over the Earth's ability to support the expected increase. Nicknamed "cornucopians," they argue that the international market will always find a substitute product or a new technology to circumvent shortages of particular resources. A case in point is copper: in the 1970s, some environmentalists predicted that the metal would be in short supply in the 1990s. Instead, there is a glut of copper and prices have plummeted because fibre-optic cable and plastic piping have replaced copper in many uses.
>
> As for crowded slums and food shortages in the developing world, the cornucopians point out that couples tend to have fewer children as their incomes rise. Economist Michael Walker of the Fraser Institute, a conservative Vancouver think-tank, says that the key is to increase the productivity of farmers . . . by protecting property rights so that farmers can take out loans and invest in tools and crops.
>
> ...
>
> But while the general optimism of the cornucopians is comforting, it conflicts with the rough consensus emerging among most demographers, scientists and policy analysts involved in population and resource research. Their view is that a high percentage of the planet's peoples are doomed to live with poverty and violence unless population growth

is dramatically reduced. That was the conclusion of a yearlong study by researchers at Cornell University's department of ecology and systematics. Interestingly, their report [released in February 1994] does not point to the depletion of nonrenewable resources like oil as the problem. Rather, they say, the Earth's biosphere can only produce enough renewable resources—food, fresh water and fish—to sustain [only] two billion people at a standard of living equal to that in Europe.

Another study by Cornell's David Pimentel, a professor of insect ecology and agricultural sciences, and Nobel-winning physicist Henry W. Kendall draws on statistics from the United Nations' Food and Agriculture Organization (FAO). Pimentel and Kendall state that even if the United Nations' population target of 7.8 billion were met, world food production would have to triple in the next 55 years for every inhabitant to have an adequate diet. That prospect is at best remote, they add, because less than half of the Earth's land is suitable for agriculture, and almost all of that is already exploited. Moreover, many of the benefits of the Green Revolution, which boosted crop yields with irrigation, fertilizer and pesticides, have already been realized—along with such unwelcome side-effects as nutrient depletion, pollution and water shortages.

Two long-term environmental problems, largely created by the industrialized countries, could also lower crop yields: increased ultraviolet radiation due to the thinning ozone layer and reduced precipitation because of global warming. Said Pimentel: "While the number of mouths to feed has increased, grain production has actually been declining since 1981."

According to yet another study [released in August 1994], by the Washington-based Worldwatch Institute, the solutions invoked by cornucopians are unlikely to stave off disaster. While Western countries helped avert large-scale famines in nations like India in the 1960s with Green Revolution aid programs, there is no new biotechnology or high-yield seed currently in development that will significantly boost world grain harvests. At the same time, marine biologists at the FAO report that all 17 of the world's major ocean fisheries are being fished at, or beyond, capacity. Nine of them are in decline or have been shut down—as in the case of Canada's Atlantic cod fishery.

Perhaps the most provocative research on the consequences of the looming gap between resources and population is being done by Thomas Homer-Dixon, head of the University of Toronto's Peace and Conflict Studies program. Homer-Dixon foresees a 21st century in which overpopulation, unequal distribution of wealth and environmental degradation combine to produce tribal warfare, mass migrations and the breakup of countries around the globe "with a speed, complexity and magnitude unprecedented in history." That bleak scenario may not be farfetched in a world where more than one billion people already go hungry and about two billion lack basics like running water or electricity.[37]

Clearly, some action must be taken to avert these crises. In 1968, Garret Hardin, a American philosopher, published a paper on a phenomenon of human nature observed about 140 years earlier called "the tragedy of the commons." Hardin applied this analogy to the problem of overpopulation and observed that although our society has an ethical precept that everyone born has certain inalienable rights, it is impossible to divide finite resources among an indefinitely large number of people. Hardin concluded that the only way we can preserve the most precious freedoms is by the draconian measure of regulating the freedom to procreate, as China has already done. Several American groups, most notably Population Connection, are presently advocating similar solutions to the population problem.[38]

Engineers and geoscientists are not in a position to alleviate the problems of overpopulation, or even to affect it significantly. What we can do is ensure that we are not contributing to the inefficient use of resources or to the creation of environmental hazards. In the long term, ethical actions are always in our own self-interest.

OPPORTUNISM AND THE TRAGEDY OF THE COMMONS

The tragedy of the commons is often raised in environmental discussions. It was first described by William Forster Lloyd, a political economist at Oxford University, in 1832. Lloyd observed the recurring destruction of common pastures, or "commons," typically found at the edge of small towns in England. Lloyd observed that the cattle which grazed on the common were smaller and stunted, compared with cattle on privately owned pastures. The privately owned pastures were obviously better kept and healthier environments for the cattle. His conclusions were republished in Hardin's 1968 paper[39] and explained as follows:

Imagine a pasture open to all local cattle owners. Each owner, quite naturally, wants to increase his or her herd and thus tries to graze as many cattle as possible on the commons. Each animal added to the herd yields a significant benefit to the owner but slightly reduces the food available to other animals. This process of increasing herd size may be sustainable for many years, especially if the commons is very large. Eventually, however, adding more animals will reduce the quality of the commons below an acceptable level. Yet even at this point, there is no way to stop the process. As Hardin says:

> Therein is the tragedy. Each man is locked into a system that compels him to increase his herd without limit—in a world that is limited. Ruin is the destination toward which all men rush, each pursuing his own best interest in a society that believes in the freedom of the commons. Freedom in a commons brings ruin to all.[40]

Hardin applied this concept as a motivating force for population limitation. However, the tragedy of the commons also applies to other environmental effects. For example, the inconsiderate citizen who dumps garbage along the roadway, or the large corporation that spills toxic or radioactive waste, or who emits heat, noise, or noxious fumes into the air, is like the cattle owner who adds one more animal to the commons. These selfish acts degrade the environment.

Whether we classify these acts as vandalism or simply as human nature, they are the philosophy of utilitarianism in reverse. Utilitarianism supports a small sacrifice by the general population in order to aid the public good. In the tragedy of the commons, a selfish individual imposes a general degradation upon society in order to reap a personal benefit. Moreover, without intervention, human nature makes the outcome inevitable. As Hardin puts it:

> The rational man finds that his share of the cost of the wastes he discharges into the commons is less than the cost of purifying his wastes before releasing them. Since this is true for everyone, we are locked into a system of "fouling our own nest," so long as we behave only as independent, rational, free enterprisers.[41]

Through thoughtful laws, regulations, and taxes, we can avoid the tragedy of the commons. The environment must be policed by the true owners—the public—and degradation must be linked to the cause. In other words, the polluter must pay. This monitoring creates the feedback loop that is required to protect the environment. This will not be an easy task, but as Case History 10.1 (below) shows, the cost of ignoring waste dumping is far more expensive than the cost of regulating it.

CASE HISTORY 10.1

TOXIC POLLUTION: LOVE CANAL, MINAMATA, BHOPAL, SUDBURY

The improper disposal of toxic waste, whether a deliberate act or simple incompetence, is professional misconduct. The following case history reviews four cases of improper disposal of harmful waste, in four different countries. The cases are so well-known that the names—Love Canal, Minamata, Bhopal, and Sudbury—are now synonymous with toxic pollution.

Love Canal, New York—Dioxin

Love Canal is named after one of the early residents of New York state. The canal was originally excavated for boats and barges, but since it was never completed for navigational purposes, "canal" is a misnomer. Between 1942 and 1953 the Olin Corporation and the Hooker Chemical Corporation saw it as a convenient hole for burying waste chemicals. More than 18,000 tons of chemical waste, including dioxins, were buried, and eventually the "canal" was again a flat piece of land. Only the people who had buried the chemicals knew what lay beneath the soil.

In 1953 the Hooker Chemical Company donated the land to the local board of education, but without saying anything about the chemicals buried there. Over time, the land became a residential area, and homes, playgrounds, and a school were built on it. In 1976, after several seasons of heavy rains, people began to notice a terrible stench of chemicals. Homes reeked, children complained of chemical burns, and pets died or became sick. Yet these problems were minor when compared with the miscarriages, birth defects, and cases of cancer that occurred in the Love Canal area more and more frequently as the years passed. Residents soon demanded some action, on the basis that these problems were far too frequent.

In 1978, a government study of the area exposed some remarkable—and frightening—statistics. More than eighty different chemicals had been detected in the air, some of which were carcinogenic. The chemical pollution in the air was 250 to 5000 times safe levels. There was an unusually high rate of miscarriages (almost 30 percent). Of seventeen pregnant women in the Love Canal area in 1978, only two gave birth to normal children.[42]

New York state authorities recognized the serious health threat posed by the buried chemicals and moved a few hundred families out of the area. The school was closed and surrounded by barbed wire. The area over the buried chemicals became a ghost town, with derelict houses, empty streets, and No Trespassing signs. Neighbouring residents who lived only a small distance from the chemicals were concerned about their health, but faced a dilemma: they wanted to move, but since "Love Canal" was now a synonym for hazardous waste, their homes were worthless. In 1980, residents' demands forced the government to carry out further testing. The tests showed high levels of genetic damage among the neighbouring residents. The area was declared a disaster area by the U.S. president, and 710 families were relocated. Many of the abandoned homes were demolished, and the chemical wastes were excavated for treatment and proper disposal. The total cost of the cleanup was estimated at $250 million.[43]

The Love Canal tragedy revealed the questionable ethics of the Hooker Chemical Corporation, which buried much of the chemical waste and then donated the land to the board of education. The company did not warn the board about the chemical contamination, even though the company was aware of the problem and concerned about it. The agreement to transfer the land contained several clauses that protected Hooker against future claims for liability. The company took credit for its public generosity, yet it kept the danger of the buried chemicals secret. Hooker guarded that secret even after the area was developed for residential use. Documents would later show that the company knew that children had suffered chemical burns from playing there. After the extent of the disaster became public, chemical industry spokespeople ridiculed the residents as hypochondriacs.[44] This toxic secret brought tragedy to many families who risked their life savings in worthless homes, whose children were born deformed, and who even now may live in fear of contracting cancer.

Love Canal heightened awareness of the need for ethical conduct and environmental regulations. In the years following Love Canal, the U.S. Environmental Protection Agency (EPA) discovered between 32,000 and 50,000 other toxic waste dumps scattered across the United States. Of these, as many as 2,000 may pose significant risks to the public.[45]

The resulting furor over Love Canal led to stricter laws and more severe penalties for improper disposal of waste. Unfortunately, the public awakening to the danger of improper and unethical dumping was too slow in coming. Lois Gibbs, the resident who took the leadership role in drawing public attention to the environmental disaster at Love Canal, recently stated: "As a society, we begin with this toxic thing and say: 'How much can we put in the environment before somebody is harmed?' This risk-assessment approach means there is a subset of our society that will always be sacrificed."[46] Dioxin was being released into the environment at Love Canal for years, and when barefoot housewives discovered that dioxin caused birth defects in their children, it still took thirteen years for scientists to verify the facts.

Minamata Bay, Japan—Mercury Poisoning

Since 1953, thousands of residents of the Minamata Bay area of Japan have fallen ill as a result of organic mercury poisoning. Mercury (also called quicksilver) is the liquid metal often used in thermometers and barometers. Whether as a pure liquid element or in compound form, mercury can cause serious renal and neurological dysfunction. Mild cases often mimic amyotrophic lateral sclerosis (ALS, also called "Lou Gehrig's disease"). The symptoms of severe poisoning include clumsiness, a stumbling gait, severe mental or behavioural problems, and partial or complete loss of speech, taste, and hearing.

The Chisso company, a nitrogen fertilizer company in Minamata, the main city on Minamata Bay, first began producing acetaldehyde in 1932. Mercury was required as a catalyst for this process and for other chemicals the company would produce later, such as vinyl chloride. The mercury was used in liquid form, and during the production process, a portion of it was lost—washed into Minamata Bay with the waste water. In the bay, microbes acted on the mercury and converted it into an organic (methyl or carbon-based) mercury compound. The organic mercury was absorbed by shellfish; since mercury is not excreted by mammals, it becomes more concentrated as it moves up the food chain. Over time, the concentrations are sufficiently large that the toxic effect becomes apparent. The first humans to be affected were fishermen and their families, who had a diet rich in fish, including shellfish. The seriousness of the problem was first recognized by physicians and health officials in the Minamata area around 1956, although cases were later traced back to 1953, about twenty years after mercury first began washing into the bay.[47]

The medical director of the hospital associated with the Chisso company became sufficiently concerned about the problem in 1956 that he began a series of tests on cats. Since family cats ate fish, they were the first to exhibit this curious behaviour. He identified the manufacturing plant effluent as the cause of the problem, but when he reported his results to his superiors at Chisso, he was ordered to stop his tests and forbidden to report his findings to the local health authorities.[48]

It was not until 1959 that medical authorities requested help from the Kumamoto medical school and an investigation was begun. By 1962, they were certain that the problem was caused by organic mercury and that the source of the mercury was the Chisso effluent.[49] Initially it was estimated that about 2,900 people had contracted Minamata disease, as the debilitating neurological syndrome is now known. However, a 1995 estimate places the number between 7,000 and 8,000 people.[50] The government indicted Chisso, but it took years for the various cases to work their way through the legal system. The first decision was made in 1970, and some compensation was also awarded by the government. However, as of 1995, some 100 of the victims were still pressing lawsuits for compensation.

The ethical issues in this case are clear. The Chisso company's actions—stopping the medical tests, suppressing knowledge of the problem, and continuing to permit mercury to be dumped in Minamata Bay—were unethical and inexcusable. Even though the damaging effects of mercury were not certain in 1932, the company knowingly inflicted personal tragedy on thousands of unwitting people.

An outbreak of Minamata disease occurred in Grassy Narrows, Ontario, in 1970. Members of two Ojibwa bands living near the Wabigoon River began to show the debilitating symptoms of mercury poisoning, and the source of the pollution was traced to the Reed Paper company in Dryden, just upstream from the Ojibwa reserves. The provincial government ordered Reed to reduce mercury usage, and the pollution was eliminated. Although compensation was eventually paid to the two bands, the economic and social effects were devastating.[51]

Bhopal, India—Methyl Isocyanate

In the early morning hours of December 3, 1984, a poisonous cloud of methyl isocyanate gas escaped from the Union Carbide plant in Bhopal, India. It killed thousands of people up to 6 km away, many while they were asleep in their beds. This was probably the worst industrial accident in history, and its social and economic impact on Bhopal was devastating. It is estimated that between 3,000 and 12,000 people died in this catastrophe; around 30,000 more suffered permanent injuries, 20,000 temporary injuries, and 150,000 minor injuries. And even these horrific numbers are disputed by victims' rights organizations, which say the real numbers were even higher.[52]

The Union Carbide plant was established in 1969 as a mixing factory for pesticides. Methyl isocyanate, which is used in large quantities in the production process, is highly volatile as well as highly toxic—estimated to be ten times deadlier than phosgene gas that the Germans used as a weapon during the First World War. Methyl isocyanate reacts vigorously with many common substances and must be maintained at very low temperatures to prevent uncontrolled reactions. The precise cause of the disaster is not known, but it is most likely that an employee closed a valve on a piping system so that a filter connected to the pipe could be washed. A metal disc should have been inserted to make certain that the valve could not leak, but this was not done. During the washing process, water leaked past the closed valve, entering piping that was connected to the methyl isocyanate holding tank. The water reacted with the methyl isocyanate, generating intense heat. The pressure in the tank increased dramatically and pushed past pressure relief valves into the atmosphere. Safety measures were either inadequate or did not work. Over the next ninety minutes, about 40 tonnes of methyl isocyanate and other reaction products were released into the atmosphere. The vapour, which is heavier than air, filled low-lying areas, crept into houses through windows and doors, and asphyxiated thousands of people while they slept.

An inquiry was held after the disaster. The construction, operation, and maintenance of the Bhopal plant were examined, as well as management decisions that permitted such a potentially dangerous plant to operate in such

The Union Carbide factory (in the background) looms over relatives and friends carrying a victim's body to cremation in Bhopal.

Source: CP Photo/Associated Press.

an unsafe manner, in an urban area, with no suitable emergency plan. The Indian government charged the company management with negligence, brought murder charges against its chief executive, and demanded $3.3 billion (USD) to settle victims' claims. Then in 1989 the Indian Supreme Court announced a settlement of all claims for $470 million, conditional on the dropping of criminal charges.[53] Shortly after the settlement was announced, a new Indian government disallowed the claim and sought to reinstate the criminal charges.[54] While the litigation continues, the Indian government is supporting the survivors of the Bhopal disaster.

Sudbury, Ontario—Sulphur Dioxide

Canada is the world's second-largest producer of nickel, and the mines around Sudbury, Ontario, are the main source of this metal. Nickel was first discovered in the area in 1856, but it was not until the Canadian Pacific Railway reached Sudbury in 1883 that anyone realized the full extent of the ore body.[55] The nickel is in the form of sulphide ore, which cannot be converted directly into metallic form. It must first be smelted—that is, burned to remove the sulphur and convert the ore to nickel oxide, which can then be reduced to pure nickel. In the early 1900s the first conversion step was typically done in huge, open "roasts," where layers of timber were interspersed with layers of ore. The roasts burned around the clock and emitted a toxic cloud of sulphur dioxide.[56] The ecological impact of this process was not given much thought a century ago, when the Sudbury area was sparsely populated. As a result, few people realized that the sulphur dioxide, when dissolved in water, created acid rain. In 1928 the federal government became aware of the problem and banned the use of open roasts.[57] However, enclosed smelting is still done.

The problem was partly solved by the erection of taller smokestacks, which spread the pollutants over a wider area, thus reducing their concentration. The largest of these superstacks, at Sudbury's Copper Cliff mine, was built in 1972 and is about 380 m high.

The environmental effects of acid rain on the Sudbury region have been severe. In the area around the smelters, trees are stunted and sparse, lakes are devoid of fish, and only the hardiest species of birds can survive (where there are any birds left at all). In 1978 the Regional Municipality of Sudbury began an ambitious program to restore 10,000 hectares of barren land. By experiment, it was discovered that in most locations a combination of fertilizer, agricultural lime (to neutralize the acidity), and a seed mixture of grass and legumes would generate a healthy grass cover. Two years after grass began growing, crews would return to plant trees and shrubs. Fifteen different tree species were planted, and by 1995 many of these trees were more than 3 m tall. Although the soil acidity is still very high, the levels of heavy metals have been reduced. Populations of insects, birds, and small mammals have increased, and successful tree growth has averaged 70 percent across all species. About 3,000 hectares have been reclaimed so far.[58] As a result of the extensive environmental damage and the efforts being made to remedy that damage, Sudbury has become one of the most closely studied ecological areas in the world.

Conclusion

These environmental disasters were not of equal magnitude, nor were they equally unethical. Thousands of deaths and injuries were directly linked to Bhopal—the world's worst environmental disaster—and to Minamata. Hundreds of illnesses, miscarriages, and cancer cases were related to Love Canal, but no such human tragedies have been directly atrtributed to the Sudbury sulphur dioxide emissions, although the pollution is evident in the lakes and vegetation downwind of the smelters.

CASE HISTORY 10.2

NUCLEAR SAFETY: THREE MILE ISLAND AND CHERNOBYL

Nuclear-generated electrical power is a major source of environmental concern, even though American and British studies show that power generation by coal is 250 times more hazardous than nuclear power and that generation by oil is 180 times more hazardous. Only natural gas poses fewer hazards than nuclear power to workers and the general population.[59] A few serious radiation releases occurred in the early experimental days of Canadian and American nuclear development, and a few deaths in North America have been attributed to power reactor accidents. However, the American reactor accident at Three Mile Island in 1979 and a later accident at Chernobyl in Ukraine (then part of the Soviet Union) in 1986 involved partial or complete meltdown of the nuclear reactor cores, and were therefore much more serious than any other accidents before or since. The public was horrified to learn that meltdowns—the most unthinkable nuclear accident—could occur. Ever since, the public has viewed nuclear power with fear, uncertainty, and distaste. These accidents have, however, fostered a greater emphasis on operating safety at nuclear plants.

Three Mile Island

Three Mile Island is located on the Susquehanna River in southern Pennsylvania. Construction on its two-unit nuclear power plant began in the late 1960s, and the second unit (unit TMI-2) was completed, tested, and brought on line in December 1978. The two units were designed to produce a maximum of 880 megawatts of electrical power. The TMI-2 unit experienced many minor problems during its commissioning and early operation. It had been operating for only three months when it was the source of North America's worst nuclear accident.[60]

The Accident and Its Causes

The problem began shortly after 4 a.m. on March 28, 1979, and recovery efforts lasted for a month. The accident's impact is still being felt today by the American nuclear power industry. The initial cause of the accident was a blocked feedwater line, which caused pumps to stop. Operator errors and other minor malfunctions then magnified the problem. James Carter, the American president at the time (and himself a nuclear engineer), commissioned an inquiry into the accident. The inquiry report describes the first few minutes of the accident as follows:

> In the parlance of the electric power industry, a trip means a piece of machinery stops operating. A series of feedwater system pumps supplying water to TMI-2's steam generators tripped on the morning of March 28, 1979. The nuclear plant was operating at 97 percent power at the time. The first pump trip occurred at 36 seconds after 4:00 a.m. When the pumps stopped, the flow of water to the steam generators stopped. With no feedwater being added, there soon would be no steam, so the plant's safety system automatically shut down the steam turbine and the electric generator it powered. The incident at Three Mile Island was 2 seconds old.
>
> ...
>
> When the feedwater flow stopped, the temperature of the reactor coolant increased. The rapidly heating water expanded. The pressurizer level (the level of the water inside the pressurizer tank) rose and the steam in the top of the tank compressed. Pressure inside the pressurizer built to 2,255 pounds per square inch, 100 psi more than normal. Then a valve atop the pressurizer, called a pilot-operated relief valve, or PORV, opened—as it was designed to do—and steam and water began flowing out of the reactor coolant system through a drain pipe to a tank on the floor of the containment building. Pressure continued to rise, however, and 8 seconds after the first pump tripped, TMI-2's reactor—as it was designed to do—scrammed: its control rods automatically dropped down into the reactor core to halt its nuclear fission.
>
> Less than a second later, the heat generated by fission was essentially zero. But, as in any nuclear reactor, the decaying radioactive materials left from the fission process continued to heat the reactor's coolant water. This heat was a small fraction—just 6 percent—of that released during fission, but it was still substantial and had to be removed to keep the core from overheating. When the pumps that normally supply the steam generator with water shut down, three emergency feedwater pumps automatically started. Fourteen seconds into the accident, an operator in TMI-2's control room noted the emergency feed pumps were running. He did not notice two lights that told him a valve was closed on each of the two emergency feedwater lines and thus no water could reach the steam generators. One light was covered by a yellow maintenance tag. No one knows why the second light was missed.

With the reactor scrammed and the PORV [relief valve] open, pressure in the reactor coolant system fell. Up to this point, the reactor system was responding normally to a turbine trip. The PORV should have closed 13 seconds into the accident, when pressure dropped to 2,205 psi. It did not. A light on the control room panel indicated that the electric power that opened the PORV had gone off, leading the operators to assume the valve had shut. But the PORV was stuck open, and would remain open for 2 hours and 22 minutes, draining needed coolant water—a LOCA [loss of coolant accident] was in progress. In the first 100 minutes of the accident, some 32,000 gallons—over one-third of the entire capacity of the reactor coolant system—would escape through the PORV and out the reactor's let-down system. Had the valve closed as it was designed to do, or if the control room operators had realized that the valve was stuck open and closed a backup valve to stem the flow of coolant water, or if they had simply left on the plant's high pressure injection pumps, the accident at Three Mile Island would have remained little more than a minor inconvenience. . . . [61]

The combination of the initial blockage in the feedwater lines, plus the overlooked lights (warning that the valves were closed from the emergency feedwater pumps), plus the relief valve that was stuck open, created a confusing combination of effects that would lead to incorrect corrective actions, exposure of the reactor core, and release of radioactive water. The series of events would end with the reactor destroyed and millions of people in fear for their health and safety.

In the days and weeks after the initial incident, the nuclear plant owners, the news media, and politicians from the local mayor to the president were involved in a debate over the need for evacuation and the correct information to be released to local residents. The risk of creating panic needed to be balanced with the risk to health should a major release of radioactive materials occur. Control was not fully regained for another month. The plant went into a cold shutdown on April 27, 1979. The full report of the incident is available on the Internet and is a chilling story of confusion and fear.[62]

The Resulting Damage

During the accident, the loss of coolant caused the reactor core to be exposed. The extreme heat melted about one-third of the core, rendering the reactor useless—a billion-dollar loss. The site cleanup likely cost another billion dollars. A small amount of radioactive material was released from the reactor into the environment, raising the fear of of radiation-induced health effects, principally cancer, in the neighbourhood of the reactor. Over the next eighteen years the state government would trace the health of 30,000 people who had been living within five miles of Three Mile Island at the time of the accident. Fortunately, no unusual trends were found, and the registry was discontinued in 1997.[63]

Although the TMI-2 nuclear plant was a total loss, the adjacent TMI-1 plant survived. Its safety measures and training protocols were greatly strengthened, and it is still productive. The TMI-2 accident is cited as a warning to proponents of nuclear energy, yet it pales in comparison with the tragedy that was to occur at Chernobyl seven years later.

Chernobyl

The Chernobyl nuclear power plant is located about 100 km north of Kiev, the capital of Ukraine. In 1986 the plant comprised four reactors constructed between 1977 and 1983. At full capacity, the plant generated 4,000 megawatts of electricity. The Chernobyl reactors were a Soviet design, designated by the acronym RBMK, which is a pressurized water reactor design that uses ordinary water as a coolant and graphite as a moderator. This type of Soviet reactor was intended for plutonium generation (for weapons) as well as electrical power production. No other power reactor design uses this combination of graphite and water.[64]

At 1:23 a.m. on April 26, 1986, the No. 4 nuclear reactor in Chernobyl exploded, releasing huge, sinister clouds of radioactive plutonium, cesium, and uranium dioxide into the atmosphere. It was the worst nuclear accident in history involving a nuclear generating plant. Ukraine was then part of the USSR, and regrettably, the Soviet authorities were slow to issue a warning or to release any details about the accident. In fact, the first information about the accident was released only two days later, by Swedish experts who had noticed the nuclear fallout over Scandinavian countries.

The Accident and Its Causes

The reactor explosion had three basic causes: poor reactor design, inadequately trained reactor personnel, and unsafe operating procedures that permitted tests to be carried out at a low and unstable power. The reactor design flaws included an inadequate containment shell, poorly designed graphite control rods, and a feature called a "positive void coefficient," which refers to the tendency for voids (or steam pockets) to form in the cooling water. Although graphite is the main moderator, the cooling water also has a moderating effect. However, when steam pockets in the water get too large, then the moderating effect of the water begins to fluctuate, and the reactor may experience rapid and uncontrollable power surges.

The RBMK reactors were known to have a positive void coefficient, so other control features were in place to prevent the instability from occurring. However, these features were overridden by the inexperienced operators. By the time the problem was recognized, the heat had deformed the channels and the control rods could not be reinserted. The WNA website describes the Chernobyl accident as follows:

> On 25 April [1986], prior to a routine shut-down, the reactor crew at Chernobyl-4 began preparing for a test to determine how long turbines would spin and supply power following a loss of main electrical power supply. Similar tests had already been carried out at Chernobyl and other plants, despite the fact that these reactors were known to be very unstable at low power settings.
>
> A series of operator actions, including the disabling of automatic shutdown mechanisms, preceded the attempted test early on 26 April. As the flow of coolant water diminished, power output increased. When the operator moved

to shut down the reactor from its unstable condition arising from previous errors, a peculiarity of the design [the positive void coefficient] caused a dramatic power surge.

The fuel elements ruptured and the resultant explosive force of steam lifted off the cover plate of the reactor, releasing fission products to the atmosphere. A second explosion threw out fragments of burning fuel and graphite from the core and allowed air to rush in, causing the graphite moderator to burst into flames.

There is some dispute among experts about the character of this second explosion. The graphite burned for nine days, causing the main release of radioactivity into the environment. A total of about 12×10^{18} Bq of radioactivity was released. . . . Some 5000 tonnes of boron, dolomite, sand, clay and lead were dropped on to the burning core by helicopter in an effort to extinguish the blaze and limit the release of radioactive particles.[65]

The Resulting Damage, Injuries, and Deaths

The WNA states that "30 people were killed, and there have since been up to ten deaths from thyroid cancer due to the accident." However, these numbers are challenged, and do not reveal the truly massive disruption of life that took place. A full list of damage and disruption was included in a 1988 UNSCEAR report. It describes:
- hundreds of direct injuries and deaths,
- evacuation of several cities, involving hundreds of thousands of people,
- extensive radiation monitoring,
- repeated decontamination of buildings and destruction of some buildings,
- creation of a prohibited area within a 5-km radius around Chernobyl,
- creation of a restricted movement area within a 30-km radius around Chernobyl,
- destruction of poisoned food and movement of thousands of cattle, *and*
- millions of protective medical treatments, radiation tests, and so on.[66]

Aftermath

Unit 4 at Chernobyl was enclosed in concrete shortly after the accident, to permit the other three reactors to continue operating. However, this concrete structure is not durable, and more repairs may be necessary. Much money was spent on improving the safety of the remaining reactors, and in view of the desperate need for electricity, they were kept running. Unit 2 was shut down after a turbine hall fire in 1991, and Unit 1 was shut down at the end of 1997. Unit 3 continued operating until December 2000.[67]

Controversial ideas and new secrets about the Chernobyl disaster were revealed by Grigori Medvedev in *The Truth about Chernobyl*,[68] first published in the Soviet Union in 1989 and translated into English in 1991. The following brief summary of the book was written by Joseph Schull of *Maclean's* magazine and is reproduced with permission.

Medvedev brings expertise and bitter experience to the tale. He worked as deputy chief engineer at Chernobyl in the 1970s, at the time of its construction. Later, he watched in frustration as the project fell into the hands of administrators who won prestigious and well-paid positions despite their lack of experience in the nuclear field. By 1986, Medvedev was deputy director of the Soviet energy ministry's department of nuclear power-plant construction in Moscow. He returned to Chernobyl in the days following the disaster to investigate its causes. His research has yielded a meticulous reconstruction of events on a minute-by-minute basis laced together with first-person accounts from people involved in the plant's operation and management.

The first truth that emerges in Medvedev's book is that the Soviet nuclear industry was run by incompetents from top to bottom: officials in charge of the construction and management of nuclear power stations simply had no training in the field, while their underlings at Chernobyl were no better prepared. Meanwhile, secrecy surrounded the industry and fostered utter ignorance about its potential dangers. Information about previous nuclear mishaps, including the 1979 accident at Three Mile Island, was reserved for high-placed officials unable to draw the appropriate lessons. A state bureaucracy that acknowledged successes but not setbacks was equally damaging. Attention to safety implied the possibility of accidents, and that could only mean that errors might be committed—a possibility that nearly everyone, from minister to technician, wanted to deny. Failure was not in the Soviet vocabulary.

The disease that led to the explosion continued through its aftermath, according to Medvedev. Several costly hours were wasted as the plant's managers denied incontrovertible evidence that the reactor had exploded, passing on to Moscow the reassuring myth of a minor explosion in an emergency water tank. A day and a half passed before the nearby town of Pripyat was finally evacuated; Chernobyl itself did not follow until May 5. The accident merely triggered a chain reaction of human errors.

The coverup continues even now. Soviet authorities have admitted to only 31 deaths in the immediate aftermath and have kept secret the numbers who have died since then. But Vladimir Chernousenko, the scientific director now in charge of the 32-km exclusion zone surrounding the Chernobyl power station, recently estimated that fatal casualties to date number between 7,000 and 10,000. . . . [69]

Conclusion

It should be noted that nuclear power engineering has matured significantly in the decades since the accidents described above, and that regulatory procedures are more rigorous. Also, it should be observed that the Canadian CANDU reactor is much safer than the Soviet RBMK design and even safer than the American light water reactor (LWR) design. Canadian nuclear power plants use heavy water, not graphite, for moderation. The heavy water is needed for the fission to proceed. Loss of heavy water causes the fission process to stop automatically.[70]

Nevertheless, the Three Mile Island and Chernobyl accidents warn us of the immense potential energy of nuclear plants. The possibility of disaster may be infinitesimally small, but it is not zero. Controlling this energy requires continual alertness, and such responsibility must never be treated casually. These tragic accidents also expose the danger of making design decisions for political or military reasons, and the danger of secrecy and information management in engineering. Competence, clear communication, and high professional and ethical standards are obviously essential in the design and operation of such facilities.

NOTES

(Further reading suggestions are included in Appendix CD-E on the CD-ROM disk included with this text.)

1. Commission for Environmental Cooperation (CEC), *Taking Stock: North American Pollutant Releases and Transfers*, published April 2003 (2000 data) by the Communications Department of the CEC Secretariat, Montreal, 2003. Available from CEC website <www.cec.org/> (October 23, 2003).
2. CEC, "Taking Stock of the CEC's Taking Stock," in *TRIO—The Newsletter of the North American Commission for Environmental Cooperation* (CEC), Summer 2003. <www.cec.org/trio/> (October 23, 2003).
3. R.W. Jackson and J.M. Jackson, *Environmental Science: The Natural Environment and Human Impact*, Longman, Harlow, Essex, 1996.
4. Ibid., p. 330.
5. Ibid., p. 334.
6. Ibid., p. 311.
7. R.E. Munn, "Air Pollution," *The Canadian Encyclopedia*, 2003 Edition, Historica Foundation of Canada, CD-ROM version. Also available at <www.canadianencyclopedia.ca> (October 23, 2003).
8. Jackson and Jackson, *Environmental Science*, p. 312.
9. Munn, "Air Pollution," ibid.
10. Ibid.
11. Jackson and Jackson, *Environmental Science*, p. 315.
12. H.L. Ferguson, "Acid Rain," *The Canadian Encyclopedia*, 2003.
13. Ibid.
14. Ibid.
15. D.E. Carr, *Death of Sweet Waters*, Berkley, New York, 1971, p. 41.
16. A.H.J. Dorcey, "Water Pollution," *The Canadian Encyclopedia*, 2003.
17. P.H. Abelson (editorial), "Carbon Dioxide Emissions," *Science*, November 25, 1983.
18. Jackson and Jackson, *Environmental Science*, p. 318.
19. "The Politics of Climate," *EPRI Journal*, June 1988, pp. 4–15.
20. M.F. Meier, "Contributions of Small Glaciers to Global Sea Level," *Science*, December 21, 1984, pp. 1418–21; and *Carbon Dioxide and Climate: A Second Assessment*, National Academy Press, Washington, DC, 1982.
21. Jackson and Jackson, *Environmental Science*, p. 320.
22. M. Smith and M. Vincent, "Tanking a Killer Coolant," *Canadian Geographic*, September–October 1997, pp. 40–44.

23. S. Tripp and B. Whiting, "The Illegal Trade in Chemicals That Destroy Ozone," in *TRIO—The Newsletter of the North American Commission for Environmental Cooperation* (CEC), Spring 2003. <www.cec.org/trio/> (October 23, 2003).

24. E. Titterton, "Nuclear Energy: An Overview," in H.D. Sharma, ed., *Energy Alternatives: Benefits and Risks*, University of Waterloo, Waterloo, ON, 1990, p. 146.

25. Jackson and Jackson, *Environmental Science*, p. 249.

26. Titterton, "Nuclear Energy," ibid.

27. Jackson and Jackson, *Environmental Science*, p. 257.

28. WNA <www.world-nuclear.org/info/inf01.htm>.

29. Allan Kupcis, Chairman, Canadian Nuclear Association (CNA), "Get Real: Nuclear Is in Your Future," *Globe and Mail*, July 25, 2003, A11. Also available on the Internet: <www.globeandmail.com/servlet/ArticleNews/TPStory/LAC/20030725/CONUKES25/TPComment/TopStories> (October 28, 2003).

30. G. Wasteneys, "CANDU Reactor Is Outstanding Canadian Feat," *Ottawa Citizen*, October 17, 2003. <www.canada.com/> (October 28, 2003). Reproduced with permission.

31. "Letters: Chernobyl Public Health Effects," *Science*, October 2, 1987, pp. 10–11.

32. Jackson and Jackson, *Environmental Science*, p. 140.

33. Population Connection (formerly Zero Population Growth—ZPG), 1400–16th Street NW, Washington, DC 20036. <www.populationconnection.org/> (October 26, 2003).

34. Jackson and Jackson, *Environmental Science*, p. 140.

35. "End of the Population Explosion?" *Discover*, July 1997, p. 14.

36. Jackson and Jackson, *Environmental Science*, p. 163.

37. P. Kaihla, C. Erasmus, J. Edlin, and B. Bethune, "Apocalypse When? A United Nations Plan to Limit Global Population Growth Triggers an Acrid War of Words," *Maclean's*, September 5, 1994, p. 22. Reprinted with permission.

38. *Population Connection*, ibid.

39. G. Hardin, "The Tragedy of the Commons," *Science* 162 (1968), pp. 1243–48.

40. Ibid.

41. Ibid.

42. L.G. Regenstein, "Love Canal Toxic Waste Contamination: Niagara Falls, New York," in Schlager, *When Technology Fails*, pp. 354–60.

43. Ibid., p. 357.

44. Ibid., p. 358.

45. Ibid., p. 358.

46. L. Baker, " Be Safe! Lois Gibbs' New Campaign Urges Caution on Toxic Chemicals," *E-Magazine*, Vol. XIV, No. 4, July–August 2003. <www.emagazine.com/july-august_2003/0703curr_gibbs.html> (October 28, 2003).

47. J. Larson, "Mercury Poisoning: Minamata Bay, Japan," in Schlager, *When Technology Fails*, pp. 367–71.

48. Ibid., p. 370.

49. Ibid.

50. "Approval of Plan for Minamata Disease Victims," *Environmental Newsletter*, No. 56. © 1996 the Tokio Marine and Fire Insurance Co., Ltd., October 1995. <www.tokiomarine.co.jp/e0300/html/EnvOct95-2.html#A> (October 28, 2003).

51. M. Bray, "Grassy Narrows," *The Canadian Encyclopedia* 2003.

52. L. Ingals, "Toxic Vapour Leak: Bhopal India," in Schlager, *When Technology Fails*, pp. 403–10.

53. "Damages for a Deadly Cloud," *Time* (February 27, 1989): 53.

54. "Haunted by a Gas Cloud," *Time* (February 5, 1990): 53.

55. B. Sutherland, "Nickel," *The Canadian Encyclopedia* 2003.

56. L.C. Ritchie, "Ecological Disaster: Sudbury Ontario," in Schlager, *When Technology Fails*, pp. 340–44.

57. Ibid., p. 341.
58. International Council for Local Environmental Initiatives, Region of Sudbury, *Canada: Land Reclamation*, ICLEI Project Summary Series, Project Summary #22. <www.iclei.org/LEICOMM/LEI-022.HTM> (October 28, 2003).
59. R.D. Bott, "Nuclear Safety," *The Canadian Encyclopedia*, 1st ed., Hurtig, Toronto, 1985, p. 1302.
60. D.E. Newton, "Three Mile Island Accident," in Schlager, *When Technology Fails*, p. 510.
61. *Report of the President's Commission on the Accident at Three Mile Island*, U.S. Government document, October 30, 1979. Also available at <stellar-one.com/nuclear/report_to_the_president.htm> (October 28, 2003).
62. Ibid.
63. World Nuclear Association (WNA), *Three Mile Island: 1979* (Information and Issue Briefs), WNA, March 2001, <www.world-nuclear.org/info/info.htm> (October 28, 2003).
64. WNA, *Chernobyl* (Information and Issue Briefs), WNA, March 2001. <www.world-nuclear.org/info/chernobyl/info07.htm> (October 28, 2003). Reprinted with permission of WNA.
65. Ibid. Excerpt reproduced with permission.
66. UNSCEAR 1988 Report, Annex D, *Exposures from the Chernobyl Accident*, United Nations Scientific Committee on the Effects of Atomic Radiation (UNSCEAR), Vienna, Austria, 1988. Available at <www.unscear.org> (October 28, 2003).
67. World Nuclear Association (WNA), *Three Mile Island*: 1979, ibid.
68. G. Medvedev, *The Truth about Chernobyl*, trans. E. Rossiter (New York: HarperCollins, 1991).
69. J. Schull, "A Fatal Coverup: The Deadly Lies of Chernobyl Come to Light," *Maclean's*, May 13, 1991, p. 65. Reprinted with permission.
70. J.A.L. Robertson, "Nuclear Power Plants," *The Canadian Encyclopedia*, 2003.

Chapter 11
Environmental Ethics

The environment is a public asset and resource; protecting it is everyone's duty. For example, the Kyoto Protocol, discussed in this chapter, calls on all Canadians to reduce greenhouse gas emissions. However, professional engineers and geoscientists have a special responsibility—they make many decisions that affect the environment, and irresponsible actions on their part could cause great harm. Engineers and geoscientists must therefore be aware of the environmental risks and hazards described in the previous chapter and must strive to reduce or eliminate them.

This chapter discusses the professional's duty to protect the environment, and lists some of the laws regulating environmental hazards. Several provincial Associations provide advice on environmental ethics in the form of environmental guidelines. These are reviewed in this chapter, along with some corporate initiatives for environmental responsibility.

When clients or employers place the environment at risk, conflicts may arise. A professional must prevent unsafe, unethical, or illegal practices, by informing or educating those responsible. If instruction is not effective, assertive action may be needed—the professional engineer or geoscientist may be forced to report unethical practices to the appropriate authorities. Although reporting (or "whistleblowing") is very rare, this chapter discusses when it might be justified.

THE DUTY TO SOCIETY

Engineers and geoscientists are bound by the Code of Ethics to protect the environment, and in general, they do a good job. Most projects involve low to moderate risks, so safety is ensured simply by using established methods, good judgment, and reasonable factors of safety. Perfection is not attainable, and some risk always exists. One judge in a tort liability case remarked: "Engineers are expected to be possessed of reasonably competent skill in the exercise of their particular calling, but not infallible, nor is perfection expected, and the most that can be required of them is an exercise of reasonable care and prudence in the light of scientific knowledge at the time, of which they should be aware."[1]

The duty to society therefore does not require perfection. But what is "reasonable"? In environmental terms, reasonable care, prudence, and scientific knowledge means the following:

Knowledge of environmental law. Professional engineers and geoscientists should seek advice before taking any action that might contravene an environmental law, regulation, or bylaw. Compliance with environmental law is essential. A few of the more general environmental laws are listed later in this chapter.

Adequate technical knowledge. Before releasing any substance into the environment, professional engineers and geoscientists must have an adequate knowledge of the effects of the release, especially if the substances are suspected to be toxic. This information is too voluminous for inclusion in this text, but advice is immediately available through the Internet, as explained later in this chapter.

Thorough analysis. In any new process, and in large or dangerous projects such as a chemical plant or nuclear facility, a "cradle to grave" systems approach is necessary to ensure that hazards are controlled. The designers must foresee the problems of handling, storing, and disposing of hazardous substances. They must also consider the decommissioning and disposing of the plant itself, even though this may be fifty years in the future. The designers must also investigate all of the possible ways that the plant operations might go astray. Sophisticated failure analyses such as failure modes and effects analysis (FMEA), event tree analysis, and fault tree analysis are essential for evaluating operating hazards. The designers must try to find every conceivable mode of failure, evaluate the probability that it will occur, and devise a remedy for combatting any harmful results.

Insistence on high ethical standards. Obviously, high ethical standards are essential. The guidelines for the environment, discussed below, are a good start.

THE DUTY TO THE EMPLOYER (AND ITS LIMITS)

If you are an employee, you have an obligation to your employer. In rare cases, however, an employer may instruct an employee to carry out acts that are contrary to the welfare of society. For example, an engineer may be asked to design a factory cooling system that takes in water from a nearby stream and dispels polluted waste water into the city sewer system, without the knowledge of authorities; or a geoscientist may be asked to falsify ore records for a mine to suggest that lower than actual quantities of toxic chemicals are being disposed in the tailings. Fortunately, the situations suggested by these examples do not arise often. The actions requested are harmful to the environment, unethical, and illegal. As a professional, you must refuse to carry out such unethical activities. However, you may have to defend or explain your refusal. The following may help—it is adapted from a previous chapter.

Illegal actions. Any activity that is contrary to any environmental law or regulation (such as those listed later in this chapter) is illegal. No employer has the authority to direct an employee to break the law. A professional engineer or geoscientist must refuse to perform any activity that is clearly illegal, and is obliged to take action if other employees are observed in illegal activities. By participating in or merely by ignoring illegal acts, the professional employee becomes liable for the penalties prescribed by the law; in addition, he or she may face disciplinary action by the provincial Association. In extreme cases it may be necessary to report the illegal acts to the appropriate authority (that is, it may be necessary to blow the whistle).

Actions contrary to the Code of Ethics or environmental guidelines. A professional engineer or geoscientist must refuse to carry out any activity that, while not clearly illegal, is a breach of the Association's Code of Ethics or is a breach of the environmental guidelines developed by many provincial Associations (discussed later in this chapter). In some situations the employer may be unaware of the legal significance of the Code of Ethics and may simply need to be informed. The professional employee should advise the employer of the appropriate section of the code or the guidelines and should decline to take any action on the employer's request. If the employer is an engineer or geoscientist, he or she has a similar obligation to obey the code. (Many other professions also obey Codes of Ethics.) The employee has a legal basis for insisting on ethical behaviour—an employer cannot direct a professional engineer or geoscientist to take an action that would result in the loss of licence.

Actions contrary to the conscience of the employee. An employee may be asked to perform an act that, while not illegal and while not clearly a violation of the Code of Ethics or the environmental guidelines, nevertheless contravenes the employee's conscience or personal moral code. These are, of course, the most difficult cases. For situations such as these, the decision-making procedure described earlier (in Chapter 6) may be useful. The employee must gather all the relevant information and define the ethical problem as clearly as possible. In precisely what way does the required action offend the employee's conscience? The employee must attempt to see the problem from the employer's perspective. Alternative courses of action must be generated and examined in light of the basic ethical theories discussed earlier; the optimum course of action must then be selected.

The professional person's career may sometimes be in jeopardy. Refusing to follow an employer's instructions may result in retaliatory action or dismissal. That is, you may be fired, or pay raises or promotions may be delayed indefinitely. You need to consider the possibility of dismissal, the consequences of unemployment, and the remedies for wrongful dismissal (as discussed in Chapter 7). However, carrying out an unethical action can also have painful consequences, both legal or disciplinary. Decisions such as these must not be made lightly.

Large industrial plants such as the Stelco steel works, shown here, are essential to Canada's prosperity. But they also emit carbon dioxide, a greenhouse gas that causes global warming. As a general rule, and to meet commitments under the Kyoto Protocol, engineers and geoscientists must endeavour to reduce greenhouse gas emissions.

Source: CP Photo/Kevin Frayer.

CANADIAN ENVIRONMENTAL LAW

If your actions affect the environment, your first priority is to know the law and to follow it. The federal, provincial, and territorial governments (and many municipalities) have environmental laws and regulations that may limit or regulate your activities. Fortunately, these laws are now very easy to find on the Internet:

- **Federal, provincial, and territorial laws.** These laws are available from the government websites listed later in this chapter.
- **International summary.** A comprehensive summary of environmental laws, including equivalent American and Mexican laws, is published by the North American Council for Environmental Cooperation (CEC). It is also available on the Internet.[2]
- **Provincial summaries.** Summary publications are available in most provinces to explain the myriad of provincial environmental laws. Two examples: *Handbook of Environmental Compliance in Ontario*,[3] and *Guide to Environmental Compliance in Alberta*.[4] A simple Internet search may reveal a similar guide for your province.
- **Local laws.** Your municipal government may have specific laws that apply only locally. Contact your city hall or regional government for information.

The most relevant Canadian environmental laws are listed below, along with a web address for the legislation (as of October 2003) or for the department that administers the legislation.

Federal Environmental Acts

The Environment Canada website (<www.ec.gc.ca/>) discusses many aspects of the environment (wind, water, climate, ozone layer, etc.). It also provides links to the on-line environmental news magazine *Envirozine*. This site also links to the Canadian Environmental Assessment Agency. Environment Canada's *Green Lane* provides case studies of several interesting environmental emergencies—mainly oil spills. The federal environmental laws can be accessed through this site. The most generally applicable laws are these:

Canadian Environmental Protection Act (R.S.C. 1985, c. 16, 4e suppl.). This is the main federal law regulating the environment. It is administered by Environment Canada and is aimed mainly at preventing pollution as the best way to protect the public. Its eleven sections deal mainly with public participation, information gathering, codes of practice, pollution prevention, toxic substances, biotechnology, and waste management.

Fisheries Act (R.S.C. 1985, c. F-14). This act also has provisions to protect the environment, especially any activities that might degrade any fish habitat. The law is administered by the Department of Fisheries and Oceans. However, the sections most relevant to this chapter—those which concern placing deleterious substances in "water frequented by fish"—are administered jointly with Environment Canada. This act provides severe penalties to prevent anyone from dumping toxic materials into water that contains fish.

Canadian Environmental Assessment Act (R.S.C. 1992, c. 37). This Act applies to projects for which the Government of Canada has decision-making authority, be it as proponent, land manager, source of funding, or regulator. All projects must receive an appropriate degree of environmental assessment before they can proceed.

Provincial and Territorial Environmental Acts

Every province and territory has environmental laws and regulations. Some of these are listed below, with the web address where the legislation (or the department that administers it, as of October 2003) may be found:

- **Alberta** <www3.gov.ab.ca/env/>
  Environmental Protection and Enhancement Act (RSA 1992, c. E-13.3)
  Water Act (RSA 2000, c.W-3)
- **British Columbia** <www.legis.gov.bc.ca/legislation/index.htm>
  Waste Management Act (R.S.B.C. 1996, c.482)
- **Manitoba** <www.gov.mb.ca/conservation/>
  Environment Act (S.M. 1987-88, c.26 C.C.S.M., c.E125)

- **New Brunswick**
  Clean Air Act (R.S.N.B. 1973, c.C-5.2)
  Clean Environment Act (R.S.N.B. 1973, c. C-6)
  Clean Water Act (R.S.N.B. 1973, c. C-6.1)
- **Newfoundland and Labrador** <www.gov.nf.ca/deptnew.htm>
  Environmental Protection Act (R.S.N. 1995, c. E-13.1)
- **Northwest Territories**
  Environmental Protection Act (R.S.N.W.T. 1988, c. E-7)
- **Nova Scotia** <www.gov.ns.ca/legi/legc/>
  Environment Act (R.S.N.S. 1994-95, c.1)
- **Nunavut** Nunavut has not yet adopted environmental legislation, so the
  regulations of the Northwest Territories continue to apply.
- **Ontario**
  Environmental Protection Act (R.S.O. 1990, c.E-19)
  Environmental Bill of Rights Act (R.S.O. 1993, c.28)
  Environmental Assessment Act (R.S.O. 1990, c.E-18)
  Ontario Water Resources Act (R.S.O. 1990, c. O.40)
  Guideline for use at contaminated sites in Ontario (Brownfields), 1997.
- **Prince Edward Island** <www.gov.pe.ca/government/index.php3>
  Environmental Protection Act (R.S.P.E.I. 1988, c. E-9)
- **Quebec**
  Quebec Environment Quality Act (*Loi sur la qualité de l'environnement*)
  (R.S.Q., c. Q-2)
  Quebec Civil Code (*Code civil du Québec*) (S.Q. 1991, C-64)
- **Saskatchewan**
  Environmental Management and Protection Act (R.S.S. 1978, c. E-10.2)
- **Yukon** <www.canlii.org/yk/sta/index.html>
  Environment Act (S.Y.1991, c.5)

THE KYOTO PROTOCOL TO HALT GLOBAL WARMING

The Kyoto Protocol is an international agreement to limit the emission of greenhouse gases and thereby stop global warming and its harmful side-effects. This protocol will have an impact on the life of every Canadian, since we will all have to reduce consumption in some way. In particular, engineers and geoscientists will have to devise ways to reduce greenhouse gas emissions.

Few people nowadays challenge the conclusion that global warming is happening; the science behind it has been validated many times over. However, the Kyoto Protocol is controversial for a number of obvious reasons: it is a far-reaching attempt to reverse human nature; it calls for immediate personal sacrifices; many people (and governments) are wary of its implementation costs, which are as yet unclear; and the proposal to grant emission credits may be open to political abuses. Moreover, global warming is a gradual process, and many people are selfishly resistant to making sacrifices now for the sake of future generations. (This is a climate version of the tragedy of the commons, discussed in the last chapter.) So in spite of the consequences of global warming, there is no guarantee that the Kyoto Protocol will succeed.

The Greenhouse Effect

As explained in the last chapter, gases in the earth's atmosphere trap much of the sun's solar energy and radiate it back to the earth's surface. This greenhouse effect increases the earth's temperature and maintains it at a level suitable for life. In fact, the temperature of the earth is 33°C higher than it would be without this insulating effect. However, the greenhouse effect is very delicately balanced, and greenhouse gases emitted by human activities can influence this balance. For the past 150 years, the concentration of greenhouse gases in the atmosphere has been increasing rapidly. The trend began with the Industrial Revolution, when people first began to burn coal for heat and power, and it has accelerated over the past century, with the proliferation of internal-combustion engines, coal-fired electrical generating plants, and similar processes, all of which generate greenhouse gases. The main greenhouse gases are carbon dioxide (CO_2), methane (CH_4), nitrogen dioxide (NO_2), ozone (O_3), and the chlorofluorocarbons (CFCs).[5] These are the gases that we must limit. (Some of these gases are more damaging to the atmosphere, so their quantities are usually expressed in the equivalent amount of carbon dioxide that would cause the same damage.)

Negotiating the Kyoto Protocol

Suspicions that global warming was affecting the climate were raised in Geneva in 1979, at the first World Climate Conference, which was sponsored by the World Meteorological Organization. During the 1980s the evidence grew stronger that human activities were affecting the world's climate (this effect was discussed in the previous chapter). The UN General Assembly established an intergovernmental panel on climate change, which produced a 1990 report asserting that global warming was probably occurring. At the 1992 Earth Summit in Rio de Janeiro, a voluntary agreement was reached to reduce greenhouse gas emissions. This agreement was titled the UN Framework Convention on Climate Change, and it was signed by 165 nations, including Canada and the United States. The goal that the signatories set for themselves was to reduce greenhouse gas emissions to 1990 levels by 2000. (This goal was not achieved.) In 1995 the parties to the convention decided that developing countries should be relieved of the hardship of some obligations, and that developed countries should do more.

In December 1997, more than 160 countries met in Kyoto, Japan, to negotiate specific emission targets. More than eighty countries agreed to reduce their emissions of greenhouse gases to an average level of 5.2 percent below

1990 levels by the year 2010 (or by some date in the five years between 2008 and 2012). Each country was allotted a different target. The Kyoto Protocol will not have legal status until it is ratified by at least fifty-five countries that emit in total at least 55 percent of the total greenhouse gases (as measured in 1990). Many meetings were held to negotiate arrangements that would be fair to developing countries, but many disagreements resulted, in particular, over issues such as credits for carbon dioxide "sinks" such as forests (which absorb carbon dioxide by natural means), the extent to which countries should be allowed to apply economic measures (that is, pay credits instead of reducing emissions), and the role of developing nations. In March 2001 the United States announced that it would no longer participate in the Kyoto Protocol.[6] In December 2002, Canada's Parliament voted to endorse this country's ratification of the Kyoto Protocol, so we need to get moving to meet our target.

Canada's Kyoto Target

The three largest emitters of greenhouse gases per capita are Australia, the United States, and Canada. Canada's target is to reduce its emissions of greenhouse gases to 570 Mt (megatonnes) of carbon dioxide and equivalents by 2010. This amount is 6 percent lower than Canada's total greenhouse gas emissions in 1990. However, emissions have increased since 1990; in fact, the predicted output for 2010 is about 810 Mt. Thus, to satisfy the protocol we will have to reduce emissions by 240 Mt by 2010. Fortunately, Canada has already adopted policies that will account for about 50 Mt of this, and another 24 Mt will be accounted for by the fact that the international community has agreed to give Canada credit for the carbon dioxide stored annually in sinks. This reduces the amount to 166 Mt, which will have to be removed from emissions over the next few years.[7] This goal is going to be difficult to achieve, especially since our population is growing, as is our economy. A serious cooperative effort by all Canadians will be essential.

Implementing the Kyoto Protocol

The Kyoto Protocol suggests a number of ways to reach the target:
- **Reduce domestic emissions.** Reduce energy used within the country, or switch to less-polluting fuels.
- **Increase carbon dioxide sinks.** A change of management practices in forests and on farms could increase the carbon dioxide sinks.
- **Implement measures jointly with another country.** Governments could cooperate on projects that reduce emissions or increase sinks. This could be especially beneficial if clean development is carried out in developing countries that are not signatories to the protocol.
- **"Trade" emissions.** The joint measures suggested above could evolve into a sort of emissions-trading system among all nations. In effect, some countries would pay other countries to reduce their emissions or create sinks.[8]

The book *Stormy Weather*[9] offers more than one hundred innovative and effective ideas for reducing global warming. Also, the David Suzuki Foundation has published a useful paper on practical and affordable methods for achieving the goals of the Kyoto Protocol. This paper, which was submitted to the federal Standing Committee on Finance, lists many ideas for tax incentives and government programs that would make energy conservation a very attractive option.[10]

Concluding Comments

Although it may entail some sacrifice, Canada can probably achieve the goals set for it under the Kyoto Protocol. It is debatable, however, whether other countries will be able to do the same.

Proponents of Kyoto, such as the David Suzuki Foundation, are very hopeful that the protocol will succeed. A recent report argues that Canada can afford to achieve the goals of the Kyoto Protocol, that the Canadian economy would continue to grow, and that the emission reductions would in fact yield net economic benefits for Canada over and above those expected. These benefits include net economic savings; an additional 52,000 jobs (because consumer savings on fuel and electricity costs would be directed toward other goods, services, activities, and investments); a significant gain in average annual household incomes arising from the new jobs; and a $2 billion increase in national GDP beyond that expected from business as usual. The reduced emissions would also generate significant health and environmental benefits such as better air quality, improved public health, and reduced damage to natural ecosystems, infrastructure, and private property.[11]

However, political manoeuvring may block the protocol's ratification. Under the treaty, countries responsible for 55 percent of emissions must ratify the protocol before it comes into force. The United States has rejected it. Russia must approve the treaty in order for the 55 percent rule to be satisfied and for the protocol to be binding on the signatories. (As this text goes to press, approval is uncertain.)

Moreover, emissions trading will be an obstacle to success, even if the protocol is approved. The definition and role of carbon dioxide sinks is highly controversial. These sinks are not easily measured. Some cash-starved countries may sell their emission credits and then pollute anyway. European countries that were once heavily dependent on coal and that are now switching to other fuels as a matter of efficiency may reap a windfall in emissions credits. And finally, since the United States—Canada's largest trading partner—is no longer part of the protocol, how will Canada compete with American corporations, which feel no obligation to reduce emissions to satisfy the Kyoto Protocol? The protocol has created some complex questions for Canada.[12]

Whether the Kyoto protocol is approved or not, the Code of Ethics, the guidelines for the environment (discussed below), and simple common sense tell us that efficient energy use and the reduction of waste are admirable goals. However, engineers and geoscientists face a stiff challenge to meet the 2008 to 2012 deadline in the Kyoto Protocol.

ENVIRONMENTAL GUIDELINES FOR ENGINEERS AND GEOSCIENTISTS

Although the care of our environment is everyone's responsibility, professional engineers and geoscientists must assume a key role because their decisions often have a great impact on the environment. Specific environmental guidelines are published by several Associations and by other agencies, as listed below.

Guidelines for Environmental Protection

A guideline on environmental practice was developed by the Association of Professional Engineers, Geologists and Geophysicists of Alberta (APEGGA).[13] This guideline has since been adopted (with some changes to the explanatory instructions) by Professional Engineers Ontario (PEO).[14]

These guidelines should be treated as complements to the Code of Ethics. They commit professional engineers, geologists, and geophysicists to protecting the environment and safeguarding the public's well-being. The original documents contain a wealth of explanatory information. The nine precepts of these guidelines are listed below, with permission:

Professional engineers, geologists, and geophysicists:

1. Shall develop and maintain a reasonable level of understanding of environmental issues related to their field of expertise.
2. Shall employ appropriate expertise of specialists in areas where the member's knowledge alone is not adequate to address environmental issues.
3. Shall apply professional and responsible judgement in their environmental considerations.
4. Shall ensure that environmental planning and management are integrated into all their activities which are likely to have adverse environmental impact.
5. Shall include the costs of environmental protection and/or remediation among the essential factors used for evaluating the life-cycle economic viability of projects for which they are responsible.
6. Shall recognize the value of waste minimization, and endeavour to implement the elimination and/or reduction of waste at the production source.
7. Shall cooperate with public authorities in an open manner, and strive to respond to environmental concerns in a timely fashion.
8. Shall comply with legislation, and when the benefits to society justify the costs, encourage additional environmental protection.
9. Are encouraged to work actively with others to improve environmental understanding and practices.[15]

An environmental guide has also been developed by the Association of Professional Engineers and Geoscientists of British Columbia (APEGBC).[16] The APEGBC guide emphasizes the importance of sustainability—a philosophy that links a viable economy to protection of the environment and to social well-being. These APEGBC guidelines are advisory and are intended to help members maintain a state in which these features flourish indefinitely. The APEGBC guidelines on sustainability specifically do not create any legal duty or obligation by any member to any person.

ENVIRONMENTAL GUIDELINES FOR CORPORATIONS

The CERES Principles

In addition to the above guidelines, which apply to individuals, a set of environmental principles has evolved for corporations. These principles have been developed over the past thirteen years by the Coalition for Environmentally Responsible Economies (CERES), an American coalition of institutions—mainly environmental, public interest, and community groups, as well as investors, advisors, and analysts.

The CERES environmental principles were developed in the wake of the *Exxon Valdez* oil spill, an environmental disaster that polluted the Alaska shoreline in March 1989. The *Exxon Valdez* ran aground in a bay on the Alaskan coast. The accident was attributed to navigational errors; also, it is possible that the captain was drunk. The resulting 38,800-tonne oil spill killed wildlife and coated the shoreline up to 750 km from the accident site. It was one of the worst environmental disasters in North American history.[17]

In the autumn of 1989, CERES published the Valdez Principles (later renamed the CERES Principles), a ten-point code of corporate environmental conduct. CERES asks industrial corporations to support the CERES principles by adopting them as a corporate Code of Environmental Ethics. The ten principles are as follows, with permission of CERES:

Protection of the Biosphere. We will reduce and make continual progress toward eliminating the release of any substance that may cause environmental damage to the air, water, or the earth or its inhabitants. We will safeguard all habitats affected by our operations and will protect open spaces and wilderness, while preserving biodiversity.

Sustainable Use of Natural Resources. We will make sustainable use of renewable natural resources, such as water, soils and forests. We will conserve non-renewable natural resources through efficient use and careful planning.

Reduction and Disposal of Wastes. We will reduce and where possible eliminate waste through source reduction and recycling. All waste will be handled and disposed of through safe and responsible methods.

Energy Conservation. We will conserve energy and improve the energy efficiency of our internal operations and of the goods and services we sell. We will make every effort to use environmentally safe and sustainable energy sources.

Risk Reduction. We will strive to minimize the environmental, health and safety risks to our employees and the communities in which we operate through safe technologies, facilities and operating procedures, and by being prepared for emergencies.

Safe Products and Services. We will reduce and where possible eliminate the use, manufacture or sale of products and services that cause environmental damage or health or safety hazards. We will inform our customers of the environmental impacts of our products or services and try to correct unsafe use.

Environmental Restoration. We will promptly and responsibly correct conditions we have caused that endanger health, safety or the environment. To the extent feasible, we will redress injuries we have caused to persons or damage we have caused to the environment and will restore the environment.

Informing the Public. We will inform in a timely manner everyone who may be affected by conditions caused by our company that might endanger health, safety or the environment. We will regularly seek advice and counsel through dialogue with persons in communities near our facilities. We will not take any action against employees for reporting dangerous incidents or conditions to management or to appropriate authorities.

Management Commitment. We will implement these Principles and sustain a process that ensures that the Board of Directors and Chief Executive Officer are fully informed about pertinent environmental issues and are fully responsible for environmental policy. In selecting our Board of Directors, we will consider demonstrated environmental commitment as a factor.

Audits and Reports. We will conduct an annual self-evaluation of our progress in implementing these Principles. We will support the timely creation of generally accepted environmental audit procedures. We will annually complete the CERES Report, which will be made available to the public.[18]

Registration Under ISO 14001

Of course, when corporations are serious about demonstrating their commitment to responsible interaction with the environment, the most visible route to follow is registration under the ISO 14000 series of Environmental Management System Standards, established in 1996—in particular, ISO 14001 (discussed earlier in this text). ISO 14001 registration requires a commitment from senior management, a review of all of the applicable environmental laws, an audit of the environmental impact of the corporation's operations, development of environmental policies, establishment of measurement techniques and methods for recording the measurements, preparation of a procedures manual to define who does what, training of employees, full communication within the corporation, and regular audits to ensure that the system is working and achieving its goals. Obviously, registration indicates that the corporation has made a serious commitment to act responsibly in environmental matters. Moreover, since the ISO 14001 standard is recognized internationally, registration should be an aid in international trade.

THE DUTY TO REPORT—*WHISTLEBLOWING*

An engineer who observes unsafe, unethical, or illegal practices must take action. Personal contact and informal communication usually yield the best results. With this idea in mind, provincial Associations have often served as mediators to help engineers who believe that clients, colleagues, employers, or employees are involved in unsafe, unethical, or illegal practices. The Association can play a useful role by helping define the ethical issues involved, advising the engineer, communicating the concerns to the client or employer in an unbiased way, and generally mediating as informally as possible.

In all probability, there is no single procedure that will work every time. Also, an engineer must take a much more aggressive approach in situations where human life is at risk. Finally, as an engineer you will have to decide whether it is the situation or the conditions that must be reported, or the individual. Each case is different.

For example, a concern about unethical billing practices is not usually urgent and can usually be resolved informally. A situation where workers' lives are at risk is clearly more urgent and might be referred immediately to the authorities (including the police) if a delay in acting might cause injury or death. Failure to take immediate action to protect human life is professional misconduct.

Professional Engineers Ontario (PEO) has defined the procedure for reporting even more clearly in the publication *A Professional Engineer's Duty to Report—Responsible Disclosure of Conditions Affecting Public Safety*. An excerpt follows:

The Reporting Process

Engineers are encouraged to raise their concerns internally with their employers or clients in an open and forthright manner before reporting the situation to PEO. Although there may be situations where this is not possible, engineers should first attempt to resolve problems themselves.

1. If resolution as above is not possible, engineers may report situations in writing or by telephone to the Office of the Registrar of PEO. In reporting the situation to PEO, engineers must be prepared to identify themselves and be prepared to stand openly behind their judgements if it becomes necessary.
2. The Office of the Registrar will expect the reporting party to provide the following information:
   a) the name of the engineer who is reporting the situation;
   b) the name(s) of the engineer's client/employer to whom the situation has been reported;
   c) a clear, detailed statement of the engineer's concerns, supported by evidence and the probable consequences if remedial action is not taken;
3. The Office of the Registrar will treat all information, including the reporting engineer's name, as confidential to the fullest extent possible.

4. The Office of the Registrar will confirm the factual nature of the situation and, where the reporting engineer has already contacted the client/employer, obtain an explanation of the situation from the client/employer's point of view.

5. Where the Office of the Registrar has reason to believe that a situation that may endanger the safety or welfare of the public does exist, the Office of the Registrar will take one or more of the following actions:

   a) report the situation to the appropriate municipal, provincial and/or federal authorities;

   b) where necessary, review the situation with one or more independent engineers, to obtain advice as to the potential danger to public safety or welfare and the remedial action to be taken;

   c) request the client/employer to take steps necessary to avoid danger to the public safety or welfare;

   d) take such other action as deemed appropriate under the circumstances;

   e) follow up on the action taken by all parties to confirm that the problem has been resolved.

Wherever possible, the Office of the Registrar shall maintain accurate records of all communications with the reporting engineer, any authorities involved and the client/employer.

In Summary: The Office of the Registrar will cooperate with any engineer who reports a situation that the engineer believes may endanger the safety or welfare of the public. Wherever possible, the confidentiality of reporting engineers and the information they disclose will be maintained. The Office of the Registrar will emphasize in all dealings with the engineer's client/employer and the public the engineer's duty to report under the Act and Regulations, and will provide the reporting engineer with an endorsement of the performance of his/her duty, provided that the Registrar has determined that the engineer has acted properly and in good faith.[19]

The above PEO policy is orderly and impartial. It clearly places public welfare first. However, not all Associations have implemented a similar policy.

THE ETHICAL DILEMMA OF WHISTLEBLOWING

Whistleblowing always involves an ethical dilemma. Every Code of Ethics requires engineers and geoscientists to consider their duty to society as paramount. However, every code also stipulates duties to clients, employers, colleagues, and employees. At what point does the duty to society override these other duties? For example, the code prohibits disclosure of confidential information concerning a client or employer, but the code requires disclosure of a situation that may endanger the health or safety of the public. Obvious, these duties may conflict.

Whistleblowing is a controversial act, so we must define the term clearly. A good definition is provided by Connie Mucklestone in her article "The Engineer as Public Defender":

Whistleblowers are people (usually employees) who believe an organization is engaged in unsafe, unethical or illegal practices and go public with their charge, having tried with no success to have the situation corrected through internal channels.[20]

True whistleblowing is rare because, as the above quote emphasizes, a true whistleblower must be concerned about "unsafe, unethical or illegal practices" and such valid complaints can usually be resolved simply by communicating the facts to the people in charge. Personal complaints or disputes are not a proper basis for whistleblowing. It follows that although whistleblowing is sometimes necessary, it is rather rare.

A whistleblower is different from a trouble-maker in two important ways: one difference is the motive of the person involved, the other difference is the methods used to protect the public. These points are illustrated in the following quote:

Engineers must act out of a sense of duty, with full knowledge of the effect of their actions, and accept responsibility for their judgement. For this reason any process which involves "leaking" information anonymously is discouraged. There is a basic difference between "leaking" information and "responsible disclosure." The former is essentially furtive, and selfish, with an apparent objective of revenge or embarrassment; the latter is open, personal, conducted with the interest of the public in mind and obviously requires that engineers *put their names on the action and sometimes their jobs on the line.* [Italics added.][21]

The whistleblower must be aware that the process may involve public exposure and scrutiny and may place his or her career in jeopardy. Obviously, whistleblowing should not be done casually, unknowingly, or wantonly. The provincial Association should be contacted, and its reporting process should be followed.

In summary, before using the reporting process described in the previous section, an engineer or geoscientist should reflect on the following three points:

- **Informal resolution.** It is extremely important to try to resolve problems informally and internally in an open and professional manner. In the vast majority of cases, clear communication is all that is required. While the Association may give guidance and provide mediation, these steps represent a greater degree of formality. A professional must be certain that an informal internal solution cannot be obtained before resorting to whistleblowing, and must assume the responsibility and consequences of any harm that results from a frivolous accusation.
- **Confidentiality**. Professionals must always report unethical cases to the appropriate regulating body and not to the news media. The goal is to remedy a problem, not to embarrass individuals.
- **Retaliation.** Where it has been necessary to report an unethical, illegal, or unsafe act to public authorities, the employer may attempt to retaliate by firing the engineer. Professional engineers and geoscientists should

know that a warranted reporting does not constitute just cause for dismissal (this was explained in Chapter 8). Furthermore, a dismissed person can file a lawsuit to recover lost wages and costs.

A DISSENTING VIEW OF THE DUTY TO SOCIETY

The first or second clause of every Code of Ethics says that professionals must consider their duty to society to be paramount. This obligation seems clear and unequivocal. However, one well-known expert on engineering ethics, Samuel Florman, spoke against this clause as a general guide, because it does not have a precise meaning. His comments are quoted, in part, below:

> If this appeal to conscience were to be followed literally, chaos would ensue. Ties of loyalty and discipline would dissolve, and organizations would shatter. Blowing the whistle on one's supervisors would become the norm, instead of a last and desperate resort. It is unthinkable that each engineer determine to his own satisfaction what criteria of safety, for example, should be observed in each problem he encounters. Any product can be made safer at greater cost, but absolute freedom from risk is an illusion. Thus, acceptable standards must be specifically established by code, by regulation, or by law, or where these do not exist, by management decision based upon standards of legal liability. Public-safety policies are determined by legislators, bureaucrats, judges, and juries, in response to facts presented by expert advisers. Many of our legal procedures seem disagreeable, particularly when lives are valued in dollars, but since an approximation of the public will does appear to prevail, I cannot think of a better way to proceed.
>
> ...
>
> The regulations need not all be legislated, but they must be formally codified. If we are now discovering that there are tens of thousands of potentially dangerous substances in our midst, then they must be tested, the often-confusing results debated, and decisions made by democratically designated authorities—decisions that will be challenged and revised again and again.
>
> ...
>
> This is an excruciatingly laborious business, but it cannot be avoided by appealing to the good instincts of engineers. If the multitude of new regulations and clumsy bureaucracies has made life difficult for corporate executives, the solution is not in promising to be good and eliminating the controls, but rather in consolidating the controls themselves and making them rational. The world's technological problems cannot even be formulated, much less solved, in terms of ethical rhetoric: especially in engineering, good intentions are a poor substitute for good sense, talent, and hard work.[22]

Florman's comments are thought-provoking and refreshing. However, it must be remembered that they were spoken in the American context: in that country, licensing regulations are much less restrictive than Canadian laws. Florman's recommendation for developing standards and regulations based on solid research cannot be refuted—such standards would be useful to the practising engineer, especially where dangerous chemicals are concerned. However, many well-known regulations and standards are already in print, yet some companies and individuals still do not follow them because of ignorance, inertia, or unethical attitudes. Developing more regulations will not change the attitudes of unethical people, and the professional engineer will still encounter situations where the public good must be weighed against the benefit of client and employer. Instead of discarding the public safety clause in the Code of Ethics, as Florman suggests, it might be better to provide more mediation between whistleblowers and their employers (as some provincial Associations are now doing), and to provide protection against retaliation for engineers who, after exhausting all other routes of action, report unethical practices.

In summary, Canadian laws for the profession provide an intermediate position that does not exist in the United States—the Code of Ethics is backed by law. Nevertheless, Florman's proposal to develop more safety regulations and standards is extremely valuable, and would permit the Canadian system to work even better.

CASE HISTORY 11.1

THE LODGEPOLE WELL BLOWOUT

Introduction

Residents of the Drayton Valley area of Alberta will long remember the autumn of 1982. At 2:30 p.m. on October 17, the Amoco Lodgepole oil well being drilled near Drayton Valley encountered sour gas (that is, gas laden with hydrogen sulphide) and blew out of control. Over the next two months, while specialists fought to regain control of the well, residents living within a 20- or 30-km radius were twice exposed to the rotten egg smell of hydrogen sulphide (H_2S) and the threat of H_2S poisoning. The first H_2S exposure period lasted sixteen days; the second lasted twelve days. During attempts to cap the blowout, two employees were overcome by H_2S and died, and the well was twice engulfed in flames. About twenty-eight people were voluntarily relocated to avoid the H_2S, and several homes were ordered evacuated during especially heavy H_2S concentrations on October 29 and November 17 to 24. Even people living far from the well were subjected to noxious and unpleasant odours, depending on the prevailing winds.

The well was not capped successfully until December 23. In January 1983, a Lodgepole Blowout Inquiry Panel was convened to investigate the causes of the blowout, the actions taken to prevent it and to regain control, the hazard to human health, and the impact on the environment, and to recommend what should be done to avoid future blowouts at wells in Alberta. The panel issued a comprehensive report in December 1984.[23]

Events Leading up to the Blowout

The Amoco Lodgepole oil well, known officially as Amoco Dome Brazeau River 13-12-48-2, is located about 140 km west of Edmonton (about 40 km west of Drayton Valley). The well was named after the nearby hamlet of Lodgepole, which is situated about halfway between the well and Drayton Valley. The Amoco Canada Petroleum Company obtained a licence to drill the Lodgepole oil well from the Alberta Energy Resources Conservation Board (now superseded by the Alberta Energy and Utilities Board). The well was "spudded" (started) in August 1982. Drilling proceeded to a depth of about 3000 m without problems. An intermediate casing was then installed, and the drilling crew began coring operations to examine the strata prior to drilling into the oil-bearing formation. Two cores were obtained without apparent problems. On October 16, the crew was obtaining a third core when they realized that fluid was entering the well from the oil- and gas-bearing formation.

The drill crew stopped the coring operations to deal with this problem, which is known as a "kick" because the entry of reservoir fluids into the wellbore forces the drilling mud out of the well. For the next sixteen hours, the crew fought to regain control of the well, but eventually the drill pipe "hydraulicked" up the hole and the kelly hose was severed, at which point the well was out of control. The intense pressure caused a continuous, uncontrolled flow of mud and sour gas into the atmosphere. The exact flow rate is unknown; however, during the inquiry it was estimated at 1.4 million m^3 of gas per day. Later tests indicated that the flow could have been even greater.[24]

Emergency Measures

Amoco immediately implemented its Major Well-site Incident Response Plan, and key Amoco personnel were notified of the blowout. People and equipment were dispatched to the site, including safety personnel, paramedics, ambulances, helicopters, and firefighting equipment (including breathing apparatus). Hydrogen sulphide monitoring equipment was ordered for both on-site and off-site monitoring. The company immediately hired specialists to cap the well, and special equipment to do so was ordered.

Over the next two months, several plans for capping the well were tried without success. On November 1 a failed attempt resulted in a fire that engulfed the well. A new control plan was developed, and on November 16 the fire was extinguished with explosives preparatory to implementing the plan. Two days later, while attempting to execute the plan, an accident occurred that resulted in the deaths of two employees, who were overcome by H_2S. On November 25, the well was again on fire. It was later determined that this fire probably resulted from an undetected underground muskeg fire that had been smouldering for some time. Amoco decided to try capping the well while it was still ablaze; however, the well specialists declined to attempt this procedure, which had seldom succeeded with other blowouts. On December 1, new well-capping specialists were hired, and by December 23 they had installed a blowout preventer (BOP) over the stub of the intermediate casing. Lines were then connected to flare off the gas and pump mud into the well. Over the next five days, 96 m^3 of mud were pumped into the well, the pressure was stabilized, and the crisis was brought to an end.

What Went Wrong at the Lodgepole Well?

The blowout occurred basically because Amoco personnel were unable to control the hydrostatic pressure in the well. This is a critically important and delicate balancing procedure. The control strategy—usually called the well control plan—sets out the basic principles and procedures that must be followed to ensure that a well will not blow out during drilling, completion, or production operations.

The well control plan is rarely a single document. Rather, it is the sum total of all drilling program documents, special instruction bulletins—including those posted at the site—company procedure manuals, and other books, manuals, and written and verbal instructions that guide drilling procedures. To help you understand the importance of the control plan and how it applies to the critical balancing of hydrostatic pressure in the well, the following explanatory note is reproduced from the inquiry report, with permission:

> The drilling fluid (mud) system has a dominant position in the general well control plan for any well. The plan requires that the hydrostatic pressure in the wellbore be greater than the formation pressure. Hydrostatic pressure depends on the height of the column of drilling mud and the density of the mud. A reduction in either or both of these will reduce the pressure that results from the column or head of drilling mud. If the hydrostatic pressure is too low, a state of underbalance exists and fluids from the formation, such as gas, may flow into the wellbore. Unless this flow is properly controlled, a blowout will result. On the other hand, if the hydrostatic pressure is too great and a state of excessive overbalance exists, the drilling mud may flow into the formation. This is referred to as "lost circulation" and will result in a loss of hydrostatic pressure which can also lead to a blowout.
>
> … In developing a drilling program, an operator must therefore consider formation pressures encountered at other wells in the general vicinity of the well being planned for and select a drilling mud density which will ensure a modest overbalance. To ensure that neither a state of underbalance nor of excessive overbalance develops as the well is drilled, close attention must be given to:
>
> (a) mud density,
>
> (b) any contamination of the drilling mud that will change its effective density, such as by drill cuttings (increase) or by air introduced during tripping (decrease),
>
> (c) maintaining a full mud column,
>
> (d) the rate of lowering the drill pipe,

(e) the rate of hoisting the drill pipe, and

(f) pumping rate and pressure.

Should the hydrostatic pressure from the mud column prove to be insufficient, and as a consequence, formation fluids such as gas enter the wellbore, a kick condition would exist. Procedures have been developed such that those fluids may be controlled within the wellbore using the BOP [blowout preventer] system mounted on the well casing. The gas is flared at the surface and control operations are continued until the flow is progressively restricted and stopped. If the kick has occurred and efforts to control and contain the in-flowing formation fluids have failed, an uncontrolled flow or blowout results and re-establishing control may be both technically difficult and dangerous.

The general well control plan must include the procedures to be used to circulate out the kick, and it must also ensure that proper equipment will be available should a kick occur. This includes the BOP system but additionally, casing and drill pipe design and selection are important components of the plan.

In order to carry out the general well control plan, the detailed drilling plan must provide for the integrity of the drilling fluid system throughout the drilling operation regardless of the circumstances encountered. The operator should have on site, at critical times, experts in geology and mud properties. The operator must also ensure that the drilling and well equipment, particularly the BOP, is functioning properly. Finally, for the general well control plan to be implemented effectively, on-site supervisors and the drilling crew must be properly trained, regularly briefed, and always prepared to act promptly in carrying out prescribed kick control procedures.[25]

Evaluating Amoco's Actions

Amoco assisted the inquiry panel by providing complete documentation on its drilling plan, drilling mud program, rig, and well equipment, and on the qualifications of the well-site crew and supervisors. It also provided a detailed chronology of events leading to the blowout and expert witnesses to explain the events surrounding the blowout. An especially important point was the density of the drilling mud used. Obtaining the right mud density is a key part of the balancing act: it requires a knowledge of the formation pressures and careful monitoring of the hydrostatic pressure. The problem is explained in the inquiry report as follows:

> [I]t is necessary to use a mud density which is neither too low, thus allowing an influx from the formation, nor too high, which could result in lost circulation. This means there is a range of mud density within which operations must take place. The closer one is to the upper or lower limit, the more careful drilling procedures must be. If the mud density is within the range but towards the "high limit," care must be taken to avoid lost circulation. For example, the mud volume must be carefully monitored and the crew must be ready to add lost

circulation material. If the mud is closer to the "low limit," drilling must proceed slowly, the potential for a kick must be carefully monitored, and plans to quickly weight up the mud must be in place.

... In developing its drilling plan, Amoco reviewed information concerning formation pressures encountered at a large number of other wells in nearby areas. These indicated that the pressures in the formation of interest at the [Lodgepole] well would ordinarily be around 33 000 to 35 000 kPa. In isolated cases, pressures as low as 22 430 kPa and as high as 46 540 kPa had been reported. Amoco decided to design the drilling mud density to meet a pressure of 33 600 kPa with provision for increasing the density if higher pressure was encountered.

Amoco also indicated that its normal practice was to design mud density to provide a mud column overbalance of some 1500 kPa above the expected formation pressures. During drilling or coring operations, mud pump pressures would add further to this, resulting in an overbalance which should avoid the possibility of excess fluid head and lost circulation but at the same time prevent influx of reservoir fluid.

Calculations by the Panel indicate that, at the predicted Nisku reef depth of 3035 m, the planned mud density of 1150 kg/m^3 would result in an overbalance of some 630 kPa relative to the expected pressure of 33 600 kPa ...

... The range of appropriate mud densities varies for each situation, and at the [Lodgepole] well, because of the high reservoir pressure with very good permeability, [the range] was likely relatively narrow. The planned pressure overbalance of 630 kPa plus or minus the effects of operations such as pumping or pulling out of the hole, was less than the 1500 kPa normally used by Amoco ...

... In summary, the Panel concludes that the planned mud density for the [Lodgepole] well was on the low side and therefore extra care should have been specified during the critical period of drilling into the Nisku zone. Although substantial seepage losses were reported and these might have been interpreted by on-site personnel as an indication that the mud was on the heavy side, an analysis of the situation indicates that the reported losses were likely due to errors. This is an indication that the drilling practices being used were less than satisfactory.[26]

Assigning Responsibility

Although the Lodgepole Inquiry Panel concluded that "no single element in the chain of events was the sole cause of the blowout," the panel's examination of the events led it to conclude that the initial kick occurred mainly because "drilling practices during the taking of cores were deficient." When combined with the marginally adequate mud density being used, this deficiency permitted the entry of reservoir fluids into the wellbore.[27]

Amoco expected the Lodgepole well to find "sweet" oil, but the company's control plan definitely recognized and accounted for the possibility of encountering sour gas. The panel accepted this testimony and did not believe that "the expectation of sweet oil played a direct role in the cause of the blowout. However, it may have influenced planning for the well and may have led to less caution in the drilling operations than might have been the case if the well was being drilled specifically for sour gas."[28]

The panel also concluded that in all likelihood the kick was not controlled because the drilling crew did not immediately recognize the problem and therefore did not immediately apply and maintain standard kick-control procedures. With regard to contributing factors, the panel noted that several pieces of vital equipment did not function properly, and also that supplies of mixed drilling mud were not adequate during the kick-control operations. As the panel wrote in its report:

> The unexpected entry of reservoir fluids into the wellbore was probably due to a combination of Amoco not adhering to sound drilling practices and only marginally adequate mud density. If the degasser had operated effectively, the initial kick might have been circulated out of the system, and subsequent kicks would likely have been avoided. Even with the failure of the degasser, control might have been maintained if there had been sufficient and properly weighted mud on hand to pump into the well. Additionally, if the casing pressure instruments had been operational from the outset, the crew might have recognized the kick at an earlier stage and implemented standard kick-control procedures when fluid influx was still relatively small. If the decision had been made to use hydrogen sulphide (H_2S)-resistant pipe for the full drill string, the pipe might not have parted and the succession of kicks might still have been successfully circulated out. And finally, if the travelling block hook latch had not failed, it may have been possible to retain control of the well by "top kill" methods.[29]

The panel was satisfied that Amoco applied reasonable judgment in selecting the type of drilling rig, the degasser, and the type of drill pipe even though, in retrospect, other choices might have been better. However, the panel concluded that

> Amoco's actions were deficient with respect to:
>
> (a) drilling practices during coring operations (cores No. 2 and 3),
>
> (b) implementation of standard kick control procedures,
>
> (c) ensuring adequate mixed drilling mud was available at all times, and
>
> (d) maintaining equipment in satisfactory operating condition (casing pressure instruments).
>
> It appears to the Panel that the fundamental problem was that Amoco did not apply the necessary degree of caution while carrying out operations in the critical zone. Amoco did not appear to be sufficiently aware of the potential for problems that could occur when coring into the Porous zone and thus the need

to be fully prepared in the event of a fluid influx. Consequently, when a kick developed, there were delays in responding to it. Then, when equipment problems occurred and supplies of mixed mud were inadequate, Amoco was forced into further delays of precious time in implementing kick-control actions.[30]

In the second phase of the inquiry, the panel made a series of recommendations for reducing the possibility of future blowouts.[31]

Conclusion

Oil drilling is a demanding, uncertain, and dangerous job. The Lodgepole blowout, which resulted in two tragic deaths and millions of dollars in financial losses, illustrates this point. There are other losses that are harder to put a price on, such as the threat of H_2S poisoning, the disruption of life, and the inconvenience caused to almost all the residents within a 30-km radius of the well. It is also difficult to evaluate the magnitude of loss suffered by Amoco and its technical staff as a result of the negative publicity resulting from the blowout. Safe, standard procedures on drill sites are essential if these dangers are to be avoided.

DISCUSSION TOPICS AND ASSIGNMENTS

(Additional assignments are found in Appendix CD-E on the CD-ROM included with this text.)

1. You have graduated from university and have been working for five years as a plant design and maintenance engineer for a pulp and paper company in northern Canada. The company is a wholly-owned subsidiary of a large multinational conglomerate. When you received your P.Eng. licence, you were promoted to chief plant engineer. You work directly for the plant manager, François Bédard, who reports to the head office, which is not in Canada. The company employs about 150 people—most of the adult population of the nearby village—either directly as employees or indirectly as woodcutters.

   In the course of your work you have become aware that the plant effluent contains a very high concentration of a mercury compound that could be dangerous. In fact, since the plant has been discharging this material for twenty-five years, water in the nearby river downstream from the plant is thoroughly unfit for drinking or swimming. You suspect that a curious new illness in an Aboriginal village about 40 km downstream is really Minamata disease, the classic symptoms of which are spasticity, loss of coordination, and, eventually, death. You also suspect that the fish in the river have been contaminated with the mercury and have spread the contamination to all the downstream lakes.

   Remedying these problems would involve drastic changes to the plant that would cost at least $1 million. So far you have told no one of your suspicions except Bédard, with whom you have discussed the problem at length. Bédard, who is not an engineer, has confided that the head office considers the plant only marginally profitable and that an expenditure of this magnitude is simply not possible. He has also told you that the head office would close down the plant, causing massive unemployment in the area and probably forcing the workers to abandon their homes to seek work elsewhere. What should you do?

2. The greenhouse effect and ozone depletion are man-made problems—that is, the carbon dioxide and chlorofluorocarbon gases that cause these problems are mainly the result of human activity. It has been suggested that man-made solutions could reverse these problems. For example, propane in large quantities could be released from aircraft at high altitudes; there, it would combine with the chlorine that is causing ozone depletion and turn it into a harmless salt (HCL). Another plan proposes to reduce the amount of carbon dioxide in the air through photosynthesis, a chemical reaction whereby water and carbon dioxide combine to form glucose and oxygen (photosynthesis is critically important in restoring oxygen to the atmosphere). Stimulating the growth of algae in the ocean would, through photosynthesis, simultaneously reduce greenhouse gases and increase the oxygen content of the atmosphere.

    Using the information resources available to you, investigate and evaluate these proposals and any others you may discover. Prepare a brief summary that answers the following questions:

    • Are these schemes chemically feasible?
    • How much material would be required to stop the current trends?
    • How much material would be required to reverse the trends and restore appropriate levels?
    • How would the processes work in practice?
    • How much would they cost, and who should pay this cost?
    • Are there any foreseeable side effects?
    • How could the proposals be tested before a full commitment to apply them is made?

3. At some point in the near future, as a result of population growth, the world is going to run out of oil and gas reserves. The consequences for humanity will be severe unless alternative energy sources are brought on line before then. Various sources put the "run-out year" between 2020 and 2050—that is, well within the professional lives of the engineers and geoscientists reading this text. Using the information resources available to you, investigate and evaluate these predictions. Prepare a brief summary of your evaluation, one that states which alternative energy sources (solar, wind, wave, geothermal, fission, fusion, etc.) are the best replacements for gas and oil. Is it likely that we will be able to maintain our standard of living well into the future?

4. Between 40,000 and 50,000 people are killed every year in car accidents in North America, yet people apparently consider driving a car to be

worth the risk. Fewer than ten people have been killed in North America in nuclear reactor accidents, yet many people are afraid of nuclear power. Many people die every year while producing food (even farming is dangerous), and many thousands of miners have been killed in the twentieth century (coal mining is especially dangerous). Clearly, there is a discrepancy between perceptions of danger and probability of death (as reflected in safety statistics). Yet perceptions often have the stronger impact on the public when it comes to public support of engineering projects. Using the information resources available to you, complete the following assignments:

a. Examine the risks associated with the various energy sources (solar, wind, wave, geothermal, fission, fusion, etc.), and develop a fair method for comparing the risks and benefits of each. That is, find the statistics for the probability of injury or death per unit of energy produced. Compare this with automobile travel on the basis of risk per unit of energy consumed.

b. Using the concepts provided in this chapter and in Chapter 6, state in one or two pages an ethical guideline for deciding when construction of a dangerous facility (such as a nuclear power plant) or production of a dangerous chemical (such as a pesticide) is morally justified. Include financial, engineering, and political arguments in your answer as well as ethical concepts.

5. We North Americans consume more energy and resources per capita than any other people on earth. By making even small lifestyle changes, we could reduce our consumption significantly; yet as a society we resist doing so for reasons unknown. The following is a brief list of simple ways that we could reduce our consumption of resources. Can you add to this list? How would you go about convincing the general public to "do the right thing" in each of the following cases?

a. Although many cities have Blue Box recycling programs, some residents insist on discarding bottles, cans, and plastics with garbage. The recyclable materials are dumped in landfills, which exceed their capacities more quickly. The value of the recyclable materials is lost, and costly new landfill sites must be found.

b. In many homes, the cellar drainage sump (which should be connected to the municipal storm drain system) is actually connected to the municipal sewage system pipe, which is usually closer. The sump typically collects rainwater from the house's perimeter drain, so this connection permits rainwater to flow into the sewage treatment plant, which must process the otherwise clean rainwater along with the sewage. After a heavy rain, the sewage plant may not be able to cope with the flow. The overflow, which is now polluted with sewage, is usually released into a stream or lake, fouling the environment.

c. Some car owners who change their own oil do not take the old oil to a gas station for recycling. Instead, they simply dump the used oil into a storm drain or septic sewer. Yet even a small amount of oil can seriously harm the environment—most obviously, by killing aquatic life. The oil may even enter the municipal drinking water system.

d. Many trailers and recreational vehicles have self-contained toilets that must be emptied regularly. Some people who use these vehicles pollute the environment by dumping the toilet contents in parks, fields, or storm sewer systems, rather than into septic systems that would carry it to sewage treatment plants.

NOTES

1. D.L. Marston, *Law for Professional Engineers: Canadian and International Perspectives*, 3rd ed., McGraw-Hill Ryerson, Whitby, ON, 1996, p. 34.
2. Commission for Environmental Cooperation (CEC), *Summary of Environmental Law in North America*, CEC, 2003, <www.cec.org/pubs_info_resources/> (October 29, 2003).
3. *The Handbook of Environmental Compliance in Ontario*, McGraw-Hill Ryerson, Toronto, ON, 2003.
4. *Guide to Environmental Compliance in Alberta*, Hazard Alert Training Inc., 4940 — 87 Street, Edmonton AB, 2003.
5. R.W. Jackson and J.M. Jackson, *Environmental Science: The Natural Environment and Human Impact*, Longman, Harlow, Essex, 1996, p. 317.
6. D. MacDonald et al., *Ratification of the Kyoto Protocol: A Citizen's Guide to the Canadian Climate Change Policy Process*, Institute for Environmental Studies, University of Toronto, September 21, 2002, p. 34 <www.torontoenvironment.org/smog/Kyoto_citizens_guide.pdf> (November 2, 2003).
7. Ibid., p. 2.
8. M. Jaccard, J. Nyboer, and B. Sadownik, *The Cost of Climate Policy*, UBC Press, Vancouver, BC, 2002, p. 18.
9. G. Dauncey and P. Mazza, *Stormy Weather: 101 Solutions to Global Climate Change*, New Society Publishers, Gabriola Island, BC, 2001.
10. J. Fulton, *Implementing the Kyoto Protocol: Practical, Affordable and Achievable Solutions*, David Suzuki Foundation, Vancouver, BC, November 4, 2003 (revised copy) <www.davidsuzuki.org/Publications/Climate_Change_Reports/> (November 4, 2003).
11. Ibid., p. 4.
12. Jaccard et al., *The Cost of Climate Policy*, p. 19.
13. Association of Professional Engineers, Geologists and Geophysicists of Alberta (APEGGA), *Environmental Practice: A Guideline*, APEGGA, Edmonton, June 1994 <www.apegga.ca/publications/guidelines.html> (November 2, 2003).
14. Professional Engineers Ontario (PEO), *Guideline to Professional Practice*, PEO, Toronto, 1988, revised 1998 <http://www.peo.on.ca/>, (November 3, 2003).
15. APEGGA, *Guideline for Environmental Practice* and Professional Engineers Ontario (PEO), "16.2 Role of the Engineer with Respect to the Environment," *Guideline to Professional Practice*, PEO, Toronto, 1988, revised 1998, pp. 27–30. <www.peo.on.ca> (November 6, 2003).
16. APEGBC, *Guidelines for Sustainability*, Association of Professional Engineers and Geoscientists of British Columbia (APEGBC), Vancouver, BC, May 1995 <www.apeg.bc.ca/library/library/guidelines/guides_sustainability.pdf> (November 2, 2003).
17. Exxon Valdez Oil Spill Trustee Council <www.oilspill.state.ak.us/facts/qanda.html> (October 29, 2003).

18. Coalition for Environmentally Responsible Economies (CERES) website: <http://www.ceres.org/our_work/principles.htm> (October 29, 2003). Reprinted with permission of CERES.

19. PEO, *A Professional Engineer's Duty to Report: Responsible Disclosure of Conditions Affecting Public Safety*, PEO, Toronto, 1996. <http://www.peo.on.ca/> (November 3, 2003). Reprinted with permission of PEO.

20. C. Mucklestone, "The Engineer as Public Defender," *Engineering Dimensions*, vol. 11, no. 2, March–April 1990, p. 29.

21. PEO, *A Professional Engineer's Duty to Report: Responsible Disclosure of Conditions Affecting Public Safety*, PEO, Toronto 1996 <www.peo.on.ca/> (November 3, 2003), p. 2. Reprinted with permission of PEO.

22. Samuel C. Florman, "Moral Blueprints," *Harper's Magazine* (October 1978). Copyright © 1978 by *Harper's Magazine*. All rights reserved. Reproduced from the October issue by special permission.

23. *Lodgepole Blowout Inquiry: Phase 1—Decision Report*, Report to the Lieutenant Governor in Council with Respect to an Inquiry Held into the Blowout of the Well: Amoco Dome Brazeau River 13-12-48-12, Energy Resources Conservation Board, Calgary, AB, December 1984, p. 5–2. Reprinted with permission of the Alberta Energy and Utilities Board.

24. Ibid., p. 7–17.

25. Ibid., p. 5–2.

26. Ibid., p. 5–5.

27. Ibid., p. 1–1.

28. Ibid., p. 5–30.

29. Ibid., p. 1–2.

30. Ibid., p. 1–2.

31. *Lodgepole Blowout Inquiry: Phase 2 Report—Sour Gas Well Blowouts in Alberta; Their Causes, and Actions Required to Minimize Their Future Occurrence* (bound as Appendix 5 of Lodgepole Blowout Inquiry: Phase 1—Decision Report), Energy Resources Conservation Board, Calgary, AB, April 1984.

Chapter 12
Computer Ethics

This chapter discusses computers in professional practice, the ethical dilemmas they create, and the Code of Ethics that guides software professionals. The question of liability for computer-generated errors—an important practical issue—is reviewed, with perspectives from several Associations. Computers create new ethical problems, from hacking, cracking, and vandalism to viruses and piracy. Professionals must be alert to these problems and to the issues of copyright infringement and possible computer disaster. This chapter addresses all of these practice and ethics problems.

THE IMPORTANCE OF COMPUTERS IN ENGINEERING AND GEOSCIENCE

In the past forty years, computers have drastically changed every design office. Tedious work is now done by hardware and software, freeing designers to be more inventive. Designs are now created at workstations that permit ideas to be visualized, simulations to be run, and alternatives to be analyzed, in the very earliest stages of the design process. Recent graduates may not realize how extensively the design process has changed. Calculations that were once tediously prepared by slide rule (and later, by calculator) are now displayed instantly, and engineering drawings that were once laboriously hand-drawn are now plotted in minutes. Conversely, the older generation can barely conceive the enormous productivity that is now possible using computer-aided design and analysis, automated drafting, dynamic simulation, project management, spreadsheets, word processing, Internet access, file transfer, e-mail, web pages, and so on. The use of computers in engineering is irreversible and will continue to grow.

Computers permit unbelievable visual power in design and incredible speed in analysis, and these capabilities have removed barriers to inspiration and eliminated the tedium of routine calculations and repetitive drawing. Yet computers are also creating new responsibilities and problems for professionals relating to security, data backup, copyright infringement, incorrect file transfer, unreliable data from the Internet, the possibility of engineering errors caused by flaws or "bugs" in computer programs, and the growing vandalism carried out by hackers and crackers, who sabotage productivity. Engineers and geoscientists must be alert to these problems.

THE CODE OF ETHICS FOR SOFTWARE ENGINEERS

Each provincial and territorial Association publishes a Code of Ethics, which is enforced under the Act, as explained in earlier chapters. In addition, most engineering societies publish voluntary Codes of Ethics to give discipline-specific guidance to their members. The SCM/IEEE Code of Ethics was developed for software engineers but is of interest to all engineers and geoscientists, because computers are essential to our work. In fact, few professions depend on computers more than engineering and geoscience. The short version of the code summarizes the key ideas; the full version (included in Appendix CD-C of this textbook) provides examples of how the code should influence the work of software engineers. Although the SCM/IEEE code is not legally binding, it provides good advice.

Software Engineering Code of Ethics and Professional Practice

as recommended by the ACM/IEEE-CS Joint Task Force on Software Engineering Ethics and Professional Practices (Version 5.2):

PREAMBLE

The short version of the code summarizes aspirations at a high level of the abstraction; the clauses that are included in the full version give examples and details of how these aspirations change the way we act as software engineering professionals. Without the aspirations, the details can become legalistic and tedious; without the details, the aspirations can become high sounding but empty; together, the aspirations and the details form a cohesive code.

Software engineers shall commit themselves to making the analysis, specification, design, development, testing and maintenance of software a beneficial and respected profession. In accordance with their commitment to the health, safety and welfare of the public, software engineers shall adhere to the following Eight Principles:

1. **PUBLIC**—Software engineers shall act consistently with the public interest.

2. **CLIENT AND EMPLOYER**—Software engineers shall act in a manner that is in the best interests of their client and employer consistent with the public interest.

3. **PRODUCT**—Software engineers shall ensure that their products and related modifications meet the highest professional standards possible.

4. **JUDGMENT**—Software engineers shall maintain integrity and independence in their professional judgment.

5. **MANAGEMENT**—Software engineering managers and leaders shall subscribe to and promote an ethical approach to the management of software development and maintenance.

6. **PROFESSION**—Software engineers shall advance the integrity and reputation of the profession consistent with the public interest.

7. **COLLEAGUES**—Software engineers shall be fair to and supportive of their colleagues.

8. **SELF**—Software engineers shall participate in lifelong learning regarding the practice of their profession and shall promote an ethical approach to the practice of the profession. Copyright (c) 1999 by ACM/IEEE. Reproduced with permission.[1]

It is interesting to observe that the SCM/IEEE code contains all of the seven guides to personal conduct (as specified in Chapter 6 of this textbook), including duties to society, to employers, to clients, to colleagues, to subordinates, to the profession, and to himself or herself. (Although the duty to subordinates is not mentioned specifically, it is defined more clearly in the long version of the code, which is included in Appendix CD-C.) The SCM/IEEE code is therefore very similar to provincial and territorial Codes of Ethics, although it lacks the enforceability of these Canadian codes.

LIABILITY FOR SOFTWARE ERRORS

Although the ACM/IEEE Code of Ethics requires software engineers to aspire to the "highest professional standards possible," computer programs often have hidden flaws (or "bugs") that generate incorrect calculations. The dangers associated with faulty software were clearly illustrated in the United States in 1978, when the Hartford Arena collapsed—an engineering misfortune that is generally considered the first computer-aided failure (see Case History 12.1). The arena's design was based on an erroneous stress analysis program; as a consequence, the structure collapsed, costing millions of dollars. Fortunately, no deaths or injuries resulted. The question is, who is to be held liable if decisions based on erroneous software cause loss or damage?

Commercial computer programs usually include a disclaimer stating clearly that in the event of malfunction, the manufacturer and supplier will not be liable for any damages whatsoever arising from the program's use, and specifically denying payment of direct or indirect damages for personal injury, loss of business profits, business interruption, and similar losses. In effect, this limits the manufacturer's liability to the price paid for the program. This shifts the responsibility to the user—a fact confirmed by the provincial Associations. For example, the APPEGA (Alberta) *Guideline for Relying on Work Prepared by Others* states: "Members are responsible for verifying that any results obtained from computer programs are reliable and valid. Professional members should: examine and understand the methodologies and input parameters, as well as the limitations of the results obtained; and verify, where appropriate, new software releases against a standard certified for general use."[2] The PEG-NL (Newfoundland) software guideline similarly states: "Members are responsible for verifying that results obtained by using software are accurate and acceptable."[3]

The PEO (Ontario) guideline *The Use of Computer Software Tools by Professional Engineers and the Development of Computer Software Affecting Public*

Safety and Welfare defines the engineer's responsibility a little more specifically. Under the heading "Use of Computer Software Tools by Professional Engineers," the guideline states:

> The engineer must have a suitable knowledge of the engineering principles involved in the work being conducted, and is responsible for the appropriate application of these principles. When using computer programs to assist in this work, engineers should be aware of the engineering principles and matters they include, and are responsible for the interpretation and correct application of the results provided by the programs.
>
> Engineers are responsible for verifying that results obtained by using software are accurate and acceptable. Given the increasing flexibility of computer software, the engineer should ensure that professional engineering verification of the software's performance exists. In the absence of such verification, the engineer should establish and conduct suitable tests to determine whether the software performs what it is required to do.[4]

Clearly, all of these guidelines hold the user responsible for verifying that the software is operating properly, and require testing or similar verification of the software before the computer output is applied to engineering design. Such tests, typically called validation tests, require hand calculations (on a calculator or independent computer). Validation tests will vary with the type of analysis and on whether the software was developed in-house (by the user) or was commercially purchased. (Typically, source code is not available for commercial software.) Let us consider these cases separately.

Software Development

When you develop software—be it for internal company use, or as a software contract, or for sale to others—it is important that you negotiate clear contracts and follow accepted guidelines for accuracy, reliability, robust operation, documentation, and testing. The demand for skilled and knowledgeable people in this important field is precisely why software engineering is now recognized as an engineering discipline. When life, health, or public welfare is placed at risk, the provincial and territorial governments have a duty to regulate the discipline. University courses are now offered in software engineering, and several Associations have developed guidelines for software use, as mentioned earlier.

NEGOTIATING CONTRACTS

The PEO guideline is presently the most comprehensive provincial guideline for software development. It offers the following advice for negotiating contracts:

> An engineer embarking on the development of engineering software for a client runs the risk of liability if the software does not perform according to the client's requirements, or if its use causes harm to the client or the public. A well-drawn legal contract, which contemplates the development of engineering software for a client and its use by the client, can minimize the engineer's exposure to liability.

It can also define the contractual rights and obligations between the parties to the contract . . . [P]rovisions addressing at least the following concerns should be included in such a contract:

- what is to be developed;

- deliverables;

- scope of use of deliverables;

- representations and warranties;

- ownership;

- limitation of liability;

- contract price, and

- maintenance and escrow.[5]

The contract terms for limiting liability are especially important. In the unlikely event that the contract should be breached, a clause limiting liability will be honoured, provided it is a thoughtful and reasonable estimate of the damages likely to result from the breach.[6] Obviously, you should get expert legal advice whenever you negotiate a contract with complex terms.

SOFTWARE TESTING

The PEO software guideline includes a lengthy discussion of several reviews and tests that should be followed during the software development. The following are suggested as a minimum:

- software requirements review;

- software design review;

- code review;

- unit testing;

- system integration testing, *and*

- validation testing.[7]

The early reviews are important because they can save much development time. Also, the final test—that is, the validation—is especially important because it is the final verification step before the software is turned over to the user. The PEO software guideline defines validation as "testing the integrated system to ensure that it meets functional and conceptual design requirements."[8]

Software developers must ensure that their work follows a guideline such as the one published by PEO, or similar documentation for their province or specific discipline.

Using Commercial Software

Many engineers and geoscientists are moving out of software development and acquiring large, commercially sold software packages, which include highly developed and highly specialized analysis methods. Errors may exist even in commercial software; however, it is far more likely that the users will introduce errors because they don't understand the software. For example, the user may:

- use an incorrect format or units for data input,
- apply the program to the wrong type of problem—one that is unsupported by the program theory (such as using a program intended for planar analysis in a 3-D application),
- set erroneous parameters (such as integration parameters) that result in incorrect computational accuracy,
- not understand the output display, *and/or*
- much more. In fact, users have long been notorious for misunderstanding software.

To avoid such misuse of commercial software, you must read and understand the documentation. Introductory tutorials provided by the software developers should be attended religiously, and their "help desks" should be consulted whenever problems arise.

Also, you should always perform a series of validation tests to verify the software. As a professional, you must never assume that commercial software "must be right." If you use the software in a project, and if that project ultimately fails because of software problems, the first question the opposing lawyer asks you during discovery is going to be, "What tests did you perform to ensure that the software was operating properly?"

An old engineering adage says that no important decision should ever be based on a single calculation. In other words, important calculations should always be double-checked. This adage, which dates back to slide-rule days, also applies to computer output. As the user, you must validate even commercial software before applying its output to make key decisions.

If at all possible, validation should involve running at least the first three of the following tests, which are discussed roughly in order of increasing effort or complexity. If the software fails any of the following tests, ask why!!!

Dummy runs. Run the software with nominal entries, such as zeroes or known values, to get an obvious answer. This is a basic check on the program's computation. A simple test is easy to imagine: if zero loads are input to a stress program, the stresses output should also be zero. If a file of identical numbers is input to an averaging program, the calculated mean, median, and mode must equal that number, and the standard deviation must be zero. Obvious software tests such as these are simply common sense.

Approximate analytical checks. Simplify the computer model and then apply analytical calculations to find the boundaries within which the computer output must fall. This is a standard engineering check. The results are approximate, but the test is reassuring. For example, a finite-element model of a complex structure can almost always be decomposed and approximated by simple beam and column equations. Take a most optimistic estimate and a least optimistic estimate and apply the analytical equations to each. The computer output should lie between these boundaries.

Independent theoretical checks. Make analytical computations using an independent theoretical basis. Where this test is possible, it is very reassuring. For example, dynamic simulations use numerical integration, but the integration can be checked by applying the laws of conservation of energy and momentum to the initial conditions and final answers.

Advanced methods. More advanced validation tests are easily developed by clever and creative analysts.

Complete duplication. For a really convincing validation, you should arrange a full-scale duplication of the computation, using different software, hardware, and input files. This test should be carried out by independent employees or consultants, if possible. This check is expensive, but it validates almost everything—input data, theory, and computation. If you have any doubts about critical analysis software, you should carry out this independent validation before making major expenditures. It is always cheaper to duplicate a computer calculation at the early stages of a project than to explain the omission to a board of inquiry after the project fails.

In summary, computer software is like any other measuring tool: it must be used properly, and like any equipment, it must be calibrated. Calibration of computer software typically means that it must be independently validated, as described above, before the output data is used for critical decisions.

PREVENTING SOFTWARE PIRACY

One of the most flagrant conflicts of interest today involves the copying of software, which is usually referred to as software piracy. Copying is so easy that whether through ignorance or intentionally, the practice is widespread. When you purchase a computer program, you are not buying the right to duplicate that program, except for backup. Some programs—especially large ones—are leased, and these may include a monitor program that checks the date given by the computer and then disables the program when the lease has expired if no extension has been obtained (usually in the form of a secret password provided by the leasing supplier). It is contrary to Canada's Copyright Act to duplicate programs for personal use or to disable the password protection on leased programs.

There are many good reasons why professional engineers and geoscientists should never use copied or pirated software.

• **Illegality.** The first and most obvious reason is that copying software is illegal. It violates the Copyright Act, which allows copying only for backup purposes. The act forbids activities such as reverse engineering; here it differs from American law, which permits research on computer programs (including reverse engineering) under the "fair use" provisions. Proposals are being made to extend Canada's Copyright Act to include these permissions.[9] Generally speaking, however, software copying is illegal.

- **Breach of contract.** The use of pirated software could result in a breach of any contract for which it is used. For example, an engineering analysis contract may be breached if the client discovers that you are using pirated software.
- **No product support, documentation, updates, or patches.** Product support, documentation, updates, and patches are usually not available for pirated software.
- **Unprofessional image.** Obviously, trying to run a professional practice with pirated software will create problems with internal morale. Furthermore, the image you project to the public is very unprofessional.
- **Penalties and embarrassment if caught.** In 1990 the software industry established the Canadian Alliance Against Software Theft (CAAST). CAAST's main goal is to reduce and eliminate software piracy. It estimates that Canada's economy lost more than $400 million to software theft in 2002. Its website enables visitors to report allegations of software piracy.[10]

Because of the extent of software piracy, this issue is addressed very pointedly in Ontario's *Guideline to Professional Practice*:

12. COPYRIGHT IN COMPUTER PROGRAMS

Under recent amendments to the Copyright Act, the uncertainty about copyright in computer programs has been eliminated by expanding the definition of "literary work" to include computer programs, which are broadly defined to include all computer programs, whether in source code or object code, regardless of how they are stored. Two exceptions under these amendments will allow certain uses to be made of computer software, which would otherwise be an infringement of copyright.

The first exception provides that it shall not be infringement for a person in lawful possession of a copy of a computer program to modify, adapt, or convert a reproduction of the copy into another program to suit that person's needs, provided:

- the modified program is essential for the compatibility of the computer program with a particular computer;
- the modified program is used only for the person's own needs
- not more than one modified copy is used by the person at any given time; and
- the modified copy is destroyed when the person ceases to be entitled to possession of the copy (i.e. upon expiry of a software licence).

The second exception provides that a person who is in lawful possession of a copy of a computer program, or of a modified reproduction of a computer program, may make a single backup copy of the program, provided the backup copy is destroyed when the person ceases to be the owner of the copy of the computer program.

The intention of these exceptions is to give the authorized software user a limited right to change the software, to ensure compatibility of the software with the authorized user's computer system, and to allow for the protection and security of the original program.[11]

The Alberta guideline defining *Illegal Copying & Use of Computer Software* contains similar information:

Legal Considerations

Computer software is covered under the Canadian Copyright Act which provides for a financial penalty as well as a jail sentence for violation. The Copyright Act protects authors' legal rights and privileges to their creative works. It should be noted that a copyright in a work exists as soon as the work is created and there is no requirement to publish the work or to affix any special notice thereto.

In addition to copyright considerations, usage of commercial software is also generally governed by contract law under the agreements of the software purchase contract and/or licence.

Ethical Considerations

The Code of Ethics establishes the duty of APEGGA members to enhance the dignity and status of the professions. APEGGA members shall conduct themselves with fairness and good faith toward other professional members and the public in the area of computer software usage to avoid conduct which would detract from the image of the professions.

In consideration of the Code of Ethics, APEGGA members must guard against any violations, real or apparent, of the Canadian Copyright Act and contract laws and the resulting legal and ethical consequences.

Guidelines

All purchased/licensed computer software is subject to the full provisions of the agreements connected with the acquisition of the software and manuals associated therewith. All APEGGA members should be aware of the agreement provisions and abide by the terms of the agreements with particular regard to copying restrictions.

The use of copies of computer software or manuals that have been obtained in violation of copyright or trade secrets or in any other fraudulent manner is deemed unprofessional conduct on the part of an APEGGA member.

In addition to exposure to possible criminal prosecution, violation of copyrights or misappropriation of trade secrets associated with computer software by our members may result in disciplinary action by APEGGA.[12]

HACKERS, CRACKERS, VANDALS, AND VIRUSES

This section provides an overview of the threats inherent in computing, which seem to be getting more serious and more threatening (and more annoying) every year. The people who cause these problems are typically called hackers, crackers, spammers, vandals (or cyber-vandals), or outright terrorists. These terms refer to different levels of threat.

In general, the term "hacker" merely applies to people who have an interest in and knowledge of computers. Most hackers are harmless; they simply know more than the average person about computers and computing. Although hackers sometimes stray into undesirable activities, they should not be confused with "crackers," who are much more malicious. Crackers use the Internet to attack an organization; they obtain access to its computers with the goal of using its resources, copying its information, or disrupting its operations. "Spammers" are crackers who gain access to unprotected computers on the Internet and then send out e-mail advertisements from those computers, typically to sell some product. Spamming amounts to a modern form of the tragedy of the commons, which was discussed in an earlier chapter. The tragedy is that thoughtless and selfish people are degrading e-mail service, which is an Internet property held in common; as spamming proliferates, at some point e-mail communication will cease to function effectively.

Vandals are like crackers but even more malicious. Typically, they set out to cause disruption, usually by destroying or modifying data and/or creating denial-of-service (DoS) attacks. The latter involve using the Internet to flood a computer with transmissions; this overwhelms the computer and effectively denies its services to legitimate users. Terrorists are much like vandals but even worse, because their goal is to inflict maximum damage.

To combat these threats, every professional office must now maintain both of the following:

- **Firewall.** This type of software guards a computer's "gates." That is, guards your computer's connections to the Internet, and blocks or admits data transmissions according to the access rules you have set.
- **Antivirus software.** This software detects, identifies, and removes any viruses that have succeeded in breaching your firewall and entering your computer.

Firewalls and antivirus software will protect your computer from a wide range of threats. A full glossary of these threats is reprinted below with permission from Symantec, a world leader in Internet security software.

- **Adware.** Programs that secretly gather personal information through the Internet and relay it back to another computer, generally for advertising purposes. This is often accomplished by tracking information related to Internet browser usage or habits. Adware can be downloaded from Web sites (typically in shareware or freeware), email messages, and instant messengers. A user may unknowingly trigger adware by accepting an End User Licence Agreement from a software program linked to the adware.

- **Dialers.** Programs that use a system, without your permission or knowledge, to dial out through the Internet to a 900 number or FTP site, typically to accrue charges.

- **Hack Tools.** Tools used by a hacker to gain unauthorized access to your computer. One example of a hack tool is a keystroke logger—a program that tracks and records individual keystrokes and can send this information back to the hacker.

- **Hoax.** Usually an email that gets mailed in chain letter fashion describing some devastating, highly unlikely type of virus. Hoaxes are detectable as having no file attachment, no reference to a third party who can validate the claim, and by the general tone of the message.

- **Joke Programs.** Programs that change or interrupt the normal behaviour of your computer, creating a general distraction or nuisance. Harmless programs that cause various benign activities to display on your computer (for example, an unexpected screen saver).

- **Remote Access.** Programs that allow another computer to gain information or to attack or alter your computer, usually over the Internet.

- **Spyware.** Stand-alone programs that can secretly monitor system activity. These may detect passwords or other confidential information and transmit them to another computer. Spyware can be downloaded from Web sites (typically in shareware or freeware), email messages, and instant messengers. A user may unknowingly trigger spyware by accepting an End User License Agreement from a software program linked to the spyware.

- **Trojan Horse.** A program that neither replicates nor copies itself, but causes damage or compromises the security of the computer. Typically, an individual emails a Trojan Horse to you—it does not email itself—and it may arrive in the form of a joke program or software of some sort.

- **Virus.** A program or code that replicates; that is, infects another program, boot sector, partition sector, or document that supports macros, by inserting itself or attaching itself to that medium. Most viruses only replicate, though many do a large amount of damage as well.

- **Worm.** A program that makes copies of itself; for example, from one disk drive to another, or by copying itself using email or another transport mechanism. The worm may do damage and compromise the security of the computer. It may arrive in the form of a joke program or software of some sort. Most worms are spread as attachments to emails.[13]

All of the activities listed above are harmful to some degree; in addition, they are either selfish, vicious, or pointless. Obviously, all four ethical theories discussed earlier in this text—Mill's utilitarianism, Kant's duty ethics, Locke's rights ethics, and Aristotle's virtue ethics—would condemn these activities as unethical. Furthermore, these activities contravene the first clause in every provincial and territorial Code of Ethics—to place the public welfare first. These activities also contravene all of the Codes of Ethics of the engineering societies. For example, the ACM/IEEE code reproduced earlier in this chapter states clearly: "Software engineers shall act consistently with the public interest."

However, an ethical explanation is unnecessary if you have ever been a victim of these activities. If you have ever had to reformat your hard drive because of a malicious virus, or had purchases charged to your credit card (or, worse, been a victim of identity theft) because a cracker stole your personal information from a database, or if you have ever simply been overwhelmed by greedy spammers, who flood the Internet with self-serving advertisements, then you know these activities are unethical. The crackers, spammers, and other vandals who perpetrate these acts are beneath contempt. Some form of Internet regulation is essential to protect users from vandalism.

So how are we to balance the open freedom of the Internet with the potential for abuse, which is now getting out of control? A similar form of abuse occurred in the early days of radio. Radio frequencies are now rigidly controlled by national laws and international treaties. But they weren't in the beginning, which meant that broadcasters could simply select their own frequencies and then increase their transmitting power until they drowned out the competition. This was unfair and unethical, and had to be remedied, and was. Society (and software engineers in particular) must now develop rules to bring fairness to the Internet. The Internet is too valuable to be merely a playground for crackers and vandals.

COMPUTER DISASTER RECOVERY

Professional engineers and geoscientists have an enormous investment in computer hardware and software—in many practices these are the major assets—so the manager must be alert to any risks, problems, or damage that threaten this investment. The most obvious responsibilities are to maintain the equipment and to provide for alternative facilities in the event of failure. The manager should have a recovery plan for the possibility, however remote, of computer disaster—the complete malfunction or destruction of the computers and/or corruption or loss of the programs and data. The first line of protection is to have critical data and programs duplicated on backup disks and stored in a safe, secure location. A more sophisticated approach, however, is to develop a disaster plan.

Every professional practice should estimate the cost and impact of a computer disaster on the practice, and how long it would take to recover. A plan should be developed that includes the following:

- identifying the possible threats,
- averting the most common forms of disaster by maintaining proper data backup,
- installing duplicate facilities, where needed, or arranging for suitable temporary operations, should they be required,
- preparing a documented plan that can be understood by everyone,
- preparing the software needed to restore the service, and
- testing the disaster recovery plan.

Many books are available that provide further advice on this subject.

CASE HISTORY 12.1

HARTFORD ARENA ROOF COLLAPSE: A COMPUTER-AIDED FAILURE

The Hartford Arena was a monumentally huge structure when it was completed in 1973. The arena housed a basketball court and seating for 5,000 spectators to watch the games. To minimize obstructions for spectators, the roof was supported by only four columns, each placed near a corner of the building. The roof, called a "space-frame," was a three-dimensional truss structure about 3 m (10 ft) deep, and approximately 91 m by 110 m (300 ft by 360 ft) in plan size, suspended about 27 m (90 ft) above the floor.

The Roof Collapse

At 4:15 a.m. on January 18, 1978, during a heavy snowfall, the huge roof suddenly and violently collapsed onto the central court, with the corners of the roof pointing up into the air. Fortunately, the collapse occurred in the middle of the night. Earlier in the evening the arena had been packed with thousands of spectators, all of whom missed death or injury by a matter of a few hours.

The Cause of the Failure

During the investigation, the snow load at the time of the collapse was estimated to be less than half the rated load for the roof. Attention shifted to the design. Several gross errors were found in the detail design of the structural steel, as described very well in the case study by Rachel Martin, which is readily available on the Internet.[14] However, the basic cause of the collapse was, as Henry Petroski stated, an "oversimplified computer analysis."[15] The Hartford Arena involved one of the earliest applications of computers to the analysis of complex space-frame structures, and the designers made a fateful error. Martin explains:

> The engineers for the Hartford Arena depended on computer analysis to assess the safety of their design. Computers, however, are only as good as their programmer and tend to offer engineers a false sense of security. The roof design was extremely susceptible to buckling which was a mode of failure not considered in that particular computer analysis and, therefore, left undiscovered.[16]

In other words, the stress analysis software overlooked the key idea that structural rods in compression buckle at a stress far lower than the yield strength of the steel, which is typically the limit for rods in tension. Any engineer could easily have discovered this error, at the earliest stages of the project, by comparing the stress calculated by the computer against the well-known Euler buckling equation—a simple calculation that can be performed in minutes.

The Hartford Arena was constructed in 1973, and housed a basketball court and seating for 5,000 spectators. On January 18, 1978, during a heavy snowfall, the huge roof suddenly collapsed, only hours after a well-attended game. The collapse was traced to an "oversimplified computer analysis." The arena is known as the first computer-aided failure.

Source: © Bettmann/CORBIS/Magma.

Ethical Implications

Why the engineers neglected to perform such a simple, obvious check of their computer output is a mystery. Moreover, the design engineers had a very strong incentive to double-check their calculations. During the construction, the truss was assembled on the ground and hoisted into place. Large deflections were immediately apparent, and the engineers were informed. In fact, as Kaminetzky reports, the deformations were so much larger than expected that contractors could not fit the windows designed to be inserted below the girders. Even the ironworkers reported that the deformations were unreasonably large.[17] Nevertheless, the engineers ignored these warnings and did not double-check their work.

It is obvious that the actions of the design engineers were negligent or incompetent. The Hartford Arena engineers failed to validate the computer output adequately and subjugated their judgment to the computer. Computer program validation should be routine due diligence. The engineers compounded their negligence when they ignored the excessive deflection of the truss—a warning sign that something was wrong.

The details of the case were never revealed in court. After six years of legal preparation, an out-of-court settlement was reached, and a probing discussion of the causes was therefore precluded.

An engineer or geoscientist cannot guarantee that every project will succeed, just as a surgeon cannot save every patient, and a lawyer cannot win every lawsuit. However, what the engineer, geoscientist, surgeon, and lawyer must all guarantee is that they possess adequate knowledge, and that they will exercise reasonable skill, care, and expertise, appropriate to the profession, to carry out the client's wishes. In the case of computer-aided design, reasonable care requires validation of computer software.

DISCUSSION TOPICS AND ASSIGNMENTS

(Additional assignments are in Appendix CD-E on the CD-ROM included with this text.)

1. One of the main advantages of the Internet is the rapid, unfettered dissemination of ideas, which are a vital stimulant to research and development. However, the Internet abounds with misuses, abuses, and negative influences, as this chapter discusses. Discuss the ethics of regulating the Internet (assuming that this could be done). In your opinion, would Internet regulation be positive or negative? Would regulation impede free speech, or assist it by reducing the "noise" on the Internet?

2. You are co-op student working in a communications company. You have access to electronic switching source code and you show it to a fellow student. Later, you suspect that your colleague may have logged on with your password and looked at the software. Although the material is read-only, you believe he may have copied the code onto a CD. Although he is a great friend, you suspect that he may be trying to sell the code or break it for illicit purposes. What should you do?

3. Computers have been used to tabulate the votes in elections for many years. Recently, however, computers are being used in new ways. Public opinion corporations write computer programs that contain the voting histories of electoral ridings; the programs then weigh that information with public opinion polls held before the election and with "exit polls" held on voting day, in which people are asked how they voted as they exit the poll. The computer programs can predict (or "project") the eventual outcome fairly accurately. Television networks then display the predictions in a way that may confuse some viewers into believing the election is over. During federal elections, viewers in Alberta and British Columbia are typically dismayed to see television commentators announce the predicted result when the polls close in Quebec and Ontario, even though the polls are open for two or three more hours in the West. Does this use of computers and television interfere with our right to fair elections? Is it fair that some citizens know the predicted outcome before they vote? Do the predictions influence the desire or determination of citizens to vote? What abuses could occur if programs were deliberately skewed to favour one particular party? Is this really an ethical problem, or is it an overreaction by technically unsophisticated people? What could be done to alleviate the problem as you understand it?

4. Assume that you are a software engineer, working as a full-time employee. You have an obligation to use your ability to achieve your employer's legitimate goals. Consider the following cases. What would your position be, ethically, if the employer obtained contracts and asked you to participate in the development of software for the following?

   - A video game that will have ultra-violent graphical images.
   - A video game in which the user is required to commit crimes such as auto theft, robbery, murder, or rape.
   - Sales accounting in small retail stores, but your employer wants you to add a feature that will permit the user to shield certain sales from tabulation, thus hiding the income (and the taxes) from the federal government.
   - An Internet casino, which will be run from a server in a foreign country.
   - E-mail communication, which will have features to permit designated super-users to track the usage and read the e-mail of all other users.
   - A website to provide pornographic images to paying users, which will be run from a server in a foreign country.
   - A utility program to filter spam very effectively, which will be distributed free, but your employer wants you to add a feature that will send secret e-mail reports back to your company listing all websites that the computer user visits.
   - A utility program to filter spam (as described above), which will be distributed free, but users will be informed that it contains a feature to send e-mail reports back to your company, and users will have to agree to this feature before the software will operate.

5. Write an essay predicting how the engineering and geoscientific professions will be altered by the widespread use of specialized, commerical technical software. Such software is increasingly developed by a few eminent specialists, and has features that no single practising professional may understand fully. Does this software increase the capability of the professional using it, or does it relegate the professional to the role of an input technician, who knows how to operate the computer but cannot comment on the validity of its output? Is the future dim for the average professional, who will merely operate tools developed by increasingly specialized foreign software experts? What can be done to ensure that this dismal prediction does not come true?

NOTES

1. ACM/IEEE Code of Ethics, Copyright (c) 1999 by the Association for Computing Machinery, Inc. and the Institute for Electrical and Electronics Engineers, Inc. This code may be published without permission as long as it is not changed in any way and it carries the copyright notice.

2. Association of Professional Engineers, Geologists and Geophysicists of Alberta (APEGGA), *Guideline for Relying on Work Prepared by Others*, APEGGA, Edmonton, June 2003, p. 13 <www.apegga.org> (Dec. 7, 2003).

3. Professional Engineers and Geoscientists of Newfoundland and Labrador (PEG-NL), *Guideline for the Use of Computer Software Tools by Professional Engineers and Geoscientists*, PEG-NL, July 28, 1995 (1 page).

4. Professional Engineers Ontario (PEO), *The Use of Computer Software Tools by Professional Engineers and the Development of Computer Software Affecting Public Safety and Welfare*, PEO, Toronto, 1993, p. 4 <www.peo.on.ca> (November 6, 2003). Reprinted with permission of PEO.

5. PEO, *The Use of Computer Software Tools*, p. 9.

6. D.L. Marston, *Law for Professional Engineers*, 3rd ed., Canadian and International Perspectives, McGraw-Hill Ryerson, Whitby, ON, 1996, p. 153.

7. PEO, *The Use of Computer Software Tools*, p. 7.

8. PEO, *The Use of Computer Software Tools*, p. 8.

9. Industry Canada and Canadian Heritage, *A Framework for Copyright Reform, Industry Canada and Canadian Heritage*, Ottawa, 2001. Available at government Strategis website: <strategis.ic.gc.ca/epic/internet/incrp-prda.nsf/vw GeneratedInterE/Home> (November 9, 2003).

10. Canadian Alliance Against Software Theft (CAAST) <www.caast.org/> (November 9, 2003).

11. Professional Engineers Ontario (PEO), "12. Copyright in Computer Programs," *Guideline to Professional Practice*, PEO, Toronto, 1988, revised 1998, p. 20 <www.peo.on.ca> (November 6, 2003). Reprinted with permission of PEO.

12. Association of Professional Engineers, Geologists and Geophysicists of Alberta (APEGGA), *Illegal Copying & Use of Computer Software—A Guideline*, APPEGGA, Edmonton, AB, 1990 <www.apegga.ca/publications/guidelines.html> (November 9, 2003). Reprinted with permission of APEGGA.

13. Symantec, *Glossary* <www.symantec.com/avcenter/refa.html#f> (November 9, 2003). Copyright © 2003 Symantec Corporation. Reprinted with permission.

14. R. Martin, *Hartford Civic Center Arena Roof Collapse*, Dept. of Civil and Environmental Engineering, University of Alabama at Birmingham (case study) <www.eng.uab.edu/cee/REU_NSf99/hartford.htm#asce> (March 19, 2003).

15. H. Petroski, *To Engineer Is Human: The Role of Failure in Successful Design*, St. Martin's Press, New York, NY, 1985, p. 199.

16. R. Martin, *Hartford Civic Center Arena Roof Collapse*, ibid. Excerpt used with permission.

17. D. Kaminetzky, *Design and Construction Failures: Lessons from Forensic Investigations*, McGraw-Hill, New York, NY, 1991, p. 224.

Chapter 13
Fairness and Equity in Engineering

A profession such as engineering should attract intelligent, creative, fair people with high personal standards. This point is well expressed in the foreword to the Code of Ethics of the Association of Professional Engineers of New Brunswick, which states:

> Honesty, justice and courtesy form a moral philosophy which, associated with mutual interest among people, constitute the foundation of ethics. Engineers should recognize such a standard, not in passive observance, but as a set of dynamic principles guiding their conduct and way of life.[1]

Furthermore, everyone has a right to work in a fair and equitable workplace. Improved employee performance is one of the most obvious results of a positive work environment. Efforts by managers and employers to introduce fairness and equity will pay off by creating a more successful enterprise.

ENSURING FAIRNESS AND EQUITY IN THE ENGINEERING WORKPLACE

To achieve fairness and equity, the first steps are to recognize the special problems that women and members of minority groups may face within the organization and to ensure that effective policies are in place for hiring and promoting employees. Communicating these policies, their meaning, and their purpose to all staff is critical to their successful implementation. If some of the issues arouse anxiety or anger in a particular group, it is often because of misconceptions, which can be defused and eliminated through open and frank dialogue.

Unfair and unethical behaviour, such as discrimination or harassment based on sex, race, national origin, disability, or sexual orientation, has no place in the engineering profession—or, for that matter, anywhere in a civilized society. In Canada this type of behaviour is illegal under the Criminal

---

This chapter was written by Dr. Monique Frize, P.Eng., O.C., Former NSERC/Nortel Chair for Women in Science and Engineering (Ontario), Professor in the Faculty of Engineering at University of Ottawa and Carleton University.

Code and human rights legislation. Under Regulation 941 of the Professional Engineers Act (Ontario), harassment and discrimination are now considered professional misconduct, for which an engineer can be disciplined if found guilty. Professional Engineers Ontario (PEO) publishes a *Guideline on Human Rights in Professional Practice*, which is available on their website.[2]

Nevertheless, some forms of these unfair practices persist in the profession because they are subtle or systematic because engineers are unaware of them and of the destructive impact that harassment and discrimination can have on individuals. These practices and their underlying causes are discussed in this text so that tomorrow's engineers will know how to recognize and avoid them.

DEFINITION OF DISCRIMINATION

Dictionaries define *discrimination* as "the action of discerning, distinguishing things or people from others, and making a difference." In recent years the term has also come to be associated with *segregation*, which is defined as "the act of distinguishing one group from others, to its detriment." It is this latter, much more harmful association that this chapter addresses.

THE CANADIAN CHARTER OF RIGHTS AND FREEDOMS

Clause 7 of the Canadian Charter of Rights and Freedoms states: "Everyone has the right to life, liberty and security of the person and the right not to be deprived thereof in accordance with the principles of fundamental justice."[3] Clause 15 defines equality rights: "Every individual is equal before and under the law and has the right to the equal protection and equal benefit of the law without discrimination and, in particular, without discrimination based on race, national or ethnic origin, colour, religion, sex, age or mental or physical disability."[4] Subsection 1 of clause 15 also addresses the right to have programs of affirmative action in cases where improvement and more balance in the participation of underrepresented groups is needed. It reads: "Subsection (1) does not preclude any law or program or activity that has as its object the amelioration of conditions of disadvantaged individuals or groups."[5] An example of the application of this subsection was the creation of a chair for women in engineering in 1989, and the addition of five chairs for women in science and engineering in 1997.[6] These chairs exist for a limited time period; their mandate is to develop strategies to increase the participation of women at all levels in these disciplines. The chairholders also serve as models for women who study or work in these fields.

The Charter has a very practical significance in professional activities: all contracts, collective agreements, work protocols, and handbooks are assumed to be consistent with the relevant provincial human rights legislation, the latter also being consistent with the Canadian Charter of Rights and Freedoms. Discrimination is against the law. Contracts, including collective agreements, can be rescinded and statutes and regulations can be nullified if found to be discriminatory. The main difference between the Canadian Charter of Rights and Freedoms and other federal and provincial human rights legislation is that the former applies to all levels of government, including agencies directly controlled by governments. In contrast, provincial human rights legislation applies (although not directly) to private parties that are not under federal jurisdiction. The Canadian Charter of Rights and Freedoms is the supreme law of Canada, with priority over all other legislation.[7]

ENROLMENT PATTERNS IN ENGINEERING PROGRAMS

Figure 13.1 shows the enrolment patterns for women in various engineering undergraduate programs between 1985 and 2001. (Note: Environmental engineering started only in 1991.) As this figure shows, electrical, mechanical, and computer engineering have been attracting the fewest women. Across all disciplines, the average proportion of women in engineering undergraduate programs in 1985 was 11 percent; by 1991 it was 15 percent, increasing to 19 percent in 1995 and 21 percent in 2001.[8]

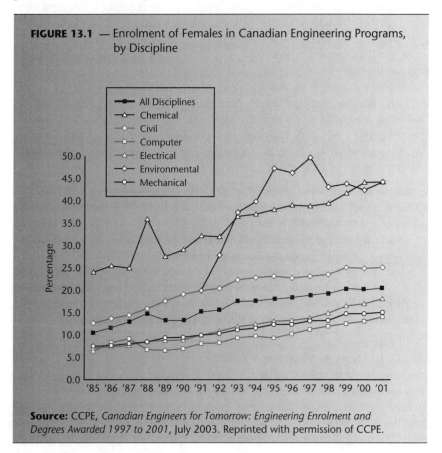

FIGURE 13.1 — Enrolment of Females in Canadian Engineering Programs, by Discipline

Source: CCPE, *Canadian Engineers for Tomorrow: Engineering Enrolment and Degrees Awarded 1997 to 2001*, July 2003. Reprinted with permission of CCPE.

In 1989, 10 percent of students in full-time engineering masters' programs were women. By 1995 the figure was 20 percent; by 2001 it was 24 percent. For doctoral programs the equivalent figures were 6 percent in 1989, 13 percent in 1995, and 17 percent in 2001. Women comprised 2.2 percent of engineering faculty members in 1991 and 7.6 percent in 2001.[9]

Universities vary greatly in their ability to attract women to their engineering programs. As Figure 13.1 shows, levels of enrolment vary greatly by choice of discipline, with higher enrolments in chemical, environmental, and industrial engineering and lower enrolments in electrical, computer, and mechanical engineering. In the past decade, other professions such as medicine, dentistry, veterinary medicine, and law have reached enrolments of 50 percent; many have even surpassed this level. The low percentages for engineering suggest that the profession is not yet fully open to women and that barriers still exist.

SOCIALIZATION IN EARLY CHILDHOOD

A major reason for women's low participation in engineering is the perpetuation of gender stereotypes—that is, the myth that certain jobs and careers are appropriate for women and others for men. For example, women are expected to be teachers, nurses, hairdressers, and secretaries; men are expected to be engineers, pilots, and firefighters and to dominate the trades. The media are an important factor in perpetuating these stereotypes; however, some teachers, parents, and guidance counsellors also condone them. A British Columbia provincial report explains how girls and boys begin to form gender-role stereotypes at a very early age and examines the effects of socialization (i.e., how children are brought up) and of self-esteem on the education and training choices of girls and women:

> Significant gender inequalities persist in Canadian society and are reflected in and reinforced through the formal and informal processes of socialization. Gender socialization begins at birth and intensifies throughout childhood and adolescence. In some ways, gender socialization continues as part of lifelong learning. From the moment we are born we begin the process of learning how to be human beings. We learn about attitudes, values, and behaviours which are acceptable in our society. We learn what is expected of us, what roles we can play, how to exercise self-control, how to live in a community. Social scientists call this learning process socialization. When this process is applied to how women and men are expected to behave, this is called gender socialization.[10]

In *Failing at Fairness*,[11] the Sadkers report how teachers tend to pay more attention to boys. Boys are reprimanded more, but they are also given more encouragement to answer questions, they are challenged more than girls, and they are praised more for their answers. In contrast, girls tend to be praised for neatness and good behaviour. It is common to hear young women describe how they were discouraged from studying mathematics and science.[12] Some guidance counsellors make a special effort to encourage girls and boys to consider nontraditional career choices; however, others—women *and* men—discourage young people from doing so. In high school, few young women study physics, which is a prerequisite for engineering studies at university.

Lupart and Cannon found that by grade 7, students have decided on careers that either include information technology or exclude it. Boys selected careers in information technology as their first choice; girls selected the same careers as their sixth and last choice.[13] In her study of enriched mini-course attendance in engineering and science by gender, McDill found that "age 13 appeared critical for girls with respect to choices made for the future . . . and little change is expected after that age. Teenage girls appear to have formed strong opinions by this point, not only with respect to general subject area, but also for specific topics within those areas."[14] These two studies indicate that girls make early career choices; clearly, for the greatest impact, intervention is needed when girls are between twelve and fifteen. In both mathematics and science, girls now show stronger academic performance than they have since the mid-1990s; even so, most of them still avoid including these fields in their career choices. For boys, the problem is different. Many boys believe that good marks are not as important as excelling in sports, and their parents often agree with this belief. This pressure obviously harms their chances of getting the marks they need for university entrance and of winning scholarships. The media, parents, and schools must all stop perpetuating stereotyped expectations for both sexes in order to ensure that all students reach their full potential. The results of gender socialization in schools are similar to those of systemic discrimination. Sadly, as a result of poor counselling, some young people end up studying subjects that seem appropriate but that are not what they would have chosen—based on their true interests and skills—had they been given full information and encouragement. Clearly, there is a need to eliminate this sort of misdirection.

The paucity of role models to inspire young women to enter engineering perpetuates the problem. Computer games and technology courses are often designed to appeal to boys, and this serves as yet another barrier against girls developing an interest in information technologies and computers.

RETENTION ISSUES

Retaining the women who have chosen engineering as a field of study is a challenge. Several factors affect the retention rates. Again, these rates vary greatly from one university to another, with some universities retaining a higher proportion of women than men, and others losing a much higher percentage of women than men from their pool of students. One factor is the masculine culture of engineering; women cope best and feel accepted when they are "one of the boys."[15, 16] Hacker,[17] Robinson and McIlwee,[18] and Sorensen and colleagues[19] describe the culture of engineering and technology as one that emphasizes the importance of technology over personal relationships, formal abstract knowledge over inexact humanistic knowledge, and male attributes over female ones. Hall and Sandler[20] describe the chilly climate for women faculty and students. These are difficult aspects to change, since women who want to be accepted will generally help maintain the status quo. The culture is also preserved through contacts with alumni/ae, faculty members, department heads, and deans, who may not understand the issue or may not want to make changes for the sake of a small group of women students.

Some improvements have been observed in recent years. For example, there is definitely more sensitivity regarding the use of inclusive language in classrooms; fewer sexist and racist remarks are being heard nowadays. That said, the learning experience of women and members of minority groups still varies from class to class and from university to university. Some inappropriate acts and behaviours still occur. For example, a professor may ignore a female or minority student in a class, or may not intervene when the behaviour of some students is demeaning to others. Whether it is inadvertent or not, this attitude on the part of the professor adds to the problem instead of solving it. Professors and instructors must do more than simply show sensitivity to social diversity. As role models, they must participate actively in advancing the concepts of fairness and equity.

Universities have made special efforts to ensure fairness and equity, but the resulting policies are not enough. If real change is to occur, those policies must be explained and enforced. Individual professors and administrators can be highly effective in bringing about improvements.

The first-year experience is the most critical, and this is when special attention must be paid to retaining students. Some students drop out because they do not receive adequate encouragement; their career expectations are thereby severely reduced. This is also a loss for the university. Some universities have increased levels of retention and morale by instituting peer mentoring programs. Scholarships based on entrance qualifications, with a number of these scholarships reserved for high-achieving women, are effective in attracting more women to engineering. Special programs will continue to be necessary until more balance is achieved, at which point it will be possible to eliminate those programs.

WHAT ENGINEERING AND APPLIED SCIENCE FACULTIES CAN DO

In addition to staging effective recruitment programs such as summer camps and various outreach activities, faculties of engineering can increase the enrolment of women beyond current levels by designing curricular content carefully and by developing teaching styles that are more appealing to women. This means including societal relevance in the content of courses and using less traditional teaching styles, such as self-learning. Attracting more women into graduate programs and hiring more women faculty members would have a positive impact. It would also be worthwhile to review the

During the Second World War, the Hawker Hurricane was a combat fighter manufactured by Canadian Car & Foundry Co. in Thunder Bay (then Fort William). The aircraft was designed in Britain but produced in Canada under the direction of Elizabeth (Elsie) MacGill, the University of Toronto's first woman electrical engineering graduate. She received a master's degree in aeronautical engineering in the United States in 1929 and became chief aeronautical engineer for Canadian Car in 1938. She designed the Maple Leaf Trainer, a small training plane, before assuming responsibility for producing the Hawker Hurricane. Total production was 2,000 aircraft—a significant contribution to the Canadian war effort.

Source: Photograph courtesy of the Canadian Aviation Museum, Ottawa.

criteria for judging achievement and success, which affect decisions in hiring, tenure, promotion, and the awarding of research chairs, for example. Universities must also create policies that allow young faculty members— female and male—to balance family and career. The criteria used for assessing faculty performance, based on decades of tradition, should be re-examined to see if they are still relevant in a changing world. This applies not only to how merit is defined and measured, but also to how awards, appointments, and promotions are given.[21] For example, examining the quality of publications (such as the number of citations) rather than their quantity, and looking at the potential of candidates instead of what they have accomplished by the time of the interview, would be fairer to candidates who have taken a career break to have children. Outdated stereotypes and biases can affect the success rates in competitions for scholarships, fellowships, awards, grants, jobs, promotions, and the allocation of research chairs.[22] These biases can be reduced through education and sensitization programs, by ensuring fair gender representation on committees making these decisions, and by making a proactive effort to find qualified women for the positions or awards.

THE BENEFITS OF ACHIEVING DIVERSITY IN ENGINEERING

Diversity encourages creativity and innovation. As yet, there is not much research on whether women and minorities bring new perspectives to the engineering profession. However, a few studies are beginning to emerge regarding this question. For a 1996 Canadian study, Ann van Beers[23] interviewed twenty female and twenty male engineers in the Vancouver area; she found that several participants thought the presence of women would bring changes to the structure of the work environment, the culture, and the practice of engineering. These responses support the view that women introduce positive values and perspectives; this is especially true for women who are established in their careers and who want to develop their own styles and approaches to problem solving. For example, women may use a more contextual approach, may have better communication and people skills, and may prefer a more consensual working relationship and flatter organizational structure to a hierarchical approach. In her study of science and engineering students at universities in British Columbia, Vickers and colleagues[24] found that a substantially larger proportion of females, compared to males, stated that making a contribution to society was an important criterion when choosing a career.

These and other studies point out that women can bring new values to the profession, provided there are sufficient numbers of them to make themselves heard and that they are allowed to be themselves. Women's affinity for a consultative style of working is very much in tune with today's management philosophy. Their verbal and interpersonal skills, when combined with a solid technical education, become a real asset, especially for smaller firms whose engineers must interact with suppliers, clients, and regulatory agencies.

Similarly, individuals from other underrepresented groups who have been raised in different cultures may bring different and innovative solutions to engineering problems. Everyone benefits when diverse, gender-balanced teams design tomorrow's products and solve environmental and technological challenges.

ELIMINATING DISCRIMINATION IN ENGINEERING

Discrimination has not yet been eliminated from Canadian society, and instances of discrimination are still evident in engineering education and practice. As mentioned earlier, Regulation 941 under the Professional Engineers Act (Ontario) now includes harassment and discrimination in the definition of professional misconduct. PEO also publishes a *Guideline on Human Rights in Professional Practice* to inform engineers and engineering students about this type of misconduct and how to avoid it.[25] A recent survey carried out for the Women in Engineering Advisory Committee of PEO (WEAC/PEO), which polled both female and male engineers across Canada, found that workplace challenges continue to exist for female engineers. Women feel they face at least some attitudinal barriers from their superiors, and a substantial proportion of men share that view. In particular, women are concerned about opportunities to network and to gain entry to executive levels in their organizations.[26]

The study concluded that "the workplace is changing in positive ways for women, but old, lingering beliefs held by even a few can act as barriers to full participation."[27]

In a large American study, the authors interviewed recipients of prestigious fellowships in science and engineering to assess the career success of men relative to that of women. This study found that the men in that survey had received faculty positions at one level higher than the women (except in biology), although all participants had been selected for their equal potential for success. It also found that the men had published slightly more (2.4 papers per year versus 1.8 for women), but that the average number of citations per paper was higher for women than for men (24 versus 14).[28] This finding points to the importance of using criteria of quality rather than just quantity when assessing academic performance.

FAIR HIRING AND PROMOTING PRACTICES

Realistic objectives for hiring people from underrepresented groups should be based on achieving better than, or at least the level of, the availability of each underrepresented group in the pool of candidates. Creating a committee to identify diversity issues can be highly effective.[29] The following advice is intended to encourage fair hiring practices:

- Jobs should be advertised widely and externally besides being posted internally.
- A strong effort must be made to encourage qualified members of underrepresented groups to apply. This means finding and contacting them, as they may be few in number and may not consult the mainstream advertising channels.
- People involved in hiring must be trained to recognize inappropriate and illegal interview questions and the importance of treating all applicants with fairness and respect.
- Unbiased interview techniques should be used. One way to test the appropriateness of a question is to ask whether everyone else—male, female, or a member of a minority group—should be asked precisely the same question. If the answer is no, it should not be asked.

Discriminatory practices in hiring are often evidenced by a predominantly female staff supervised by a predominantly male senior management. For example, in the 1940s, Canadian banks were staffed by tellers who were almost exclusively female, while branch managers and senior executives were exclusively male. Since then, women with management potential have been identified and assisted, through proactive training, to qualify for promotions. This process provides female role models for younger women and helps integrate women's values into the corporate culture. In turn, the women who are promoted must be committed both to the organization and to the goal of employment equity if real change is to occur. Unexpected new benefits and insights result from hiring women, especially when they are encouraged to introduce diversity and innovation and when they feel their attributes and values are respected.

Employment Equity

Instituting an employment equity program to achieve a fairer societal representation is quite different from applying affirmative action or quotas. A Saskatchewan government brochure defines employment equity as follows:

> Employment equity is a comprehensive pro-active strategy designed to ensure that all members of society have a fair and equal access to employment opportunities. It is a process for removing barriers that have denied certain groups equal job opportunities . . . Employment equity programs encourage employers to hire, train, and promote members of these groups.[30]

This concept is quite different from establishing quotas, which involves hiring from one group until the desired numbers are reached. Employment equity is often misunderstood to be identical with a quota system; this can lead to frustration and anger among the group that feels disadvantaged by these policies.

Some employers, especially in private industry, still fail to see the benefits of hiring women. A survey of engineering graduates from 1989 to 1992 found that most female graduates had been hired by organizations with employment equity policies.[31] Moreover, these same organizations also hired male engineers in proportion to the existing pool of male graduates. This was evidence that employment equity policies remove obstacles for underrepresented groups without adding barriers for the majority group.

Avoiding and Removing Double Standards

Stereotypes often shape our perceptions and can have a major impact on the career progress and success of women and members of minority groups. For example, research by Foschi and colleagues found that even when files for female and male candidates were created equal as to their background and skills, prior to being selected for a summer job, the male candidates were viewed by the male subjects making the hiring choices as better qualified than the females.[32] This view was even stronger when the field was a nontraditional one for women. Foschi and colleagues also found that when women subjects made the hiring choices, they rejected the stereotypes and selected close to half the women candidates for the jobs. This points to the importance of having fair gender representation on hiring selection committees. A 1997 study of postdoctoral awards in Sweden—a country known for its progressive attitudes toward gender equality—found that a woman had to be twice as productive as a man to be perceived as equally competent and be awarded a postdoctoral fellowship.[33] Whether deliberate or not, stereotypes create a double standard in subjective evaluations. It is anticipated that false assumptions about gender roles and capabilities will disappear as employers hire capable women who, through commitment and excellence, will dispel these myths.

Fairness in Employee Performance Assessment and Promotions

In a large study of American scientists and engineers in high-tech companies conducted by DiTomaso and Farris,[34] Caucasian men were rated by their managers as average for the attributes of innovation, usefulness, and promotability, and were rated a little lower on cooperativeness. These managers rated women lower on all these attributes except for cooperativeness. When the employees made a self-assessment, the Caucasian men rated themselves slightly higher than their managers' rating, but women rated themselves lower than their managers' rating for all attributes except cooperativeness. A possible interpretation of this result is that male Caucasian scientists and engineers understand the corporate culture better and interpret the feedback more accurately, since for the most part they come from a similar group as the managers. For women, self-confidence may be an issue and the uneven understanding of feedback may be a problem.

Managers had also rated foreign-born non-Caucasian men lower than the American-born non-Caucasian men. However, the foreign-born men rated themselves higher on each point than their managers did, which shows a high level of self-confidence and self-esteem.

In the same study, women rated their managers lower than the men did on the following skills: getting people to work together, letting people know where they stand, being sensitive to differences among people, and minimizing hassles with support staff. However, women rated managers more highly on communicating goals clearly, defining the problem, getting resources, and motivating commitment. This result shows that managers should put more effort into developing objective and measurable criteria for assessments. They must also focus more attention on the type of feedback they provide, communicate the rules clearly, and test whether these rules have been understood. Managers must especially work on understanding the different approaches and perspectives that women and minority groups bring to the organization, and avoid underestimating the performance of particular groups of employees. Managers who build teams with people from diverse backgrounds and perspectives can expect the organization's performance to be enriched and improved.

WHAT IS SEXUAL HARASSMENT?

Sandler and Shoop define sexual harassment as follows:

> Unwelcome sexual advances, requests for sexual favours, and other verbal or physical conduct of a sexual nature constitute sexual harassment when any one of the following is true:
>
> • Submission to such conduct is made either explicitly or implicitly a term or condition of a person's employment or academic advancement.
> • Submission to or rejection of such conduct by an individual is used as a basis for employment decisions or academic decisions affecting the person.
> • Such conduct has the purpose or effect of unreasonably interfering with a person's work or academic performance or creating an intimidating, hostile, or offensive working, learning, or social environment.[35]

The three important characteristics of sexual harassment are as follows: the behaviour is unwanted or unwelcome; the behaviour is sexual or related to the sex or gender of the person; and the behaviour occurs in context of a relationship where one person has more *formal* power than the other, or more *informal* power. Some examples of a formal power relationship: a work supervisor and an employee; a faculty member and a student; a physician and a patient. An informal power relationship exists where one peer or colleague, while nominally equal in the hierarchy, nevertheless exerts an influence over another.[36]

A 1992 study reported that sexual harassment still occurs in the workplace and that engineering is not immune from it.[37] In fact, sexual harassment may be more of a problem for women in engineering because there may be only a few women at each site or in each firm. They may feel isolated and perhaps find little support or sympathy for what may be perceived as a "boys will be boys" attitude. Companies vary greatly in their policies and procedures for dealing with harassment. Education is at the heart of the solution. To eliminate harassment in the workplace, companies need a fair investigation procedure that does not victimize the complainant and that provides checks and balances for verifying accusations.[38]

When sexual harassment occurs, providing moral support for colleagues who are complainants can do a great deal to reduce their stress levels. Most of the time, informal investigations can accomplish more than formal ones that are more confrontational. That said, in some situations a formal complaint is the only possible approach. Most organizations should consider both approaches when developing their policies and procedures.

FAIRNESS IN AWARDS AND TRIBUTES

In the past, it was rare to see a woman being nominated for an award or a prize, or for her to be invited as a keynote speaker or to serve on an expert panel in a technical society or professional engineering association. In the first decade of the twenty-first century, there are still engineering events where only men appear in visible roles. For the sake of fairness, organizations should monitor the proportion of women on executive committees, receiving awards, and being invited as speakers on specialty topics. When the proportion of women invited to participate in these activities equals or surpasses their actual proportion in the Society or the Association, then progress will be visible and real. If the quality of the candidates is really the criterion for rewards, women should be making the lists more often than is currently observed.

IMPLEMENTING CHANGES

Every organization must set its own goals based on where it currently stands and should draw up an achievable plan for creating an environment where the work is challenging and comfortable for all employees. Many workplace issues and their solutions are discussed in greater detail in the report *Women in Engineering: More Than Just Numbers.*[39] Another useful document is a 2002 publication, *Becoming Leaders: A Handbook for Women in Science, Engineering and Technology*, which provides useful advice for women about strategies for success and some pitfalls.[40] This handbook is also very helpful for employers in both the private and public sectors and for deans, department heads, and professors in university settings. It explains how they can attract and retain more women to an organization. The authors state:

> Organizations need women's leadership. With it, they will change in ways needed by women, do more things that serve women, and become better places for women to work, succeed and contribute. This diversity will enhance the organization for both women and men.[41]

This handbook discusses work–life balance, career skills and strategies, proactive diversity (for employers and deans), women's leadership styles, strategies for students, tenure strategies for new faculty, and the importance of mentors and personal networks.

CONCLUSION

This chapter focused mainly on the issue of women's participation in engineering, because women are a majority in the Canadian population and their scarceness in an important field like engineering raises obvious questions. In the end, the profession of engineering has everything to gain by integrating a diversity of values into its current practice.[42] When the culture of engineering changes from a predominantly male one to a culture that integrates diversity, many of the obstacles and challenges identified in this chapter will disappear.

The following case studies are based on real situations that have arisen in recent years. The names of the individuals and companies have been changed to protect the privacy of those concerned.

CASE STUDY 13.1—DISCRIMINATION[43]

STATEMENT OF THE PROBLEM

You are an engineer and the chief executive officer of a profitable company called the Exeter Corporation. You are contacted by Susan Smith, a highly valued sales manager at Exeter, who has been passed over for promotion to director of product development. The promotion was given to Sam Brown on the recommendation of the vice-president of marketing, Peter Young. Smith sees this as a classic case of discrimination and is threatening to sue Exeter for unfair practices. She asks you to respond to her concerns within twenty-four hours. If you don't, you will probably lose her, a valuable employee, and her lawyer will be exploring the possibility of a settlement through the Human Rights Commission or the courts.

You arrange a meeting with Young and the human resources director to ask for more information regarding why Brown was selected over Smith. You are told that the difference between the two candidates was marginal. Young's explanation for his recommendation includes both objective and subjective criteria. His first comment is that Brown's experience in terms of seniority and familiarity with the industrial sector weighed slightly in his favour. Young adds that through Brown's greater participation in company social events and in the squash ladder, he was better known to all the vice-presidents, who said that Brown "looked like a winner." They could not say the same about Smith because she was less well known.

When prodded by the human resources director, Young suggests an additional list of problems and shortcomings that he attributes to Smith:

- "Mark Tannen, vice-president of manufacturing, is thought to be having an affair with Smith, and he is pushing her for promotion."
- "If Smith was promoted, Exeter might be liable to discrimination charges placed by Brown because of Mark Tannen's push for promotion for his honey."
- "The director of product development is a man's position. Human resources—soft, person-to-person stuff—is for women. Factories are for men."
- "Exeter clients prefer to deal with men. They know how to relate to their wives, mothers, and girlfriends, but not to women product development managers."
- "Women are undependable. They get married, get pregnant, want time off, and are less committed to the job."

Young provides no evidence to support these assertions; however, it is clear that they have influenced his decision to appoint Brown. He believes his decision made good business sense.

After the meeting, you reassess the situation. According to the objective data presented, Brown and Smith were both qualified for the position. Smith has shown excellent achievement as a product line manager. The same could be said about Brown. Understandably, choosing between the two would be difficult. Ignoring Young's subjective evaluation of Smith's "shortcomings," you must make a choice between promoting a woman (Smith) to a higher management level and promoting a man (Brown) who has marginally more experience.

You review your company's existing employment equity policies and current equity situation. Although one-quarter of the employees at Exeter are women, there are no women at the executive level and none on the board of directors. Recently, you and the human resources director issued a policy stating that the company would make great efforts to ensure equity and fairness in the manner in which employees are recruited, trained, and promoted. Therefore, although you have no hard evidence, you worry that gender inequity may permeate the organization. Also, if Smith pursues her lawsuit, you wonder whether it may encourage other women to come forward and state similar experiences. You realize that if Peter Young's comments on Smith's "shortcomings" were repeated in the courts, Exeter would surely be found guilty of discrimination. Thus, the firm would experience both a financial loss and the loss of an excellent employee.

QUESTIONS

1. *What criteria should have been used to select the new director of product development? Are these the same as Young's criteria?*
2. *Based on your criteria, who should have been appointed to the job: Sam Brown or Susan Smith? Explain your answer.*
3. *Since you are the CEO, what should you do in the next twenty-four hours regarding the potential lawsuit threatened by Smith?*
4. *As CEO, what long-term issues do you face if you want to ensure employment equity at Exeter, and what steps should you take to put this equity process in place?*
5. *Does your provincial Code of Ethics address this type of issue? Does it make a difference whether Peter Young, Sam Brown, and Susan Smith are also engineers? If your answer is yes, quote the appropriate sections of the code. If not, should the code provide guidance for dealing with this case? Alternatively, could you suggest the proper wording for an appropriate clause?*

CASE STUDY 13.2—SEXUAL HARASSMENT

STATEMENT OF THE PROBLEM

Michelle Kirkland has been a mechanical engineer in a consulting firm for four years. Recently, she wrote to a senior female engineer to discuss a serious work-related problem and to ask for advice on how to solve it. Below are extracts from her letter:

In my academic years I never had any problems being a woman in a male-dominated environment, and therefore very naïvely entered the workforce with a very positive and healthy attitude toward men in engineering. Today, unfortunately, that is no longer the case. After four years of verbal abuse and three incidents of sexual harassment from my immediate supervisor, I have become so cynical about men that I no longer enjoy my work. Most men quite naturally treat women without respect and as second-class citizens without even being aware of it.

The worst part of the situation is that I feel I cannot talk to anyone about this. In our corporation, female managers are practically unheard of and men seem to stick together like glue. Their attitude is that everything seems to be my fault: "Women are more sensitive" and "Women are less reliable" are the most recent comments that I bluntly received from my manager.

I was considering leaving the profession at one point, but meeting other women engineers motivated me to fight back harder and try again. Should I transfer to another department? Should I leave the company? (But are there any better ones out there?) Should I leave the profession and let my daughter solve the problems? I really do not know what to do. Sticking it out means additional stress in an already stressful job, headaches, and more anger. On the other hand, leaving means letting "them" win.

The senior engineer has sufficient personal knowledge of Kirkland and the atmosphere in Kirkland's workplace to believe that these allegations are true.

QUESTIONS

1. *What would you recommend that Kirkland do?*
2. *How does this work atmosphere of verbal abuse and harassment within the company affect the company's effectiveness and profitability? Is this a "professional" environment? Explain and justify your answer.*
3. *Was the manager's behaviour in violation of your province's Code of Ethics? If so, quote the appropriate sections. If not, what new clauses would you add to the code to deal with specific issues of harassment?*

DISCUSSION TOPICS AND ASSIGNMENTS

(Additional assignments are found in Appendix CD-E on the CD-ROM included with this text.)

1. Prepare a display or talk for a local high school to present engineering careers at your company or engineering studies at your university. Ensure that your message conveys the idea that the career paths and opportunities are open to all, regardless of race, sex, handicap, or other irrelevant criteria. Consider using recent videos about careers in engineering and science and other means of attracting young people to these fields. You especially wish to portray how engineers and scientists apply

their knowledge for the benefit of humankind, to solve problems and to design the world in which we live and work. What will you say, and how will you say it?

2. You are the senior engineer responsible for employing and orienting new engineers in a large consulting firm. What policies would you expect the firm to have in place for interviewing, hiring, and promoting employees and for resolving internal disputes in order to ensure fairness and equity?

3. The term "employment equity," which has become the norm in Canada, is often confused with the terms "affirmative action" and "equal employment opportunity," which are used more in the United States, even though the definitions of these three terms are significantly different. Search the Internet and the human resources literature in your library to obtain clear definitions of these three terms. What are the similarities and differences between them? Does the accusation of reverse discrimination, which is sometimes made against affirmative action, apply to employment equity? Is employment equity the best policy for Canada? Explain the basis for your answer in detail.

4. You have heard a rumour that a supervisor is harassing a young person in your company. You are the senior engineer responsible for the department in which these two people work. What will you do about this situation? What measures should be in place, or should be devised, to eliminate or to deal with such a situation in the workplace? Would your actions be the same or different for the following four cases:

   • A female employee being sexually harassed by her male immediate supervisor.
   • A male employee being sexually harassed by his female immediate supervisor.
   • A visible minority employee being harassed by a white supervisor.
   • A white employee being harassed by a visible minority supervisor.

5. You are a career counsellor in a university and you must give advice to first-year engineering students and to applicants who want to be admitted into engineering. What advice would you give, or what action would you take, to deal with the following situations?

   a. A young woman in first-year engineering who is academically excellent has been told by her parents that they will not finance her education because she will just get married and quit her job when she has children, so it would be a waste of money to pay for her education.

   b. A woman who had excellent marks in high school married immediately after finishing high school, and then went to work as a secretary. Her husband was unfortunately abusive, the marriage has now ended in divorce, and she is stuck in a menial job at minimum wage. How can she afford to enrol in university to improve her skills and resume her dream of a professional career?

    c. A young woman in high school wants to enter university. Her family has encouraged her to study engineering or science, but her school guidance counsellor has advised her to study art or nursing. Her confidence has been shaken by this contrasting advice from people whom she respects.

6. Many ideas have been suggested over the years to encourage people to live up to their full potential and not be deterred by artificial barriers in education and employment. For example, mentor programs, in which young people meet role models in nontraditional occupations, create long-term support that helps overcome obstacles. Another strategy has been to organize a nontraditional career day at a junior high school. Such a project was carried out very effectively in the Yukon, where a nurse and a ballet dancer (both of whom were male) and a carpenter, a geologist, a firefighter, and an engineer (all of whom were female) organized workshops with grade 8 and 9 students. A third technique involves introducing methods of engineering problem-solving into science and mathematics courses at the secondary level and/or discussing the various engineering fields. This brings students closer to these fields at the critical time when they are making career choices.

    Conduct a brainstorming session and generate a list of at least four other techniques for encouraging people—especially young people—to ignore stereotypes and achieve their personal goals. Consider how you might act on these ideas in your university or in your local public school system.

NOTES

1. Government of New Brunswick, "Code of Ethics," *Engineering and Geoscience Professions Act*, Statutes of New Brunswick, 1999, c. 88, s. 2(1).

2. Professional Engineers Ontario (PEO), *Guideline on Human Rights in Professional Practice* <www.peo.on.ca> (July 22, 2003).

3. Government of Canada, *The Canadian Charter of Rights and Freedoms*, clause 7, <laws.justice.gc.ca/en/charter/> (August 5, 2003)

4. Ibid, clause 15(1).

5. Ibid, clause 15(2).

6. M. Frize, C. Deschenes, E. Cannon, M. Williams, and M. Klawe, *A Unique National Project to Increase the Participation of Women in Science and Engineering* (CWSE/Canada) <www.carleton.ca/cwse-on/enffound.htm>.

7. Government of Ontario, *Human Rights Code*, Government of Ontario, R.S.O. 1990, Chapter H.19 <92.75.156.68/DBLaws/Statutes/English/90h19_e.htm> (August 5, 2003).

8. Canadian Council of Professional Engineers (CCPE), *Statistics on Enrolments in Canadian Engineering Undergraduate Programs* (Ottawa: CCPE).

9. Ibid.

10. R. Coulter, *Gender Socialization: New Ways, New World*, British Columbia Ministry of Equality, Victoria, BC, 1993.

11. M. Sadker and D. Sadker, *Failing at Fairness: How America's Schools Cheat Girls*, Charles Scribner's Sons, New York, 1994.

12. W.H. Peltz, "Can Girls + Science − Stereotypes = Success? Subtle Sexism in Science Studies," *The Science Teacher*, pp. 44–49, December 1990.

13. J.L. Lupart and M.E. Cannon, "Gender Differences in Junior High School Students Towards Future Plans and Career Choices," *Proceedings of the 8th CCWEST*, St. John's, Newfoundland, 2000.

14. J.M.J. McDill, "Participation Trends in and Lessons Learned From Outreach" (WEPAN 2003 Conference Proceedings).

15. Canadian Committee on Women in Engineering (CCWE), *Women in Engineering: More Than Just Numbers*, April 1992 <www.carleton.ca/cwse-on/webmtjnen/repomtjn.html> (August 5, 2003).

16. Wharton, E. *Women into Engineering: Project and Report Summary with Proposed Next Steps*, December 2001 <www.peo.on.ca> (August 5, 2003).

17. S. Hacker, "The Culture of Engineering: Women, Workplace and Machine," *Women's Studies International Quarterly* 4, 1981.

18. J.G. Robinson and J.S. McIlwee, "Men, Women and the Culture of Engineering," *Sociological Quarterly* 32, pp. 403–21, March 1991.

19. K.H. Sorensen and A.J. Berg, "Genderization of Technology among Norwegian Engineering Students," *Acta Sociologica* 30 (2), pp. 151–71, 1987.

20. R. Hall and B. Sandler, *The Classroom Climate: Projecting the Status and Education of Women* (Association of American Colleges, Washington., DC, 1982).

21. P. Caplan, *Lifting a Ton of Feathers: A Woman's Guide to Surviving in the Academic World*, University of Toronto Press, Toronto, 1992; and Natural Sciences and Engineering Research Council (NSERC), *Towards a New Culture: Report of the Task Force on How to Increase the Participation of Women in Science and Engineering Research*, NSERC, February 1996.

22. Gender Statistics on CRC Chair awards compiled by Dr. Claire Deschenes and Dr. Monique Frize <www.fsg.ulaval.ca/chaire-crsng-alcan/publicat/publicat.shtml>.

23. A. van Beers, *Gender and Engineering: Alternative Styles of Engineering*, MA thesis, Department of Sociology and Anthropology, University of British Columbia, Vancouver, July 1996.

24. M. Vickers, H.L. Ching, and C.B. Dean, "Do Science Promotion Programs Make a Difference?" *Papers and Initiatives* (M. Frize ed.), More than Just Numbers Conference, University of New Brunswick, Fredericton, May 1995.

25. Professional Engineers Ontario (PEO), *Guideline on Human Rights in Professional Practice*.

26. Women in Engineering Advisory Committee, *National Report of Workplace Conditions for Engineers*, Professional Engineers Ontario (PEO), Toronto.

27. Ibid.

28. Sonnert and G. Holton, "The Career Patterns of Men and Women Scientists," *American Scientist* (January–February 1996).

29. Canadian Committee on Women in Engineering (CCWE), *Women in Engineering: More Than Just Numbers*.

30. Women's Secretariat, *Employment Equity (Women in the Workplace)*, brochure, Government of Saskatchewan, Regina.

31. M. Frize, ed., *Papers and Initiatives*, More Than Just Numbers Conference, University of New Brunswick, Fredericton, May 1995. Report released in October 1996. Available at <www.carleton.ca/cwse-on> (August 5, 2003).

32. M. Foschi, L. Lai, and K. Sigerson, "Gender and Double Standards in the Assessment of Job Applicants," *Social Psychology Quarterly* 17, pp. 326–39, April 1994.

33. C. Wenneras and A. Wold, "Nepotism and Sexism," *Nature*, May 1997.

34. N. DiTomaso and G.F. Farris, "Diversity in the High-Tech Workplace," *Spectrum*, pp. 21–32, June 1992.

35. B.R. Sandler and R.J. Shoop (eds.), *Sexual Harassment on Campus: A Guide for Administrators, Faculty, and Students*, Allyn & Bacon, Boston, 1997, p. 4.

36. Ibid.

37. C.M. Caruana and C.F. Mascone, "Women Chemical Engineers Face Substantial Sexual Harassment: A Special Report," *Chemical Engineers Progress*, pp. 12–22, January 1992.

38. M. Frize, "Eradicating Sexual Harassment in Higher Education and Non-Traditional Workplaces: A Model," *Proceedings*, Canadian Association against Sexual Harassment in Higher Education Conference, pp. 43–47, Saskatoon, November 1995.

39. CCWE report.

40. F.M. Williams and C.J. Emerson, *Becoming Leaders: A Handbook for Women in Science, Engineering and Technology*, NSERC/Petro-Canada Chair for Women in Science and Engineering and WISE Newfoundland and Labrador <www.mun.ca/research/2003report/publications/becoming_leaders.php> (August 5, 2003).

41. Ibid.

42. F.L. Huband, "Why Women?" Preview column in *ASEE Prism*, September 4, 1991.

43. This case is based on S. Seymour, "Case of the Mismanaged Ms.," *Harvard Business Review*, November–December 1987, pp. 77–87. See also A. Mikalachki, D.R. Mikalachki, and R.J. Burke, *Teaching Notes to Accompany Gender Issues in Management: Contemporary Cases*, McGraw-Hill Ryerson, Toronto, 1992, pp. 5–8.

Chapter 14
Disciplinary Powers and Procedures

The provincial and territorial Associations of professional engineers and geoscientists protect the public by regulating professional practice. To do so, each Association has been delegated the powers to prosecute people who practise unlawfully and to discipline licensed practitioners who are found guilty of professional misconduct, negligence, or incompetence. These infractions and the disciplinary actions that may result are explained in this chapter.

PROSECUTION FOR UNLAWFUL PRACTICE

A person who is not a member or licensee of a provincial or territorial Association but who nevertheless

- practises professional engineering (or professional geoscience), or
- uses the title Professional Engineer (or Professional Geoscientist, etc.), or
- uses a term or title to give the belief that the person is licensed, or
- uses a seal that leads to the belief that the person is licensed

is guilty of an offence under the Act. The procedure for prosecution and the penalties vary slightly, depending on the province or territory. The Association must initiate the action to prosecute offenders in the appropriate court under the authority of the Act. The trial judge then assesses the penalty, which is typically a fine proportional to the seriousness of the infraction. These prosecutions are typically administrated by Association staff. Professional engineers or geoscientists generally are not involved in these proceedings except perhaps as witnesses.

DISCIPLINE FOR PROFESSIONAL MISCONDUCT

In the case of licensed members, disciplinary action for professional misconduct or incompetence is conducted within the Association by a discipline committee formed of members of the governing council and other professional engineers or geoscientists. Under the authority of the Act, the committee has the power to discipline members for professional misconduct, as defined below.

Definition of Professional Misconduct

The various Acts have slightly different definitions of what constitutes grounds for disciplinary action. These definitions are reproduced in Appendix CD-B. Although the definitions are not identical, they are very similar. In British Columbia, for example, incompetence, negligence, and unprofessional conduct are grounds for disciplinary action.[1] In Ontario, the Act states that professional misconduct and incompetence are grounds for disciplinary action,[2] but regulations under the Ontario Act define professional misconduct in more detail, and include negligence.[3]

The provincial Acts typically identify six causes for disciplinary action: professional misconduct (or unprofessional conduct), incompetence, negligence, breach of the Code of Ethics, physical or mental incapacity, and conviction of a serious offence. Each of these terms is discussed briefly in the following paragraphs.

Professional Misconduct

Professional misconduct (or unprofessional conduct, as it is called in some Acts) is the main source of complaints to provincial Associations. In about half of the Acts, the term is not defined. This places an additional burden of proof on the Association's legal counsel in any formal hearing, since alleged misconduct must be proven both to have been committed and to constitute professional misconduct.

However, Alberta, Newfoundland, and Prince Edward Island have much more general definitions. For example, Alberta's Act defines "any conduct . . . detrimental to the best interests of the public" or that "harms or tends to harm the standing of the profession generally" as unprofessional conduct.[4] While such clauses will stand the test of time because of their generality, they are really not specific enough to serve as guidance in individual cases (although the Association's Code of Ethics may give more specific guidance).

At the other extreme, Ontario's definition of professional misconduct includes some very specific acts, such as "signing or sealing a final drawing . . . not actually prepared or checked by the practitioner."[5] In fact, the Ontario regulation is a fairly comprehensive definition of professional misconduct and may be of interest to readers whether they reside in Ontario or not (see Appendix CD-B). Such specific guidance is clear and unambiguous. However, the regulations cannot define every possible form of professional misconduct, so they contain a general clause stating that professional misconduct includes any act that "would reasonably be regarded" as unprofessional.[6] This circular definition is very general, and any complaint based on this clause would first have to prove that the person's actions were "unprofessional."

Incompetence

Incompetence is defined in several Acts as a lack of knowledge, skill, or judgment or disregard for the welfare of the public of a nature or to an extent that demonstrates the member is unfit to carry out the responsibilities of a professional engineer. Depending on the Act, undertaking work that the engineer is not competent to perform may be considered either incompetence or professional misconduct. This rather subtle distinction covers the all-too-common occurrence of an engineer practising outside the area of his or her expertise, even though the engineer may be fully competent within his or her major field of practice.

Negligence

In most Acts, "negligence" means "carelessness," or carrying out work that is below the accepted standard of care or performance. In many instances it means the omission of an activity required to ensure the proper care or safeguarding of life, health, and property. In fact, the omission of care or insufficient thoroughness in performing duties would probably be the most common complaint under this heading.

Breach of the Code of Ethics

In four provinces (Alberta, New Brunswick, Newfoundland, and Nova Scotia), a breach of the Code of Ethics is specifically defined in the Act to be equivalent to professional misconduct. These codes therefore have the full force of the Act. In other provinces (British Columbia, Manitoba, Prince Edward Island, Quebec, and Saskatchewan) and the territories, where the term "professional misconduct" is undefined or defined in very general terms, it would likely be understood to include the Code of Ethics, thus giving the code some enforceability under the respective Act.

In Ontario the Code of Ethics is not clearly enforceable under the Act. However, there is an detailed definition of professional misconduct in the regulations that contains many concepts that are in the code, such as "failure to act to correct or report a situation that the practitioner believes may endanger the safety or the welfare of the public," as well as failure to disclose a conflict of interest, and about sixteen additional clauses (see Appendix CD-B).[7] Behaviour that contravenes the Code of Ethics may fall below the ideal, but such behaviour is not punishable under the Act unless it breaches the definition of professional misconduct. In other words, the Ontario Code of Ethics describes ideal professional conduct, whereas the definition of professional misconduct sets a lower limit.

Physical or Mental Incapacity

Most Acts also include a "physical or mental condition" as a definition of incompetence, provided the condition is of a nature and extent that, to protect the interests of the public (or the member), the member no longer may practise professional engineering or geoscience.

Conviction of an Offence

The provincial Acts also permit disciplinary action against a member who is guilty of any offence relevant to the member's suitability to practise. In other words, should a member be found guilty of an offence under any other Act, and should the nature or circumstances of the offence affect the person's suitability to practise as a professional engineer or geoscientist, then the person can be found guilty of professional misconduct. Proof of the conviction must be provided to the Discipline Committee. This clause is used relatively rarely, since convictions for minor offences (traffic violations, local ordinance violations, etc.) do not affect one's suitability to practise engineering. However, convictions of serious offences such as fraud or embezzlement, which involve a betrayal of trust and questionable ethics, could be grounds for declaring a member unsuitable. This condition clearly imposes a standard of conduct on professional engineers that is somewhat higher than that expected of the average member of the public.

THE DISCIPLINARY PROCESS

Any member of the public can make a complaint against a licensed engineer or geoscientist, although most complaints are brought by building officials, government inspectors, or other professionals. Disciplinary procedures are very unpleasant, and since they can have a dramatic effect on a professional's career, they must be fair, as well as be seen to be fair. The Association's disciplinary procedures are defined in the Act in very formal legal terms. When a complaint of negligence, incompetence, or professional misconduct is made against a licensed professional, it sets in motion a three-stage process: gathering information, evaluating the complaint, and conducting a formal hearing that renders a judgment.

To ensure complete impartiality, the three stages of the disciplinary process are usually carried out by three different groups of people. No one who participates at an earlier stage is permitted to participate in the final hearing and judgment.

- The first stage is generally conducted by Association staff.
- The second stage is conducted by a Complaints Committee or an Investigations Committee (depending on the province) composed of members of the Association's governing council and other licensed members.

- The third stage is conducted by a Discipline Committee, which is composed of members of the governing council and people who have not previously been involved with the case.

The disciplinary process is typically confidential, unless the Discipline Committee, in the final stage, orders that its findings be published.

A summary of the complaints process carried out by Professional Engineers Ontario (PEO) is reprinted below, with permission. The procedure in other provinces and territories is very similar although not identical. Here's what happens when a member of the public has a complaint against an Ontario engineer. (The process is explained as advice to the complainant.)

Stage 1: Gathering evidence about your complaint

If you believe that a practitioner may have acted in an improper or incompetent manner, PEO would like to hear from you. . . . The Registrar's staff can answer your questions about the conduct expected from a practitioner, even if you do not wish to register a formal complaint. They can also verify whether someone is a licensed professional engineer or a C of A [Certificate of Authorization] holder. . . . The Registrar's staff will review your concerns with you, and advise you on the evidence you will need to provide to support allegations of either professional misconduct or incompetence, or both.

Making a complaint

1. You may submit a letter to PEO, which should include a description of your concerns, their time frame and the supporting documentation.

2. PEO staff complete a preliminary investigation, in which they examine your complaint in the context of the Professional Engineers Act and Regulation 941 [which defines professional misconduct], and help you to identify the evidence required. Staff may also engage an independent engineer to review and to comment on the work of the practitioner.

Depending on the available evidence, they will either draft the complaint for your signature, or assist you in putting your letter(s) to the Association into a complaint format.

If the evidence does not support a formal complaint, staff may suggest other means of addressing your concerns. However, you do have the right to insist that a complaint be submitted directly to the Complaints Committee.

3. Your signed complaint is filed with the Registrar for review. It is then sent to the practitioner in question, who has a period of time to respond in writing to the allegations made in the complaint.

4. As the complainant, you will be asked to comment on the practitioner's response.

Stage 2: Peer review of the complaint

At this stage, your complaint is presented in confidence to the Complaints Committee, which may:

1. refer the complaint, in whole or in part, to the Discipline Committee, as explained in Stage 3;

2. refer the complaint, in whole or in part, to the Discipline Committee via Stipulated Order [as defined below], which is a process designed to handle less serious discipline matters;

3. dismiss the complaint;

4. send a "letter of advice" [as defined below] to the engineer, or interview the engineer, without referring the case to the Discipline Committee;

5. direct staff to obtain more information, which is then considered by the Complaints Committee.

In all of the above scenarios, both the complainant and the engineer in question are sent a copy of the committee's written decision with reasons.

If you are dissatisfied with the way your complaint has been handled after the Complaints Committee has made its decision, you can request that it be reviewed by PEO's Complaints Review Councillor. The Complaints Review Councillor reviews only the procedures and process followed by the Complaints Committee to arrive at its decision, not the merits of the complaint.

Stage 3: Discipline hearings

If the case is referred to the Discipline Committee, a written notice of hearing is prepared by PEO's lawyer and served on the accused practitioner, who may also hire a lawyer. Before the hearing, PEO's lawyer usually meets with the accused practitioner or his or her lawyer to disclose the nature of PEO's case.

Discipline hearings are held at PEO's offices and follow court procedure, with a court reporter present. The hearing panel formed from among the members of the Discipline Committee comprises five PEO members.

As the complainant, you may be asked to testify at the discipline hearing. After the hearing panel has heard all of the evidence, the Discipline Committee gives a written decision to both the practitioner and you as the complainant. . . .

The practitioner has the right to appeal the decision to the Divisional Court of Ontario.

Note: The procedures followed during the discipline process are fully laid out in sections 23-28 of the [Ontario] Professional Engineers Act. . . .[8]

Letter of Advice

A letter of advice is intended to warn a practitioner about actions that are not categorized by the Complaints Committee as professional misconduct, but justify some potential concern.

Stipulated Order

A Stipulated Order process is a simpler form of disciplinary hearing for less serious cases. The Stipulated Order may be used instead of a formal hearing when the Complaints Committee has reason to believe, after reviewing the complaint, supporting materials, response of the accused engineer, and so on, that the Act, regulations, or bylaws have been breached but a formal hearing is not warranted. The written consent of the complainant and the accused engineer are required before the process can begin. The Stipulated Order process is carried out as follows:

> A single representative of the Discipline Committee will review the Complaints Committee findings and meet separately with the member and the complainant to discuss the evidence. The Discipline Committee representative will determine—with staff and legal assistance if necessary—whether the Act, Regulations or Bylaws have been breached or not, and decide on an appropriate penalty, if applicable.
>
> If the member is found guilty of breaching the Act, Regulations or Bylaws, the member will be given a Stipulated Order, comprising:
>
> • a detailed description of the alleged offence;
>
> • the finding of guilt; and
>
> • the penalty.
>
> The member [is then requested to] sign the Order. The penalty can include one or more of the penalties available to a full discipline hearing, as defined in the Act. By signing the Stipulated Order, the member agrees to guilt and accepts the penalty. There is no appeal.
>
> If the member is found not guilty, the member and the complainant will be notified in writing that the matter is resolved and no further action will be taken.
>
> If the matter cannot be resolved by way of a Stipulated Order, it proceeds to a discipline hearing. The entire Stipulated Order process is carried out "without prejudice," which means that it is not referred to during the hearing, nor can the Discipline Committee representative who participated sit on the discipline panel.[9]

Obviously, although this process is appropriate only for less serious and perhaps more clear-cut complaints, it can be highly effective in avoiding the costs of formal hearings and reducing the time needed to obtain judgements.

DISCIPLINARY POWERS

The penalties that can be meted out by the Discipline Committee are fairly general; they include fines and the payment of costs but not, of course, imprisonment. The disciplinary powers awarded under each provincial and territorial Act are summarized in Appendix CD-B. Typically, if a member or licensee should be found guilty, the Discipline Committee can

• revoke the licence of the member (or the permit or certificate of authorization, if a corporation).
• suspend the licence (usually for up to two years).
• impose restrictions on the licence, such as supervision or inspection of work.

- require the member to be reprimanded, admonished, or counselled, and publish the details of the result, with or without names.
- require the member to pay the costs of the investigation and hearing.
- require the member to undertake a course of study or write examinations set by the Association.
- have any order that revokes or suspends the licence of a member to be published, with or without the reasons for the decision.
- impose a fine (up to $10,000 in Alberta; up to $5,000 in Ontario; but only hearing costs in the Yukon).

The severity of the penalty will, of course, vary with the circumstances of the case.

Throughout the above discussion and this text, the terms licensee and member have been used interchangeably. The usage varies across Canada, and includes Temporary Licences, Certificates of Authorization, Limited Licences, Permit Holders, and (in Ontario) designation as a Consulting Engineer. The disciplinary actions described in this chapter also apply to these other categories. That is, other forms of permit or certificate may be revoked or suspended in the same manner as, or in addition to, the revocation of a member's licence.

CASE HISTORY 14.1

THE BURNABY SUPERMARKET ROOF COLLAPSE

The opening-day collapse of the rooftop parking lot of the Save-on-Foods supermarket in Burnaby, British Columbia, was tragic for everyone involved. Many people suffered injuries and financial loss. The engineers who designed the structure were the subject of litigation and disciplinary action by the provincial Association. The lessons learned were very expensive, even though they led to improvements in building safety and reliability. Sadly, the event bears some similarity to an infamous earlier disaster.

The Day of the Collapse

In April 1988, just before 9 a.m. on a rainy Saturday morning, a gleaming new Save-on-Foods supermarket was opened in the Station Square development in Burnaby. The supermarket had a parking lot built on the roof, and shoppers were asked to park their cars in the lot and descend to the main floor. About 600 shoppers, mainly senior citizens, came to the opening of the supermarket, which was staffed by about 370 employees.

After an opening ceremony attended by the Burnaby mayor and a host of local dignitaries, the shoppers milled about the store, examining the goods for sale. A sudden sharp crack was heard, and water sprayed from an overhead sprinkler pipe near a column in the produce area of the store. Startled shoppers looked up to see that the sprinkler pipe had been broken by a severely

The Save-on-Foods supermarket in Burnaby, B.C., opened on the morning of April 23, 1988. A parking lot on the roof of the supermarket provided extra convenience to shoppers. However, within minutes of opening, a main beam supporting the roof collapsed, dumping twenty automobiles into the produce section of the supermarket. Fortunately, no lives were lost. The subsequent inquiry discovered basic errors in the design, which led eventually to changes in licensing procedures for engineers.

Source: Photo by Craig Hodge. Reproduced with permission of The Tri-City News.

twisted roof beam. A cameraman attending the opening ceremony took a photograph of the beam—a photograph that would later assist in the investigation of the failure. While the water sprayed out over a nearby bulk foods area and a cheese counter, the supermarket staff immediately and efficiently directed shoppers to vacate the area. An announcement was made over a public address system, asking shoppers to leave the store. Some shoppers were reluctant to leave, but those who saw the twisted beam had no hesitation.

At about 9:15 a.m., less than five minutes after the initial bang, a huge section of the roof collapsed, dropping twenty automobiles from the roof parking lot into the produce area of the store. The remaining shoppers panicked, and people fell as they rushed outside. A pile-up occurred at the exit doors. About twenty-one people were injured, mainly from injuries (such as broken bones) sustained in the melee. One employee was pinned under the falling roof beam and suffered a crushed pelvis. The injured were rushed to hospital. Fortunately, no one died.

The Inquiry and Report

The provincial government appointed a commissioner, Daniel J. Closkey, to head an inquiry into the cause of the collapse and to suggest how similar failures could be avoided in the future. The commissioner held ten days of hearings over two months. Forty-seven witnesses testified, and the hearings were broadcast live on cable television. A final report was published in August 1988.[10]

The report's conclusions were surprising. The design engineering firm was experienced and well established, yet several basic errors had been made in the design calculations. Moreover, the calculations had been thoroughly reviewed by a second experienced engineering consulting firm. How could basic design errors have been missed by both of these experienced engineering firms? An excellent summary is presented in the 1990 research paper by Vancouver consultant C. Peter Jones and Professor N. D. Nathan.[11] Jones and Nathan were engaged as advisors to Closkey. Based on their investigation, they identified nine errors by the design engineer (in assumptions, decisions, judgments, or miscalculations) that reduced the factor of safety. Although none of these errors, by itself, would have caused the collapse, together they led to a failure that Jones and Nathan called "inevitable." The engineer who reviewed the design evidently made most of the same errors in assumptions, decisions, judgments, and miscalculations, and neglected to discover the fatal flaws. These errors are described, in detail, both in the commissioner's report and in the cited paper by Jones and Nathan, and are summarized very briefly as follows:

Errors in dead load estimation. The original design was correctly carried out. The roof consisted of two concrete slabs separated by a layer of insulation and a waterproof barrier. The bottom slab was part of a concrete/sheet-steel roof deck. During the early design stage, the top concrete slab was made thicker but the beam size was not adjusted to resist the added weight. When the concrete slabs were poured, the actual thickness was (inadvertently) even greater than specified. Finally, a concrete walkway on the roof was widened but this extra weight was ignored. These three factors increased the moment in the beam (caused by dead load) by about 55 percent, effectively eliminating the factor of safety.

Error in beam specification. The beam was originally checked for strength and deflection, and the deflection limit was the governing case. However, the design engineer concluded that a greater deflection was acceptable, and a redesign was carried out. The commissioner found no records or clear recollection of this second design. Nevertheless, subsequent issues of the structural drawings showed the beam incorrectly reduced to a smaller-weight section. The revision was not flagged on the drawing, perhaps explaining why it was overlooked when reviewed later.

Error in live load estimation. The live load estimation was also reduced by re-evaluating the area of the roof that was supported by the beam. This was a valid point, and a minuscule reduction of about 1 percent was appropriate. However, an error in judging the "tributary area" resulted in an erroneous live-load reduction of 12.6 percent.

Optimistic calculation of bending moment. We usually calculate the bending moment from the centre-line of a structural-steel support.

However, in concrete design it is acceptable to calculate a moment from the edge of the column. This optimistic convention was used, slightly reducing the expected bending moment in the beam.

Optimistic tests of beam strength. When the (erroneous) beam section was analyzed with the (erroneous) load data, it was still slightly under-strength. However, at this point the engineer received "mill certificates" from the steel supplier, showing that the steel yield strength was 25 percent greater than the strength used in computations. The beam was therefore judged adequate in strength. However, the commissioner concluded that this strength estimate, based on only a few test specimens, was unrealistically optimistic, because the test specimens were taken from the web of the beam, which typically has a higher strength than the flanges, and in bending, the flanges are more highly stressed than the web.

Lack of lateral support. To prevent lateral buckling, long, deep beams must be supported laterally, or loads must be reduced. Lateral supports at the failed column were noted on a shop drawing at one point in the design, but the supports were deleted when the engineer was informed of the extra cost. Clearly, the engineer did not study the reduced capacity of the beam caused by the lack of lateral support.

No check of column buckling strength. The commissioner concluded that there was no evidence that the load-carrying capacity of the beam/column assembly was studied.

The Cause of the Collapse

After hearing all of the testimony, the commissioner concluded that the roof failure involved two modes. These modes, illustrated in Figure 14.1, are explained as follows:

Beam failure. The primary cause of failure was that the beam was under-strength for the imposed bending loads, and the laterally unsupported lower compression-flange of the beam buckled at the supporting column.

Column failure. In addition, the vertical beam-column assembly likely buckled simultaneously, although this was probably not the primary cause.

Buckling is usually a rapid and catastrophic occurrence, but in this case, the roof did not crash down immediately. When the beam buckled, the movement snapped the sprinkler pipe, and the noise, motion, and water spray alerted the supermarket staff that total collapse was imminent. However, the roof and the automobiles on top of it were supported by an apparent "membrane action" of the roof components, acting in tension. This unexpected support delayed total collapse for about four minutes, giving the shoppers time to leave the danger area before everything came tumbling down.

In their paper, Jones and Nathan also discuss the "fragmentation" of the design process, which created communication problems, leading to inconsistencies in the design. Although one person was responsible for the design,

FIGURE 14.1 — Collapse of the Main Beam in the Save-on-Foods Supermarket

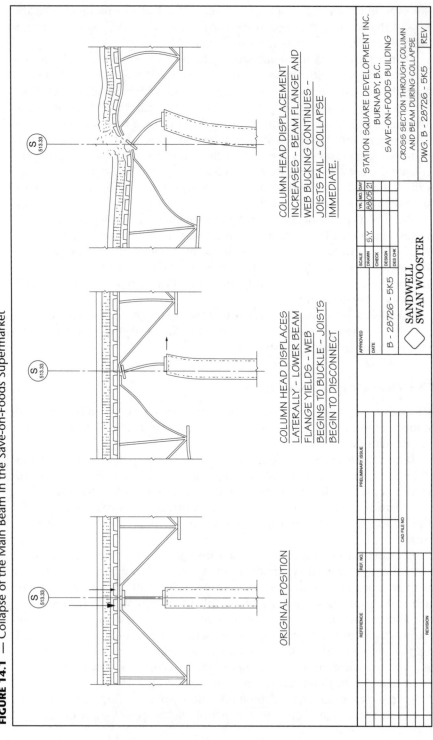

Figure 14.1—Collapse of the Main Beam in the Save-on-Foods Supermarket (on previous page). The first drawing shows the original design of the column supporting the beam. In the second drawing, the beam flange has yielded, causing the web to buckle. Fortunately, this position was retained for more than four minutes, while the supermarket was evacuated. The final drawing shows the joists disconnecting and the column collapsing.

Source: Drawing B28726-SK5, page 108 of Closkey Inquiry Report. Copyright © 2003 Province of British Columbia. All rights reserved. Reprinted with permission of the Province of British Columbia. www.ipp.gov.bc.ca.

decisions were actually made by several participating groups. Jones and Nathan conclude with some good advice for design engineers and their clients:

> In the design and construction of a structure, hundreds of calculations are made, and hundreds of items of information are communicated from one participant to another. It is certain that many errors will be made, and the process must be designed to eliminate them. In budgeting the manpower for a project, allowances must be made for careful and detailed checking at each step. The engineers checking the design must have nothing else on their minds: they must not be burdened with many other simultaneous responsibilities. They must avoid a mind-set that the design is probably good, particularly if the designer is a respected senior. The checker should cultivate an attitude of mind that "anyone can make mistakes and it is up to me to find them in this design."[12]

Recommendations

The commissioner made nineteen recommendations in his final report, directed at the provincial government, the municipalities, the engineering profession, the Canadian Institute of Steel Construction, the Canadian Standards Association, and the Canadian Sheet Steel Building Institute. All of the six recommendations (5 through 10) directed toward the engineering profession have been implemented.

Recommendations 5 and 6 suggested that companies, partnerships, and other firms be required to register under the Act and should be subject to deregistration for unethical practices. This recommendation was implemented: corporations, partnerships, and other legal entities must now hold a Certificate of Authorization in British Columbia.

Recommendation 7 suggested that APEGBC require structural engineers to satisfy higher qualifications than those required for membership. The status of Structural Engineer of Record (SER) has since been implemented, and requires additional experience and examinations. Only SERs are permitted to approve the design of a building's primary structure. (APEGBC has also established several special interest "divisions," which encourage the exchange of information among engineers in various specialties.)

Recommendation 8 suggested that structural engineers be required to carry a specified minimum of professional liability insurance. This recommendation has been partially implemented: all members, licensees, and certificate holders offering services to the public must notify clients, in writing, whether professional liability insurance is held and applies to the services offered, and must receive the client's acknowledgement before proceeding with the work. (APEGBC has also instituted a secondary liability insurance plan, as explained elsewhere in this text.)

Recommendation 9 suggested that APEGBC establish and enforce a minimum fee schedule. This recommendation resulted from the obvious conclusion that the fees paid to engineers on the Save-on-Foods building design were far too low to permit an unhurried and careful review and analysis of the structure. APEGBC has developed a comprehensive Fee Guideline for Engineering Services that sets minimum recommended fees. This guideline is available from the APEGBC website.[13]

Recommendation 10 suggested that provincial standards of practice be established for building design drawings and calculations. APEGBC now publishes several guidelines for professional practice—including structural, mechanical, electrical, and geotechnical services—with applications to building projects. All of the guidelines are available from the APEGBC website.

In the years since the Burnaby roof collapse, many other provincial Associations have examined their practices and have implemented similar rules, procedures, and guidelines.

Costs of the Roof Collapse

The fortuitous period of about four minutes between the cracking of the roof and its collapse saved the lives of those directly below it. However, the financial costs of the collapse were high. In their paper, Jones and Nathan make an approximate estimate:

- The budget of the commissioner inquiry was $250,000.
- Private legal costs during this period are not known, but 12 legal firms communicated with the inquiry and many were present at one or more hearings.
- The legal costs to the APEBC [now APEGBC] for the disciplinary inquiry amounted to $80,000. This did not include staff time.
- All defendants had legal counsel.
- All perishable stock in the supermarket was destroyed. All other stock had to be removed. The collapsed area was rebuilt and the entire store upgraded structurally at a cost understood to be approximately $5,000,000. The store finally opened in the fall of 1988.
- Some competitive advantage was possibly lost, as a competing store opened nearby, late in 1988. Six months of sales were lost. Smaller stores in the complex also suffered losses.

- Personal injuries occurred, automobiles were lost, and other damages were alleged.

- The extent of litigation is not known, but total costs are clearly very high.[14]

Conclusion

In the preface to his report, the commissioner made the following statement concerning the public safety roles of engineers, architects, municipalities, and professional Associations:

> Owners are primarily motivated by a satisfactory return on investment. Contractors and suppliers likewise are profit-oriented. The professional engineers and architects have dual responsibilities. On the one hand, by training and through professional ethics, they have a duty to maintain a high calibre of service to the public. In the context of building design and erection, this translates into the protection of public safety. On the other hand, professional consultants operate businesses in a commercial world. They, too, require a reasonable stream of revenue to survive. In the middle stand the regulators: municipalities, responsible for enforcing building standards, and the professional associations, for maintaining professional standards.[15]

This is the only statement in the commissioner's report with which I would disagree. The municipalities and professional Associations establish the bylaws, professional standards, and Codes of Ethics that regulate the profession, but it is the engineer who is "in the middle" between the pressures of the profit-making organizations and the escalating demands for guaranteed safety.

The Save-on-Foods roof collapse bears some similarity to the collapse of the Quebec Bridge more than ninety years earlier (see the case history in Chapter 1). Both collapses were caused by the buckling of a cantilever structure that had an undersized cross-section, as a result of a miscalculation of the applied load (especially the dead load). In both cases, the design engineer was a respected senior person and the engineering firm was underpaid for the work expected and the responsibility undertaken. These lessons, taught by the 1907 disaster that stimulated the regulation of the engineering profession, are still valid a century later.

DISCUSSION TOPICS AND ASSIGNMENTS

(Additional assignments are found in Appendix CD-E on the CD-ROM included with this text.)

1. Using Appendix CD-B, "Excerpts from the Provincial and Territorial Acts and Regulations," compare the definitions of professional misconduct, negligence, and incompetence that form the basis for disciplinary action under each Act. Which province or territory has the most specific definitions? Which has the most general definitions? Would you say that

the Acts are generally in agreement on the definitions? Do you see any major inconsistencies between them? Define and discuss these similarities and inconsistencies.

2. Would an infraction of the Code of Ethics in your province or territory be clearly enforceable under your provincial Act? Should all codes always be fully enforceable, or should they be purely voluntary codes of personal behaviour? Explain your answer. Regardless of your answer, check your provincial Act in Appendix CD-B. Does it agree with your own views on enforceability?

3. Using Appendix CD-B, compare the disciplinary powers awarded under each provincial and territorial Act. Which Act provides the most severe fines and penalties? Would you say the disciplinary powers in the Acts are generally similar? Are there serious inconsistencies between them? Point out and discuss these similarities and inconsistencies.

4. In your employment as an engineer, you discover that some of your fellow employees who supervise the delivery and storage of materials on the job site (and who are also professional engineers) have been involved in "kickback" schemes with suppliers. The suppliers invoice your employer for materials that have not been delivered, your colleagues validate the invoices, and the suppliers pay them a hidden commission. Obviously, these schemes violate criminal law. Which clauses in your provincial Code of Ethics have been broken by your colleagues? To what types of disciplinary action have they exposed themselves as a result? Suppose you confront them and they promise they will discontinue these schemes if you agree not to reveal them. Would your silence be consistent with your provincial Code of Ethics? If not, could any disciplinary action be brought against you? Describe the course of action you should follow. Would it be different if your fellow employees were not professional engineers?

5. You receive a registered letter from the registrar of your provincial Association stating that you are the subject of a formal complaint made by a former client or employer. The letter contains a description of the complaint, which alleges that you are guilty of incompetence because advice included in a report you wrote was faulty, and the client or employer suffered a financial loss as a result of following your advice. A preliminary investigation by the Association appears to show that the complaint has some validity. The registrar asks you to respond to the complaint. Describe the actions you would take to protect yourself.

NOTES

1. British Columbia, *Engineers and Geoscientists Act*, RSBC 1996, c. 116, s. 33(1)(c).
2. Ontario, *Professional Engineers Act*, RSO 1990, c. P.28, s. 28.
3. Ontario, Regulation 941/90, s. 72(2), *Professional Engineers Act*, c. P.28, RSO 1990.
4. Alberta, Engineering, *Geological and Geophysical Professions Act*, SA 1981, c. E-11.1 (as amended), s. 43(1).

5. Ontario, Regulation 941/90, s. 72(2)(e), *Professional Engineers Act*, c. P.28, RSO 1990.

6. Ibid, s. 72(2)(j).

7. Ibid, s. 72.

8. Professional Engineers Ontario (PEO), *Making a Complaint: A Public Information Guide, brochure*, PEO, Toronto <www.peo.on.ca/> (October 8, 2003). Reprinted with permission of PEO.

9. PEO, "Alternatives to Discipline Hearings: Stipulated Order," *Gazette* vol. 14, no. 1 (January–February 1995): 2. Excerpt printed with permission of PEO.

10. D.J. Closkey, P.Admin., *Report of the Commissioner Inquiry, Station Square Development*, Burnaby, British Columbia, Province of British Columbia, August 1988. Copyright © 2003 Province of British Columbia. All rights reserved. Reprinted with permission of the Province of British Columbia. www.ipp.gov.bc.ca.

11. C.P. Jones and N.D. Nathan, "Supermarket Roof Collapse in Burnaby, British Columbia, Canada," *ASCE Journal of the Performance of Constructed Facilities*, vol. 4, no. 3, August 1990, pp. 142–60. Reprinted by permission of the publisher, ASCE.

12. Ibid., p. 160.

13. Association of Professional Engineers and Geoscientists of British Columbia (APEGBC) website: <www.apeg.bc.ca/> (October 6, 2003).

14. Jones and Nathan, p. 158.

15. Closkey, p. x.

Part Four
Maintaining Professionalism

Chapter 15
Maintaining Professional Competence

Engineering and geoscience are continually evolving. Keeping abreast of new theories, techniques, hardware, and software is a challenge. Fortunately, most professionals meet the challenge easily and satisfy the Code of Ethics clause that requires continued competence. This chapter explains how the Associations, the engineering societies, and the universities are assisting engineers and geoscientists by providing and monitoring professional development activities.

CAREER MOMENTUM VERSUS OBSOLESCENCE

Recent graduates enter the workplace enthusiastic to learn new ideas. This positive attitude is the start of a successful career and must be sustained. Continued professional development prepares you for the next project, challenge or promotion and gives you the impulse you need to reach your career goals. In other words, engineering education must be rejuvenated periodically. Otherwise your bachelor's degree is like a radioactive mineral that decays over time. In previous decades, the half-life was about ten years, but our high-tech environment has accelerated the process. Even if you were at the top of your graduating class, you will be obsolete, eventually, without some intellectual renewal. Continued professional development permits you to remain professionally competent and productive for life.

Technical obsolescence is a special worry for older engineers, especially during bad business cycles. In good times, employers encourage professional development, retraining, and upgrading; when the economy slows down they reduce costs through downsizing (dismissal). Older engineers are seen as obstacles to less experienced, less expensive younger people, who are eager to learn and move ahead. To avoid becoming a liability, the older engineer must develop skills that are essential to the employer and that justify the higher pay.

Ideally, employers should view professional employees as corporate intellectual assets, like patents or trademarks—the employer has an investment in these assets that must be protected by encouraging professional employees to join engineering societies and participate in technical conferences, courses, and seminars (delivered in person or electronically). Formal education is likely most effective in prolonging productivity; however, it is also the most expensive, even when taken part-time. A common rule of thumb is that a professional with a postgraduate degree is productive for as much as ten years longer than a professional with a basic bachelor's degree.

LICENSING AND COMPETENCE

Maintaining professional competence is not entirely voluntary. Almost all of the provincial and territorial professional engineering Acts contain clauses (usually in the Code of Ethics) requiring continued competence. For example, the CCPE Code of Ethics (which is not a legally binding document but serves as a model code supported in principle by the Associations) states that professional engineers shall "keep themselves informed in order to maintain their competence, strive to advance the body of knowledge within which they practise and provide opportunities for the professional development of their subordinates."[1]

This requirement is clarified in the CCPE *Guideline on Continued Competency Assurance of Professional Engineers*, which advises the Associations to adopt a strategy of encouraging, monitoring, and auditing the continued professional development (CPD) of professional engineers and geoscientists.[2] For the Associations, CPD is a win/win arrangement: besides helping individual members remain productive and competitive, CPD is a quality assurance measure that helps protect the public.

Competence Assurance Programs

Most professional engineering Associations (eight of the thirteen) have mandatory competence assurance programs that follow the CCPE model and that require each member to submit an annual report of CPD activities undertaken in the previous year. APEGGA (Alberta) implemented its mandatory CPD plan in 1998. (APEGGA seems to have been the first Association to initiate the CCPE guideline.) British Columbia, Manitoba, Ontario, Quebec, and the Yukon do not have compulsory programs.[3] However, APEGBC (British Columbia) launched a voluntary program in 2001.

CPD Activities

CCPE's model for maintaining professional competence is very simple: it merely requests professional engineers to participate in learning experiences (of any relevant type), record these experiences, and report them to the Association annually. The reporting process and rules vary slightly, depending on the Association. A wide range of activities are usually acceptable as CPD experiences. Note, however, that although any of the following activities may qualify, they would be valued differently according to the weighting scheme specified by the Association:

- **Professional practice.** Practice or employment involving work that falls within the definition of professional engineering or professional geoscience.
- **Formal education.** Attendance (or teaching) at workshops, university courses, in-house instruction, or professional development programs provided by a university, college, engineering society, or internal industry educational program, for which there is typically a test or evaluation of the candidate's attendance and performance and a permanent record of the syllabus and standards.
- **Informal or self-directed education.** Attendance (or teaching) at short courses, seminars, conferences, and so on, for which there is typically a record of the event but no syllabus, formal evaluation process, or record of attendance.
- **Writing publications.** Writing or co-authoring journal papers, patents, monographs, books, codes, standards, and so forth, on engineering or geoscience topics. Note that a heavier weighting would likely be given to peer-reviewed publications.
- **Society participation.** Participation in engineering society activities, such as editing journals, reviewing articles for publication, organizing professional seminars, or presenting papers at conferences.
- **Nontechnical activities.** Nontechnical community service, mentoring a member-in-training, service on public bodies, writing for nontechnical publications, and the like. These activities might be lightly weighted, but they should certainly count toward professional development.

Time Commitment Required

Each Association typically provides (or suggests) both a weighting system for CPD activities and a recommended annual sum to be achieved. The unit of measure may be either the professional development hour (PDH) or the continuing education unit (CEU). The CEU is likely the more common unit, since it is used by the International Association for Continuing Education and Training. The following definitions permit a simple conversion:

- The PDH is equal to 1 contact hour of noncredit learning.
- The CEU is equal to 10 contact hours of noncredit learning.

APEGGA (Alberta) suggests 240 PDHs over three years, with activities in three of the various categories above per year. This implies that 80 PDH/year are recommended. APEGGA has more detailed rules for special cases, and nonpractising members are of course exempted from the CPD program.[4] The Alberta standard is followed in several other jurisdictions.

APEGBC (British Columbia) has a recently established a voluntary program, and recommends an average of only 30 PDHs per year.[5]

Reporting

CPD activities are typically reported when the member's licence is renewed. At that time, the member submits a statement listing the activities completed during the previous year. This process requires personal record-keeping, but the vast majority of engineers are already maintaining their competence, and the reporting process merely confirms this fact. In British Columbia, members may report their CPD activities electronically by connecting to the APEGBC website.

Monitoring and Auditing

Obviously, the Associations must perform some type of monitoring to ensure compliance. A random check of a small sample is proposed by CCPE, to ensure that the self-assessment is fair and honest. In other words, the Association selects a small sample of professional engineers each year and verifies that the records submitted are accurate and properly evaluated.

The CANDU (CANada Deuterium Uranium) nuclear reactor is an impressive Canadian engineering achievement. The newest CANDU installation is the Darlington Nuclear Generating Plant about 70 km east of Toronto. Darlington consists of four CANDU reactors, with a total electric output of 3,524 megawatts, which satisfies about 20 percent of Ontario's electricity requirements. The photo shows the four CANDU reactors (in the four large, windowless buildings behind the cylindrical vacuum building). The long building behind the reactors is the turbine hall, which houses the electrical generators. The heat from the reactors creates steam in the boilers atop the reactors. The steam is piped to the turbine hall, where it drives turbines coupled to the generators. The electricity produced by the generators is fed to transmission lines, which carry the power to the consumers.

Source: Courtesy of Ontario Power Generation.

MAINTAINING COMPETENCE THROUGH PROFESSIONAL PRACTICE

For most professional engineers and geoscientists, the best way to maintain competence is to stay involved in interesting and advanced projects. This method also generates income, so it is an obvious first choice.

However, to remain competitive, professionals must accept personal responsibility for finding and applying new theories, methods, and equipment. This is impossible for engineers working in isolation, so some form of collaboration or communication with colleagues is essential. Fortunately, a wealth of information is available from the engineering societies. As a matter of common sense, all professionals should join at least one society, to benefit from the knowledge, techniques, advice, and information that is distributed so freely. Engineering societies can play a key role in helping you to maintain your professional competence (as discussed in Chapter 16).

MAINTAINING COMPETENCE THROUGH CONTINUING EDUCATION

Formal courses are an excellent avenue for continuing education and can lead to a recognized advanced degree, as discussed in more detail later in this chapter. However, most employed engineers choose to enrol in short courses or seminars on a part-time basis. Several alternatives should be considered:

Short courses. Many short courses are offered by universities, colleges, and engineering societies, usually on a regular basis. If you live near a university or college, you might find such courses are very convenient. If you are not near a university or college, other options are available.

Internet courses. A new educational source is now evolving—the Internet. Some universities are already offering complete programs over the Internet. Engineering societies are also realizing that they have an important role to play in providing advanced or specialized courses. The availability and quality of Internet courses is still rather haphazard. The Internet is rife with fraud. Finding suitable educational programs in the desired specialty may take some determined digging. (A simple Internet search using the key words "online engineering education" will yield over 100,000 hits.) Even so, at some point Internet courses may eventually become the preferred method of professional development.

In-house courses. More and more employers are arranging courses in the workplace. This popular short-course format is supported by universities and engineering societies; some private educational businesses are also providing it. The obvious advantage is that engineers can walk a short distance to a classroom at their place of work, take a class, and be back at work immediately afterwards. The working engineer participates in a live classroom experience, but no time is lost commuting. Useful, specialized topics can be presented by leading authorities, in a short, convenient format.

Some engineering societies and universities now provide in-house courses electronically, using satellite TV or the Internet. Engineers at remote locations can ask questions during class by telephone or through Internet audio. Systems like this have been highly successful at many locations in Canada and the United States. As Internet capacity expands, many more opportunities for electronic education will develop.

The EIC Continuing Education Program

Engineering societies are good sources of engineering information. Most of them are organized by discipline, highly specialized, and oriented toward useful results. The Engineering Institute of Canada (EIC), in particular, has taken on the task of encouraging and coordinating continuing education activities. Engineering societies cannot award degrees, but many of them award CEUs.

Private companies are also offering CPD courses. However, to ensure that a course is of acceptable quality and that the diploma or credit will be recognized, it is important to check the provider's credentials. CEAB-accredited university courses are, of course, universally recognized in Canada. Recently, the EIC has moved into the role of coordinating both university and privately offered courses and validating their quality.

EIC's Continuing Education Program is supported by the Canadian Council of Professional Engineers (CCPE), the Association of Consulting Engineers of Canada (ACEC), and the Canadian Academy for Engineering. EIC is a member of the International Association for Continuing Education and Training (IACET) in Washington, D.C.

EIC does not offer courses; however, it advertises more than thirty course providers and provides links to their websites. More importantly, EIC has verified that the providers deliver professional development programs according to established quality guidelines. These providers are entitled to use the EIC logo as an indication that their standards have been verified by EIC. Course providers are listed on the Internet[6] and include the following:

- **Engineering societies that are members of the EIC.** EIC has nine constituent members, including the Canadian Society for Civil Engineering, Canadian Society for Chemical Engineering, Chemical Institute of Canada, Canadian Geotechnical Society, and Canadian Society for Mechanical Engineering (all discussed in Chapter 16).
- **Universities and other teaching institutions.** Many universities participate, including the Universities of British Columbia, Calgary, Ottawa, Toronto, and Windsor, as well as Concordia, Dalhousie, Ryerson, RMC, and the British Columbia Institute of Technology.
- **Industry associations with expertise in specific areas.** Several industry organizations participate, including the Canadian Nuclear Society, Canadian Dam Association, and Canadian Electricity Forum.
- **Other organizations.** Many other organizations and institutions link to EIC for specific courses or conferences.

The EIC is a useful place to start your search for appropriate CPD courses.

MAINTAINING COMPETENCE THROUGH POSTGRADUATE STUDIES

Professionals who want to specialize, carry out research, or study advanced techniques—especially computer-related topics—should consider postgraduate study. Most people do not think about postgraduate study until it is too late; that is, they delay the decision until they have occupational, financial, or family commitments that take precedence. Since this textbook is intended for recent graduates and senior undergraduates, the best time for you to consider postgraduate studies is probably right now!

Getting Information

If you are thinking about postgraduate study, you should first examine the annual catalogues for the universities that interest you. Catalogues are simple to find and view on the Internet. Admission requirements are clearly specified in the catalogue and vary slightly from university to university. If you have been out of university for three or more years, you may qualify for special consideration as a mature student. Admission requirements may be lower for mature students because they are usually more determined, more goal-oriented, and thus more effective in their postgraduate studies.

Admission Requirements—Master's Degree

To qualify for admission to a master's degree program, an engineer from an accredited undergraduate program must usually have ranked in the upper half of the undergraduate class (B average or better). However, admission requirements may be slightly different for master's programs that require a thesis (research master's) and for those that require mainly courses (course-work master's). A master's degree usually requires a minimum of one academic year, although two years may be required if the student must make up any deficiencies or if the research project is especially challenging or time-consuming. Some master's candidates may be hired as part-time teaching or research assistants. The engineering master's degree may be awarded in applied science (MASc), science (MSc), or engineering (MEng).

Admission Requirements—Doctoral Degree

Candidates for doctoral degrees must usually be ranked in the upper quartile of their undergraduate classes. To enter a doctoral program, applicants must usually have completed research at the master's level in a related area of study. Although some candidates are admitted directly into the doctoral program from the bachelor's degree in Canada, this is not common. The master's degree is usually required, although occasionally students who begin a master's degree and show exceptional ability in the first term or two may be permitted to transfer to the doctoral program, thus achieving the same result.

It is important to contact professors in your area of interest, personally, and to become familiar with their research projects. You should consider several projects (and supervisors) and discuss all aspects of the projects before making a commitment, because your doctoral research topic usually sets the direction for the rest of your career. A minimum of three years beyond the bachelor's degree (or two years beyond the master's degree) is usually stated for the doctoral degree, although the actual time is typically about a year longer than the minimum. One of the reasons for the longer time is that doctoral candidates usually work part-time as lecturers, teaching assistants, or research assistants. The usual engineering doctoral degree is the Doctor of Philosophy (PhD).

Thesis Requirements

Some universities require a thesis to be written for the master's degree, and some allow courses to be taken in place of the master's thesis, but a thesis is universally required for a doctoral degree. The master's thesis is generally written under the guidance of the supervisor and is expected to contribute to the supervisor's research goals and demonstrate the candidate's level of achievement. However, the doctoral candidate is less dependent on the supervisor, and the doctoral thesis is expected to represent an original and independent contribution to the literature of the discipline.

Benefits and Sacrifices

New employees with postgraduate degrees start at higher salaries than employees with bachelor's degrees, of course. However, since postgraduate study delays the time when those higher earnings begin, several years may pass before you break even. In the long run, the greatest benefit is that your professional life may be considerably extended, providing higher salaries in the later years. Advanced engineering degrees almost always pay off, although the break-even point varies from discipline to discipline. The greatest sacrifice in postgraduate study occurs in the early years, when family time is limited and major financial purchases, such as homes and cars, must be delayed. It is important to weigh future benefits against present sacrifices. Although you may be attracted by the challenge of postgraduate study, you should analyze this trade-off for your specific circumstances, before making a commitment. Will the sacrifice pay off for you?

CONCLUSION

Provincial and territorial Associations must be proactive in encouraging continued competence and must not view their role as merely one of setting standards for their colleagues to meet. In particular, APEGGA (Alberta) seems to have been successful in this proactive role and has set a standard for other Associations to follow.

Associations must recognize the many ways in which continuing competence can be achieved and give credit appropriately. The documentation and verification process must be simple and unintrusive, and must recognize that some forms of achievement, experience, and education may be very difficult to document.

In the engineering profession, few organizations are dedicated to assisting individual engineers, and the Associations, with their established communication links, are in a unique position to encourage, advertise, and otherwise facilitate continuing professional development programs. The Associations must assist, motivate, and encourage colleagues, even though these tasks may be at the limit of their regulatory duties.

The engineering faculties of our universities and colleges also have an obligation to provide more evening, part-time, and short courses to satisfy the need for continuing education. Rules must be more flexible for full-time professional employees enrolled in postgraduate courses on a part-time basis.

DISCUSSION TOPICS AND ASSIGNMENTS

(Additional assignments are found in Appendix CD-E on the CD-ROM included with this text.)

1. Use the Internet to examine the requirements and procedures for the CPD program in your province or territory. As an exercise, compare the CPD requirements for your province or territory with the CPD guideline published by the CCPE (reference 2) and with the CPD rules for Alberta (reference 4) and British Columbia (reference 5). (Alberta and B.C. residents may substitute any other province.) What similarities and differences do you observe in these four sets of rules? Which is most demanding, and which is least demanding?

2. Prior to the development of the CCPE model for continuing professional development, it was suggested that competence be maintained by requiring professional engineers to undergo formal examinations every five or ten years. Discuss the effectiveness of the CCPE model, based on personal assessment, with formal examinations. Which is more likely to protect the public? Compare the personal assessment in the CCPE model with the personal assessment in income tax.

3. As mentioned above, most engineers do not consider postgraduate studies until they have occupational, financial, or family commitments that limit their flexibility. Since this text is intended for recent graduates and senior undergraduates, the best time to consider postgraduate studies is probably right now! The following exercises may help you think effectively about this option:
   (a) Using the Internet, examine the postgraduate engineering opportunities of at least three universities of your choice. Choose universities in at least two countries. Make a list of the programs that most inspire your interest and curiosity. Check the admission requirements. On

graduating from your undergraduate program, did you have (or will you likely have) the required minimum academic record for admission? If so, continue with part (b). If not, read the university's rules for makeup courses and mature or probationary students. What further action could you take to qualify for admission?

(b) For your own personal situation, write down the pros and cons of getting a postgraduate degree. Consider such matters as full-time or part-time study and the effects that either of these might have on your earnings (both present and future), your family life, your career satisfaction, and any other factors that you feel are relevant. Prepare a summary on a single sheet, with advantages on the left side of the page and disadvantages on the right. Does the summary confirm that you are following the proper path regarding postgraduate studies?

4. Using the Internet, compare postgraduate programs in your discipline at five universities of your choice. Prepare a summary in chart form, and rate the programs and the universities for the following: type of research in progress, research grants per faculty member, courses provided, size of laboratories, number of books in the library, number of students in undergraduate and postgraduate programs, tuition fees, and any additional factors that you feel are relevant. (The annual *Maclean's* survey of universities may also help.) Using your results, rank the universities in terms of attractiveness to you as a potential postgraduate engineering student.

5. As discussed in an earlier chapter, many engineering graduates are interested in the MBA degree. Assuming that you have completed questions 1 and/or 2 above, consider whether an MBA would be a more appropriate degree for you. Make a summary of your decision on a single sheet, listing the advantages and disadvantages for the various courses of action as you see them.

NOTES

1. Canadian Council of Professional Engineers (CCPE), *Code of Ethics*, Document G03-97, CCPE, Ottawa, ON, clause 4. <www.ccpe.ca/e/guide_guidelines.cfm> (November 28, 2003).

2. Canadian Council of Professional Engineers (CCPE), *Guideline on Continued Competency Assurance of Professional Engineers*, Document G05-96, CCPE, Ottawa, ON <www.ccpe.ca/e/guide_guidelines.cfm> (November 28, 2003).

3. Association of Professional Engineers and Geoscientists of British Columbia (APEGBC), "Member Professional Development: Where to From Here?" *Innovation, The Journal of APEGBC*, vol. 7, no. 5, June 2003, p. 8.

4. The Association of Professional Engineers, Geologists and Geophysicists of Alberta (APEGGA), *Guideline: CONTINUING PROFESSIONAL DEVELOPMENT*, APEGGA, Edmonton, AB <www.apegga.org/members/prof_dev/cpd_bro.htm> (December 2, 2003).

5. Association of Professional Engineers and Geoscientists of BC (APEGBC), *Continuing Professional Development (CPD): A GUIDE FOR MEMBERS*, APEGBC, Vancouver, BC, January 2003 <www.apeg.bc.ca/prodev/guidelines.html> (December 4, 2003).

6. Engineering Institute of Canada (EIC), EIC course providers (website) <www.eic-ici.ca/english/cont_ed/pp1.html> (September 23, 2003).

Chapter 16
Engineering Societies

Engineering societies were formed during the Industrial Revolution to communicate newly discovered technical information; they are even more important in the twenty-first century. The original societies now include many innovative new branches, disciplines, and specialties, and new societies are formed every year. The term "engineering society" (or "society") is used here partly for historical reasons, but also to avoid confusion with "learned societies," which serve the same purpose in the arts and humanities.

Engineering societies play a key role for professionals, by linking colleagues with similar interests. New theories, techniques, and equipment usually appear first in conferences or publications sponsored by engineering societies. This chapter reviews the role of societies and explains why every professional engineer and geoscientist should be a member of at least one society.

THE PURPOSE OF ENGINEERING SOCIETIES

The principal goals of engineering societies—goals that have not changed in over 150 years—are to encourage research into new theories and methods, to collect and classify this new information, and to disseminate it to members so that it can be put to good use. Engineering societies are the most important publishers of new research results in conference proceedings and monographs; they also lead the way in developing new design standards. The world's libraries are bulging with useful publications from engineering societies, and more such publications are available through the Internet. Everyone has benefited from this free exchange of information; the benefit to society and to professional practice has been immense.

The purpose of engineering societies is totally different from that of the provincial and territorial Associations, yet many people confuse the two, perhaps because their activities sometimes overlap. Two examples follow:

- British engineering societies perform a sort of voluntary regulation by awarding the Chartered Engineer status. However, this status does not permit the holder to practise in Canada, where all professional engineers and geoscientists must be licensed by a provincial or territorial Association.

- Almost every American (and international) engineering society publishes a Code of Ethics; however, while infractions of the code may lead to expulsion from the society, such actions are very rare. In Canada, the Associations are far more effective in enforcing their Codes of Ethics and in regulating professional conduct.

THE EVOLUTION OF ENGINEERING SOCIETIES

The first technical society for engineers was the Institute of Civil Engineers, established in Britain in 1818. It was followed thirty years later by the Institution of Mechanical Engineers. Shortly after that, other societies were established for naval architects and for gas, electrical, municipal, heating, and ventilating engineers.[1]

In the United States the first engineering society was the American Society of Civil Engineers, founded in 1852. Many others were established in the 1800s—the American Society of Mechanical Engineers (1880), the American Institute of Electrical Engineers (1884), and the American Society of Heating and Ventilating Engineers (1894), to mention only a few.

In Canada the first society to be formed was the Engineering Society of the University of Toronto, in 1885. The "Society was, indeed, a 'learned society' and published and disseminated technical information . . . in addition to looking after the University undergraduates in engineering."[2] In 1882 the Canadian Institute of Surveys was formed, followed by the Engineering Institute of Canada in 1887 (although the EIC name was not adopted until 1917), the Canadian Institute of Mining and Metallurgy (1898), the Canadian Forestry Association (1900), and others.

Some Canadian engineering societies were established only recently, after engineers came to realize that one of the oldest and most prestigious societies, the Engineering Institute of Canada (EIC), could no longer maintain the diverse specialties of engineering within a single organization. For this reason, several constituent societies were established. The Engineering Institute of Canada is now a federation of nine member societies.

Agreements signed between EIC and the Canadian Council of Professional Engineers (CCPE)—which acts on behalf of the provincial Associations when requested—clearly state the roles and duties of the organizations: the provincial Associations are responsible for regulating engineering, and the role of the EIC (and its member societies) is to collect, organize, and disseminate engineering, scientific, and technical information.[3] EIC has taken on this role aggressively, and now plays a vital role in certifying and coordinating courses for continued professional development. This service is especially useful to engineers and geoscientists in those provinces which have mandatory professional development requirements (as discussed in the previous chapter).

CHOOSING AN ENGINEERING SOCIETY

Most professionals join a society that promotes their discipline. Societies often sponsor undergraduate student chapters, so you likely know the principal society in your discipline. New or more specialized societies may not be well known, but they are easily found by asking a colleague or senior engineer or through an Internet search. Every society has a website. For example, a simple search of Yahoo's Directory of Engineering Organizations reveals several hundred engineering societies, listed under sixteen disciplinary headings, from aeronautical to systems engineering.[4]

The *International Directory of Engineering Societies*, published annually, summarizes the activities of hundreds of engineering societies throughout the world.[5] This directory, which is available in most university libraries, lists the purpose, membership, address, dues, and much other information for each society. At least one society exists for every one of the following:

- *Discipline:* agricultural, chemical, civil, computer, electrical, environmental, geological, geotechnical, manufacturing, marine, mechanical, mining, nuclear, petroleum, systems design, etc.
- *Product:* abrasives, automation, automotive, computers, concrete, explosives, gears, illumination and lighting, lasers, machinery, paper, plastics, powder metallurgy, rubber, robotics, steel, textiles, vehicles, welding, wire, etc.
- *Facility:* electric power plants, highways and bridges, hospitals, radio and television, railways, shipbuilding, utilities, etc.
- *Innovation:* design, inventions, human-powered vehicles, etc.
- *Evaluation:* cost engineering, nondestructive testing, instrumentation, measurement, etc.
- *Function:* consulting, engineering education, engineering management, human resources, quality control, etc.
- *Environment:* conservation, environmental impact, glaciology, mapping, sustainable development, occupational health, remote sensing, etc.
- *Language:* English, French, German, Japanese, Spanish, Portuguese, etc., and
- *Geographical areas.*[6]

A simple Internet search will find the society that specializes in your interests. A few well-known Canadian and American societies are listed by engineering discipline in Table 16.1.

The older, larger, and better established American and British societies have greater storehouses of technical information and are usually able to offer more services to members. The smaller, newer Canadian societies are in the process of building up their reputations and memberships. However, the Canadian engineering societies are more effective in dealing with problems that are typically Canadian, and in promoting Canadian interests.

Regardless of your engineering discipline, industry, or personal interests, there is an engineering society that needs you as a member and that can help you professionally. Take advantage of this valuable source of useful knowledge.

TABLE 16.1 — A Brief List of Engineering Societies

Acronym	Society	Web address
Canadian Societies		
EIC	Engineering Institute of Canada	www.eic-ici.ca/
CGS	Canadian Geotechnical Society	www.cgs.ca/
CSCE	Canadian Society for Civil Engineering	www.csce.ca/
CSME	Canadian Society for Mechanical Engineering	www.csme.ca/
CSChE	Canadian Society for Chemical Engineering	www.chemeng.ca/
CSEM	Canadian Society for Engineering Management	www.csem-scgi.ca/
IEEE—Canada	Institute of Electrical and Electronic Engineers—Canada (formerly CSECE)	www.ieee.ca/
CNS	Canadian Nuclear Society	www.cns-snc/
CIM	Canadian Institute of Mining, Metallurgy and Petroleum	www.cim.org/
CSAE	Canadian Society for Engineering in Agricultural, Food and Biological Systems	www.csae-scgr.ca/
CMBES	Canadian Medical and Biological Engineering Society	www.cmbes.ca/
CIC	Chemical Institute of Canada	www.cheminst.ca/
MTS	Canadian Maritime Section of the Marine Technology Society	www.mtsociety.org/
American and International Societies		
ASCE	American Society of Civil Engineers	www.asce.org/
AIME	American Institute of Mining, Metallurgical and Petroleum Engineers	www.aimeny.org/
ASME	American Society of Mechanical Engineers	www.asme.org/
IEEE	Institute of Electrical and Electronic Engineers	www.ieee.org/
AIChE	American Institute of Chemical Engineers	www.aiche.org/
ASAE	American Society of Agricultural Engineers	www.asae.org/
SAE	Society of Automotive Engineers	www.sae.org/
ASEE	American Society for Engineering Education	www.asee.org/
AIPG	American Institute of Professional Geologists	www.aipg.org/

Notes: Most societies permit membership applications to be submitted electronically. Because of the value and usefulness of engineering societies, membership dues are deductible from personal income (for practising professionals, under Canadian income tax laws).

The Engineering Institute of Canada (EIC) is a federation of nine member societies: CGS, CSCE, CSME, CSChE, CSEM, IEEE-Canada, CNS, MTS, and LMO (LMO is a charitable organization of life members of EIC). EIC is also now a leading provider and coordinating body for continued professional development.

HONORARY SOCIETY—THE CANADIAN ACADEMY OF ENGINEERING

The Canadian Academy of Engineering is Canada's highest honorary engineering society. The Academy, located in Ottawa, is an independent, self-governing, nonprofit organization established in 1987 to serve the nation in engineering matters. The Fellows of the Academy are professional engineers from all disciplines, elected on the basis of distinguished service and contributions to society, to the country, and to the profession. The total number of Fellows may not exceed 250 at any one time.

The mission of the Academy is to enhance, through science and engineering principles, well-being and the creation of wealth in Canada. The Academy is self-financing and does not receive grants from government, although it sometimes conducts critical studies and surveys on a variety of engineering topics, on a contract basis. The Fellows of the Academy provide independent advice, based on their experience and expert knowledge, gained as practising members of the engineering profession in Canada.[7]

STUDENT ENGINEERING SOCIETIES

Canadian Federation of Engineering Students

The Canadian Federation of Engineering Students (CFES) traces its history back to a tumultuous inaugural congress of engineering students at McGill University in February 1969. CFES is organized into four regions: Western Canada, Ontario, Quebec, and the Atlantic region. The federation was established to improve communication between university students across Canada. Another CFES goal is to help engineering students grow culturally, morally, intellectually, academically, and economically. CFES also serves as a liaison between Canadian engineering students and the Canadian Council of Professional Engineers (CCPE), as well as the National Council of Deans of Engineering and Applied Science (NCDEAS). CFES organizes many student projects, such as the Canadian Engineering Competition. It also provides interactive on-line information about job and graduate school opportunities, and publishes *Project Magazine*, a national magazine for engineering students.[8]

Student Programs Provided by the Associations

In recent years, several provincial and territorial Associations have introduced student programs. These programs are intended to improve communication with the next generation of professional engineers and geoscientists. The Associations are to be applauded for this initiative. These programs provide a smooth transition from student status, through the internship phase (as EIT, MIT, GIT, etc.), to that of licensed engineer or geoscientist. Some provinces have also established links to faculty members in engineering universities and created networks of mentors to assist the students. If you are an engineering student, contact your Association (see the Web address in Appendix A) and inquire whether your Association has a student program.

The Canadair CL-215 Waterbomber is the only aircraft in the world specifically intended to fight forest fires. The CL-215 was designed in the 1960s and is capable of scooping 5670 kg (12,500 lb) of water from a lake, without stopping, and dumping it on a forest fire. A newer, turboprop version, the CL-415, first flew in 1993.

Source: CP Photo/AP Photo/Franco Arena.

CHARITABLE ENGINEERING SOCIETIES

Engineers Without Borders—*Ingénieurs Sans Frontières*

Engineers Without Borders (EWB), established in January 2000, is a registered charity dedicated to international development. The mission of EWB is to promote human development through access to technology. In other words, EWB seeks to narrow the technology gap between the developed world and the developing one by promoting the involvement of engineering students in development issues. EWB focuses on the role of technology in fundamental areas—water, food availability, health, energy, and communications—and tries to address basic problems in developing communities. EWB does not bring technology from the West; rather, it encourages simple technology, developed with local input and innovation. Such solutions are longer-lasting and promote sustainability and self-sufficiency. EWB is a Canadian initiative, and EWB chapters have been started at most Canadian universities with engineering programs. The group's newsletters document a startling array of successful projects. The achievements of this group, in the few years since its inauguration, speak well for the initiative and idealism of Canada's engineering students.[9]

Registered Engineers for Disaster Relief (RedR) Canada

Registered Engineers for Disaster Relief (RedR) Canada is a recently established Canadian branch of an international organization that relieves suffering in disasters by selecting and training competent personnel and providing them to assist humanitarian agencies. The founding members of RedR Canada are ACEC, CCPE, EIC, and the Canadian Academy of Engineering. In January 2001, ACEC signed a Memorandum of Understanding with RedR International that led to the founding of RedR Canada as a nonprofit organization.

RedR members provide technical assistance to restore roads, bridges, water supplies, and communication systems. They also assist, after a disaster, in managing waste, protecting the environment, and managing financial, material, and human resources. RedR Canada is an independent organization; however, all national offices work together as members of RedR International, based in Geneva.[10]

EIC Life Members Organization

The EIC's Life Members Organization (LMO) was incorporated as a charitable organization on May 28, 1967. This group of retired engineers promotes the advancement of science and engineering in Canada and supports benevolent causes. The LMO engages in a wide range of charitable works, including research projects and awards for science fairs. It also recruits skilled volunteer engineers and technicians to assist people with disabilities.[11]

THE IRON RING AND THE RITUAL OF THE CALLING OF AN ENGINEER

The Corporation of the Seven Wardens is a little-known group that has performed a vital role in Canadian engineering for many decades. The wardens arrange the Iron Ring ceremonies held on most campuses just before graduation day. This ceremony is a milestone in the engineer's education. The following account of the Iron Ring was written by J.B. Carruthers, P.Eng., and is reproduced with permission:

Most engineers in Canada wear the Iron Ring and have solemnly obligated themselves to an ethical and diligent professional career through the Ritual of the Calling of an Engineer. This Ritual is the result of efforts by the Corporation of the Seven Wardens, started in 1922 when a group of prominent engineers met in Montreal to discuss a concern for the general guidance and solidarity of the profession. These seven prominent engineers formed the nucleus of an organization whose object would be to bind all members of the engineering profession in Canada more closely together and to imbue them with their responsibility towards society.

They enlisted the services of the late Rudyard Kipling, who developed an appropriate Ritual and the symbolic Iron Ring. The purpose was outlined by Rudyard Kipling in the following words:

"The Ritual of the Calling of an Engineer has been instituted with the simple end of directing the young engineer towards a consciousness of his[/her] profession and its significance, and indicating to the older engineer his[/her] responsibilities in receiving, welcoming and supporting the young engineers in their beginnings."

The Ritual has been copyrighted in Canada and the United States, and the Iron Ring has been registered. The Corporation of the Seven Wardens is entrusted with the responsibility of administering and maintaining the Ritual, which it does through a system of separate groups, called Camps, across Canada. There are presently 20 such Camps.

The Corporation of the Seven Wardens is not a "secret society." Its rules of governance, however, do not permit any publicity about its activities and they specify that Ceremonies are not to be held in the presence of the general public.

The original seven senior engineers who met in Montreal in 1922 were, as it happens, all past presidents of the Engineering Institute of Canada. There is, however, no direct connection between the Engineering Institute of Canada and the Corporation of the Seven Wardens.

The wearing of the Iron Ring, or the taking of the obligation, does not imply that an individual has gained legal acceptance or qualification as an engineer. This can only be granted by the provincial bodies so appointed and, as a result, it should also be mentioned that the Corporation of the Seven Wardens has no direct connection with any provincial association or order.

The obligation ceremonies for graduating students are held in cities where Camps are located, and for convenience, in some cases, on the university campus itself. Such ceremonies must not be misconstrued as being an extension of the engineering curriculum. The Iron Ring does not replace the diploma granted by the University or the School of Engineering nor is it an overt sign of having successfully passed the institution's examinations.

The purpose of the Corporation of the Seven Wardens and the Ritual is to provide an opportunity for men and women to obligate themselves to the standard of ethics and diligent practice required by those in our profession. This opportunity is available to any who wish to avail themselves to it, whether they be new graduates or senior engineers. The Ritual of the Calling of an Engineer is, of course, attended by all those who wish to be obligated, along with invited senior engineers and, when space permits, immediate family members. A complete explanation of the Ritual, its obligations and history is given to every man and woman before the ceremony so that they may decide in advance whether or not they wish to take part in the spirit intended. A few people, for one reason or another, have chosen to refrain from being obligated, and so cannot rightfully wear the Iron Ring. The Corporation of the Seven Wardens feels that this in no way detracts from their right to practise in the profession and further feels that the obligation should continue to be a matter of personal choice, taken only by those who wish to take part in the serious and sincere manner intended.[12]

THE EARTH SCIENCE RING—A RITUAL FOR GEOSCIENTISTS

The awarding of the Earth Science Ring is a ceremony comparable with the Iron Ring ceremony and, in fact, adopts some of the format and wording of the Iron Ring ceremony. The following description of the Earth Science Ring and award ceremony was written by Philippe Erdmer, P.Geol., and Edward S. Krebes, P.Geoph., and is reproduced with permission:

> The Earth science ring ceremony, a ritual of welcome into the profession of newly qualified geologists and geophysicists by senior practising Earth scientists, started in Alberta in 1975. This yearly tradition for the university geoscience graduating classes at Edmonton and Calgary has spread to other provinces and jurisdictions in Canada. The ceremony carries many of the same passages written by Kipling for the Engineers' iron ring ceremony and symbolizes the commitment and responsibility that come with wearing the title of a professional.

> Like the engineer's iron ring, the Earth science ring's simplicity and strength bear witness to the calling of the geologist and geophysicist. The ring is made of silver and marked with the crossed hammer of geology and with the seismic trace of geophysics— signifying both the immediate and the remote searching out of Nature's knowledge. Without beginning and without end, it also represents for those who wear it the continuous interplay of ideas and of material realities.

> The ceremony includes a charge (speech) by senior Earth scientists and an obligation (pledge) taken by the group of newly graduated geologists and geophysicists. The charge reads in part: "We tell you here that you will encounter no difficulty, doubt, danger, defeat, humiliation or triumph in your career which has not already fallen to the lot of others in your calling. . . ." The obligation includes the words: "I will not pass . . . false

information or too casual interpretations in my work as an Earth scientist. My time I will not refuse, my thought I will not grudge; my care I will not deny towards the honour, use, stability and perfection of any project to which I may be called to set my hand. . . . My reputation in my calling I will guard honourably. . . . I will strive my uttermost against professional jealousy and the belittling of my co-workers in any field of their labour."

On a lighter note, following the obligation, new ring bearers are reminded that "From now on, we surrender to you what lies under the earth, and the tools to interpret or misinterpret. Sooner or later, you will drill the holes that bring no return, lose the vein in which lie extra riches and reputation, misinterpret the signal from the depths. This will equally baffle, bewilder and break your heart to your professional and personal education."

Receiving an Earth science ring is neither a prerequisite nor a later condition of professional membership with APEGGA. Although there is no obligation to obtain or wear a ring, it is significant that almost no one in the graduating classes willingly misses the ceremony. In addition, the ceremony is not strictly a graduation event, as it has occasionally included already practising geologists and geophysicists in Alberta who express the wish to receive a ring. Like the iron ring of the obligated engineer, the Earth science ring is a symbol of values that lie at the core of our individual beings and of the trust placed in us by society.[13]

DISCUSSION TOPICS AND ASSIGNMENTS

(Additional assignments are found in Appendix CD-E on the CD-ROM included with this text.)

1. Canadian engineers and geoscientists often debate whether to join the newer Canadian engineering societies, or whether to join foreign-based societies that have established journals, committees, and conferences. Write a brief summary debating the pros and cons of the two alternatives. Does Canada need distinct engineering societies? Discuss the implications for Canadian sovereignty if Canadian engineering societies should be absorbed into the larger, American-dominated societies. Are these societies truly nonpolitical, or do national interests influence their policies and the content of journals and transactions? Are there uniquely Canadian conditions that would justify uniquely Canadian societies? Give examples. Write your summary as if you were sending it to a federal politician who has very little engineering knowledge.

2. Engineering and geoscience students must study hard to succeed. How can they find time to devote to professional activities, such as student government, CFES, the Association's student participation program, or charitable organizations such as Engineers Without Borders? Discuss whether the CEAB (the Canadian Engineering Accreditation Board) should permit a portion of such activities to be considered as complementary (nonengineering) studies, which are required for engineering accreditation. How would these credits be awarded? What weighting would you give to the various activities?

3. If you are a student, contact your Association (see the Web address in Appendix A) and inquire whether your Association has a student program. If it does, find out the costs and benefits of joining the program.

NOTES

1. L.C. Sentance, "History and Development of Technical and Professional Societies," *Engineering Digest,* vol. 18, no. 7, July 1972, pp. 73–74.
2. Ibid.
3. EIC, "Canadian Engineers Close the Ring," *Engineering Journal,* vol. 60, no. 1, January 1977, pp. 15–19.
4. Yahoo Directory of Engineering Organizations, <dir.yahoo.com/Science/Engineering/Organizations/> (September 23, 2003).
5. American Association of Engineering Societies (AAES), *International Directory of Engineering Societies and Related Organizations* (Washington, DC: AAES, 1995).
6. Ibid.
7. Canadian Academy of Engineering, *Mission Statement,* Canadian Academy of Engineering, 130 Albert St., Suite 1414, Ottawa, ON, KIP 5G4 <www.acad-eng-gen.ca/> (September 23, 2003).
8. Canadian Federation of Engineering Students (CFES) <www.cfes.ca/> (December 1, 2003).
9. Engineers Without Borders (EWB) <www.ewb-isf.org/> (December 1, 2003).
10. Registered Engineers for Disaster Relief (RedR) Canada <www.redr.ca/> (March 22, 2003).
11. Engineering Institute of Canada, *Life Members Organization* <www.eic-ici.ca/english/tour/lmo2.html> (December 1, 2003).
12. J.B. Carruthers, P.Eng., "The Ritual of the Calling of an Engineer," *Project Magazine: The National Magazine for Engineering Students,* April 1985, p. 19. Reproduced with permission.
13. P. Erdmer, P.Geol., and E.S. Krebes, P.Geoph," "The Earth Science Ring: Made in Alberta," APEGGA <www.apegga.ca/aboutapegga/earth_ring.html> (September 28, 2003). Reproduced with permission.

Part Five
Exam Preparation

Chapter 17
Writing the Professional Practice Exam

All provincial and territorial Associations of professional engineers (including the *Ordre des ingénieurs du Québec*) require applicants to write a Professional Practice Examination (PPE). The exam ensures that the applicant is familiar with the principles of engineering practice, has a general understanding of Canadian law as it applies to engineers, and understands the laws and regulations governing the practice of engineering in the applicant's province or territory of practice.

Applicants for the geoscience licence must also write the Professional Practice Examination, in those provinces and territories where geoscientists are licensed. (Prince Edward Island and Yukon do not yet have such legislation.) In provinces such as Quebec and Nova Scotia, where legislation regulating geoscience has only recently been promulgated, the PPE (and the procedures for administrating it) are still under development as this text goes to press.

This chapter describes the PPE and shows typical solved questions from previous exams. It should be especially informative and useful to people writing the exam.

THE PROFESSIONAL PRACTICE EXAM SYLLABUS

The Canadian Council of Professional Engineers (CCPE) has published a general syllabus for the exam, which is accepted in principle by most Associations. The CCPE syllabus includes both professionalism and engineering law, defined as follows:

Professionalism

Topics to be covered by the examination in the general areas of engineering practice and ethics should include, but need not be limited to: the definition of professional engineering; the role of the association and the responsibilities associated with self-governance; professional accountability, conduct and ethics, the professional engineer's responsibility to the public and duty to report illegal or unethical engineering practice; the ethical use of the engineer's seal; continuing competence; and the social and environmental impacts of engineering on society.

Engineering Law

Topics to be covered by the examination in the area of the law as it relates to engineers and to the practice of engineering should include, but need not be limited to: the basic structure of the Canadian legal system, common law, Quebec civil law, statute law and the provincial court system; tort law, liability and liability issues; business organizations; contract law, specifications and tendering, discharge and breach of contract, bonding, estoppel and construction lien legislation; intellectual property, patents, technology transfer, copyrights, trademarks, industrial designs and trade secret; fiduciary responsibility; professional advertising, unfair competition and merchandising rights; dispute resolution, negotiation and arbitration; litigation and the engineer as expert witness; the Canadian Human Rights Act; environmental legislation; worker's compensation and occupational health and safety legislation.[1]

The format of the PPE varies. Some provinces (notably Ontario) have a three-hour, essay-type written examination, typically containing eight to ten questions, half of these devoted to professionalism (practice and ethics), and half to Canadian engineering law. In Ontario, candidates are usually provided copies of the Code of Ethics and the definition of professional misconduct (Ontario Regulation 941, sections 72 and 77), but all candidates should read the exam instructions to find out what aids are permitted during the exam.

Since 1998 the PPE developed by the Association of Professional Engineers, Geologists and Geophysicists of Alberta (APEGGA) has served as a pilot form of the National PPE. It has been adopted in many other jurisdictions, from the Northwest Territories to Manitoba to Newfoundland. The National PPE consists of a set of one hundred multiple-choice questions, administered in a two-hour, closed-book format. All questions are common to engineering, geology, geophysics, and geoscience. The pass mark is 65 percent, with no penalty for wrong answers. The grade is final and cannot be appealed. The major subject areas on the exam are as follows:

A. Professionalism (30%)

1) Definition and interpretation of professional status

2) The role and responsibilities of a professional in society

3) The role and responsibilities of a professional to management

4) Professional conduct, ethical standards and codes

5) Safety and loss management—the professional's duties

6) Environmental responsibilities

B. Professional Practice (20%)

1) Professional accountability for work, workplace issues, job responsibilities and standards of practice

2) Continuing competence

3) Quality management and standards of skill in practice

4) Business practices as a professional

5) Insurance and risk management

6) Professional and technical societies

7) Non-statutory standards and codes of practice

C. Regulatory Authority Requirements (9%)

1) Safety and loss management—regulatory aspects

2) Environmental regulations

3) Occupational health and safety

4) Workers compensation

5) Other statutory standards of practice

D. Law and Legal Concepts (25%)

1) Canadian legal system and international considerations, basics of business organizations

2) Contract Law—elements, principles, types, discharge, breach, interpretation etc.

3) Tort Law—Elements, application of principles, interpretation, liabilities of various kinds

4) Intellectual Property—patents, trademarks, software issues, copyright

5) Arbitration and Alternative Dispute Resolution (ADR)

6) Expert Witness

E. The Act (16%)

1) Definitions of the professions and scopes of practice

2) Structure and functions of a Provincial Association

3) Regulations and By-Laws

4) Registration

5) Discipline and enforcement

6) Use of seals and stamps[2]

British Columbia has a slightly different approach: the PPE administered by APEGBC is a three-hour examination consisting of a two-hour multiple-choice section and a one-hour essay question.[3]

Therefore, the first step in preparing for the PPE is to determine the format of the exam in your province or territory. Are any aids or references permitted or provided? The PPE typically will include general-knowledge questions selected from the topics above, but will also include some applied ethics and

(A) The CN Tower, built in 1976, is the word's tallest building and freestanding structure and an impressive icon on the Toronto skyline. The tower rises to a height of 553 metres (1,815 feet) and is the centre of telecommunications for Toronto. Antennae on the tower broadcast signals for many Canadian radio, television, and communication companies. The CN Tower is also one of Toronto's foremost entertainment and tourist attractions. Its construction required innovative techniques. Its concrete core and three curved supporting arms were formed using a novel "slip-form" supported by hydraulic jacks, which moved upwards, gradually decreasing in size, to create the tower's elegantly tapered shape. In 1995, the CN Tower was declared one of the Seven Wonders of the Modern World by the American Society of Civil Engineers (ASCE). (B) The design of the CN Tower required extensive analysis and testing. The photo shows a 1:500 aero-static scale model of the CN Tower being tested in the boundary layer wind tunnel at the University of Western Ontario.

Source: (Left) Corel "Toronto" CD #462019; (Right) Courtesy of The Boundary Layer Wind Tunnel Laboratory, University of Western Ontario. © Canada Lands Company (CLC) Limited. Reprinted with permission.

law questions. This textbook is intended to help you with the general-knowledge topics and the applied ethics questions, but other references should be consulted for the law and legal concepts in the PPE syllabus.

The following review of the problem-solving process should be useful for readers preparing to write the PPE. The process is especially useful for exam questions in applied ethics.

THE EGAD! STRATEGY FOR ETHICS QUESTIONS

The PPE typically tests your knowledge of ethics (and the Code of Ethics, specifically) by describing a hypothetical situation and asking you to suggest the proper course of action. The strategy for solving ethical problems, described in Chapter 6, has been simplified for easier use on the exam. This six-step strategy has been renamed the "EGAD" method and is similar to a solution method taught to law students.[4] It is remembered easily by these three words:

READ—EGAD!—WRITE

The term EGAD! (an old English exclamation of surprise) is an acronym or mnemonic for the four key steps in the solution strategy: Ethical issues, Generation of alternatives, Analysis, and Decision. These are explained as follows:

Step 1: READ—*Read the problem thoroughly and gather information*

Exam questions contain less information than real problems, so read each question thoroughly! Highlight or underline key facts, but do not copy the question into the exam book, since this wastes valuable time. Ask yourself the typical reporter's questions:

- *Who is involved?*

- *What type of harm or damage has occurred?* (or what type may potentially occur?)

- *How has this harm occurred?* (or how may it potentially occur?)

Step 2: E—ETHICAL ISSUES—*Identify the basic ethical issues*

The exam question may state the ethical problem directly. For example: "Has Mr. Smith broken the Code of Ethics?" However, some exam questions may say simply: "Explain and discuss this case." You must then imagine which ethical issues should apply, and if possible compare the similarities (and differences) of the case with previous cases. If the ethical issue is not obvious, then ask yourself: "What, exactly, is wrong in this situation? Do any actions contravene the law or the Association's Code of Ethics? What is unfair?" Once you identify the ethical problem, the proper course of action may be obvious. If so, write down your answer. However, in some problems you may have to suggest or imagine (or "generate") a proper course of action.

Step 3: G—GENERATION—*Generate (or suggest) possible courses of action*

Some exam questions simply ask: "What should you do?" You must suggest the proper action. This step requires creative thought, so it may be difficult. The creative techniques that are commonly used in engineering (such as brainstorming techniques) may be useful. You might suggest a compromise, or a totally new idea. You might also try to imagine yourself as one of the participants. What would you do in this situation? The goal is to find a new course of action that is ethically correct and that has a minimum of nasty side effects.

Sometimes the exam question involves an ethical dilemma with only two alternatives, both of which are unacceptable. In this case, you should try to suggest a third possibility that is better. In your answer, you might list the reasons why the two choices are undesirable; do not, however, assume that you must choose one of them.

Step 4: A—ANALYSIS—*Analyze the possible courses of action*

When two or more courses of action have been suggested (in the previous step), you must examine each one, to find the best. You want the simplest course of action that solves the problem without nasty side effects. You should test each course of action as follows:

- Is this course of action legal?

- Is it consistent with human rights, employment standards, and design standards?

- Does it obey the Code of Ethics and maintain the ideals of the profession?

- Can the solution be published and stand the scrutiny of your colleagues and the public?

- What benefits will result, and are they equally distributed? (Utilitarianism)

- Can this solution be applied to everyone, uniformly? (Kant)

- Does this solution respect the rights of all participants? (Locke)

- Does the solution develop or support moral virtues and/or is it a golden mean between unacceptable extremes? (Aristotle)

Step 5: D—DECISION—*Make a logical decision*

The previous step should yield at least one acceptable course of action. However, in some cases all of the alternatives may be unacceptable, or the alternatives may be so equally balanced that none is clearly superior. In this case, you may need to review the above steps. If you still face a dilemma, with two equal choices, you must decide which is better (or least negative). If the choices are equally balanced, select the course of action that does not yield a benefit to the person making the decision. This will help you to defend the decision.

Step 6: WRITE—*Write a professional summary of your answer*

Finally, you must explain your answer clearly, logically, and neatly. You can't afford to waste time, so you must practise writing good answers. Start by stating your decision, which answers the question asked by the examiner. Then explain why you came to that conclusion. It is very important to cite subsection numbers (from the Code of Ethics or regulations) for the ethics questions, just as you would cite past cases (or precedents) for the law questions. Do not copy clauses from the Code of Ethics; identify them by number. It is also important to write neatly and legibly. The examiners greatly appreciate this courtesy.

An Important Hint

An examiner who sets one of the essay-type PPEs says that candidates using the EGAD! method spend too much time on the EGA steps (ethical issues, generation of alternatives, and analysis) and not enough time on step D (explaining the decision). These short or incomplete answers get lower grades. Remember that the EGAD! process is intended merely to help you think about the problem in an orderly way. Do not write out all of the steps; your exam grade is based only on your written decision (step D), so explain it thoroughly.[5]

SOLVED QUESTIONS FROM PREVIOUS EXAMINATIONS

(Additional exam questions are in Appendix CD-E on the CD-ROM disk included with this text.)

This section contains twenty-five examination questions selected from previous PPEs in several provinces.[6] Readers are encouraged to attempt all of the following questions, regardless of their province of residence. Ethics concepts are universal, problems are similar, and answers will differ only slightly from province to province. The questions have been chosen to show the various exam formats: essay-type, short-answer, multiple-choice, and true–false. Solutions are suggested for all questions. An asterisk (*) indicates where the specific clause number(s) from your provincial Act (or Code of Ethics) should appear, if appropriate.

Essay-Type Examination Questions

In the essay-type examination used in Ontario, the applicant is typically asked to answer four or five ethics questions and is permitted about twenty minutes per question. A copy of the Code of Ethics may be provided for reference during the exam.

1. Professional Engineer A takes a job with a manufacturing company and almost immediately thereafter is given responsibility for preparing the draft of a bid for replacement turbine runners for a power corporation. While working on preparing the bid for the manufacturing company, Engineer A, as president and shareholder of his own company, which he runs privately from his home, writes to the power corporation requesting permission to submit a tender on the same project. A few days later, and while continuing to work on the bid for the manufacturing company, he receives word from the power corporation that a bid from his company would be considered. The day after learning this, he resigns his position with the manufacturing company and proceeds to finalize and submit a bid on behalf of his own company.

   Discuss Engineer A's actions from an ethical point of view.

   Suggested answer. Engineer A is clearly unethical in his actions. By running a private company in competition with his employer, he is not being fair or loyal to his employer, as required by the Code of Ethics. (*) He has taken advantage of inside information, betrayed the trust of his employer, and yielded to a conflict of interest. If his private company was unknown to his employer, then he has failed to disclose his conflict of interest as required by the Code of Ethics. (*) By his actions, he has failed to show the necessary devotion to professional integrity required by the Code of Ethics.

In his defence, it could be said that since he resigned before actually signing the contract, he did not compete with his employer, but this would be a technical point; the serious conflicts of interest occurred during the bid preparation stage. The only positive statement in his defence is that he provided an additional option for the power corporation in its selection of bids. Engineer A has exposed himself to the possibility of serious disciplinary action for conflict of interest, under the provincial or territorial Act.

2. You are a Professional Engineer with XYZ Consulting Engineers. You have become aware that your firm subcontracts nearly all the work associated with the set-up, printing, and publishing of reports, including artwork and editing. Your wife has some training along this line and, now that your children are at school, is considering going back into business. You decide to form a company to enter this line of business together with your neighbours, another couple. Your wife will be the president, using her maiden name, and you and your neighbours will be directors.

   Since you see opportunities for subcontract work from your company, you reason that there must be similar opportunities with other consulting firms. You are aware of the existing competition and the rates they charge for services and see this as an attractive sideline business. Can you do this ethically, and if so, what steps must you take?

   Suggested answer. You can do this ethically, but there is a potentially serious conflict of interest unless you scrupulously follow the Code of Ethics for your province or territory. You can undertake the sideline business provided it does not interfere with your regular employment and provided your employer is fully informed, as required by the Code of Ethics. (*) Your wife, of course, is free to use any legal name in her business affairs; however, if the sole reason for using her maiden name is to conceal your participation in the company's ownership and operations, your cooperation could be considered unethical. If you tell your employer all the details about your wife's company, then you have disclosed your conflict of interest, and no ethical problem exists. (*) However, your wife's other clients may worry about a possible loss of confidentiality. A publishing company often receives confidential reports and must not reveal the contents of those reports to others. Therefore, your wife must not allow you to see sensitive engineering information submitted to her by other clients for publishing. Obviously, confidentiality must be guaranteed to her other clients, or conflicts of interest may arise in the future.

3. Brenda MacDonald, a Professional Engineer, is manager of a chemical plant in a northern Canadian town. Early this summer she noticed that the plant was creating slightly more water pollution in the lake into

which its waste line drains than is legally permitted. If she contacts the province's environment ministry and reveals the problem, the result will be a considerable amount of unfavourable publicity for the plant. The publicity will also hurt the lakeside town's resort business and may scare the community. Apart from that, solving the problem will cost her company well over $100,000. If she tells no one, it is unlikely that outsiders will discover the problem, because the violation poses no danger whatever to people. At the most, it will endanger a small number of fish.

Should MacDonald reveal the problem despite the cost to her company, or should she consider the problem little more than a technicality and disregard it? Discuss the ethical considerations affecting her decision.

Suggested answer. MacDonald must, legally and ethically, take action to remedy this situation. She is obligated under the Code of Ethics to consider the public welfare as paramount. (*) The legal limit for pollution has been exceeded, and failure to take action could be considered professional misconduct under the Act. (*) If she has known about the excess for some time, she may already be considered negligent and therefore subject to disciplinary action under the Act. (*)

MacDonald must abide by the ministry's regulations, which would probably require her to submit a complete, factual report to inform officials about the pollution. Before sending the report, she should discuss it fully with her employer. If the employer reacts adversely, MacDonald must, nevertheless, forward the report to the ministry as required by law and by the Code of Ethics. (*) If the employer attempts to dismiss her, MacDonald may find it useful to ask the provincial Association to mediate and to inform her employer of the requirements under the Act. Should MacDonald be dismissed while acting properly and in good faith, she would have grounds for a suit against the employer for wrongful dismissal to recoup lost wages and costs. It would be advisable for her to consult a lawyer in that event.

The engineer's concern over adverse publicity and the cost to the company must not obscure the requirement to act within the law. If the situation is permitted to continue unabated, the long-term consequences will be much more serious. The pollution could ruin the neighbouring resort industry, and MacDonald could find herself subject to disciplinary action for negligence or professional misconduct.

4. You are a Professional Engineer employed by a consulting engineering firm. Your immediate superior is also a Professional Engineer. You have occasion to check into the details of a recent invoice for work done on a project for which your boss is the project manager, but on which both you and members of your staff have done work.

You are surprised to see how much of your time and the time of one of the senior engineers who reports to you have been charged to the job. You decide to check further into this by reviewing the pertinent time sheets. The time sheets show that time charged to other work has been deliberately transferred to this job. You try to raise the subject with your boss but are rebuffed. You are quite sure something is wrong but are not sure where to turn. You turn to the Code of Ethics for direction.

Which articles are relevant to this situation? What action must you take, according to the Code of Ethics?

Suggested answer. According to the Code of Ethics, you must be loyal to the employer. (*) However, the code also states that you must be fair and loyal to clients. (*) This creates an ethical dilemma. The dilemma can be resolved by observing that the deliberate transfer of charges from one job or client to another could be a form of fraud or theft, which is illegal. Therefore, it is important to obtain a clear explanation or justification for this transfer. If your superior is completely unwilling to reassure you of the reasons for this action, you must expose this unprofessional, dishonest, or unethical conduct, as required by the Code of Ethics. (*) The information should be conveyed to the client who is being overcharged.

Should your superior threaten to dismiss you, consult the provincial Association and ask it to mediate or to explain the requirements placed on you and your superior by the Code of Ethics. (*) If you are dismissed while following the requirements of the Code of Ethics in good faith, you should consult a lawyer about suing for wrongful dismissal.

5. A consulting engineering firm is preparing to submit a proposal to clean up an area contaminated by a chemical spill during a train derailment. From past experience, the engineers in the firm know the amount of work involved in doing the job properly. The experts will include people with training in ecology, water quality, ground water, soils, air pollution, and other areas. The methodology that they feel must be followed will result in an expenditure of about $5 million. Before their proposal is submitted, however, the federal government, which is the potential client, issues a news release saying that it has budgeted only $1 million for this work.

What can the consulting firm do? To reduce the level of work to one-fifth of what it thinks is necessary would infringe on the firm's perceived ethical responsibilities to the environment.

Suggested answer. The question implies that the consulting firm feels pressure to submit a bid to do a partial or inadequate job, within the $1 million limit, simply to get the work. This behaviour contravenes every Code of Ethics (either directly or indirectly), which states that the professional engineer must act competently in providing engineering services. (*) Moreover, most Codes of Ethics require the professional engineer to uphold the principle of adequate compensation for engineering work. (*) It is therefore unethical for the consulting firm to submit a bid to perform an inadequate job, or to perform a $5 million job for $1 million.

However, the question asks what the consulting firm should do. In response, recall that every Code of Ethics requires a professional engineer to explain the consequences to be expected if engineering judgment is overruled by nontechnical authority. (*) In this case, the budget has likely been set by financial officials who are unaware of the extent of the required engineering work. This fact must be communicated to the federal department that issued the call for tenders, which must be advised to correct its specifications.

If the problem cannot be resolved by simple communication, the consulting firm must evaluate the seriousness of the matter. (This requires details not provided in the question.) For example, if an inadequate job will endanger the public, the consulting firm has an obligation to put the public interest first, by publicizing the issue. (*) Furthermore, if the consulting firm submits a proper ($5 million) bid, and a competing firm obtains the job for less and thereby creates a dangerous situation, the consulting firm has an obligation to expose any unprofessional or unethical conduct by the other practitioner. (*)

6. Engineer A enters into a consulting contract with a client to provide design and construction supervision of road surfaces in a partially completed land development project. He has taken over from another consultant, who was discharged partway through the job. Before Engineer A can finish the project, his contract also is terminated. Shortly thereafter, it becomes obvious that there are deficiencies in the work done under A's supervision. Investigation shows that hastily paved road surfaces, completed under adverse late-fall weather conditions, are not up to specifications. It seems that A is aware of this. He intended to require remedial work by the contractor in the spring, but his termination occurred before that time. Engineer A did not advise his client that he was expecting to reinspect in the spring and to have deficiencies corrected, nor did he inform his client of the existing state of the roads after he was released from his contract.

   Did Engineer A act in an ethical way in his dealings with his client, even though he may feel that he was unfairly terminated? Discuss the articles of the Code of Ethics that have a bearing on this case.

   Suggested answer. Engineer A, as a professional engineer, is required to act as a faithful agent for the client, in spite of other problems that might interfere. (*) Therefore, even if Engineer A felt that he was unfairly terminated, it would be unethical for him to neglect his responsibilities, such as listing the deficiencies that were known only to Engineer A, so that the work could be continued. This is especially important if the deficiencies might lead to endangering other workers or the general public. The Code of Ethics requires the professional engineer to put public safety first. (*)

However, in his defence, the atmosphere of communication between Engineer A and his former client is obviously poor. The question states that Engineer A feels that he was unfairly terminated, and he replaced a previous engineer on this task. These facts might be relevant (but more information is needed that the question does not provide). If the client created an atmosphere in which professional communication was not possible, then Engineer A may sue the client for breach of contract. However, Engineer A must not endanger the public safety by neglecting the duty to the client.

7. An engineer enters into a contract with a public body [city or town] whereby he agrees to conduct such field investigations and studies as may be necessary to determine the most economical and proper method of designing and constructing a water supply system. He also agrees to prepare an engineering report, including an estimate of the cost of the project, and to estimate the amount of bond issue required. The contract provides that if the bond issue passes, the engineer will be paid to prepare plans and specifications and supervise the construction, and he will be paid a fee for his preliminary services. If the bond issue should fail, the public body will not be obligated to pay for the preliminary work. The public body is prohibited by law from committing funds for the preliminary work until the bond issue is approved.

May an engineer ethically accept a contingent contract under these conditions?

Suggested answer. At first glance, this may appear to be a simple entrepreneurial activity. However, the project is structured to create a massive conflict of interest for the engineer. Any public project must be carried out such that the public interest is protected (and is seen to be protected). The water supply system for a city or town is an especially sensitive matter, since the life of the community depends on a reliable water supply.

In this case, the contract proposes that the engineer will not be paid for the report if the bond issue is not approved (presumably by city and/or financial officials). Such a proposal is contrary to the Code of Ethics, which requires the engineer to uphold the principle of adequate compensation for engineering work. (*)

Moreover, the contract proposes that the engineer reap a double reward if the bond issue passes. He will be paid both for the report and for future work—to prepare plans and specifications and supervise the construction. This creates pressure on the engineer to put the bond issue ahead of the engineering quality of the project. This is a clear conflict of interest, and in this case the conflict is insurmountable, since disclosing the conflict would discredit his report in the eyes of the public, and concealing the conflict is unacceptable. Therefore, since such a conflict of interest is contrary to the Code of Ethics (*), he must not accept the contract as proposed.

The proper way to organize a project such as this is to split it into two parts. The first contract would be for preparing the report on the most economical water supply system. The engineer should be paid for this work, regardless of the success of the bond issue. The second contract— to prepare plans and specifications and supervise the construction— would be arranged only if the bond issue passed. Such a contract should be offered for tender, and payment would follow the standard procedures (lump sum, percentage, per diem, etc.) recommended by the provincial Association.

8. Engineer X, a civil engineer and an employee of ABC Consultants Ltd., signed the Ontario Application for Renewal of Certificate of Authorization (C of A) for that company as the engineer taking responsibility for seeing that the Professional Engineers Act, its bylaws, and its regulations would be complied with.

   ABC Consultants Ltd. prepared the electrical and mechanical designs for a multistorey building. Although Engineer X had very little to do with this project, he permitted his seal to be applied to the design drawings. These designs were found to be deficient in a number of respects. Contrary to the Ontario Building Code, fire walls were omitted, fire dampers were not shown, and sprinklers were improperly connected, among other things. On investigation it was found that both the electrical work and the mechanical work were done by other professional engineers working for ABC.

   What is Engineer X's ethical position in this matter?

   Suggested answer. In almost every province, corporations that practise engineering are required to identify the individuals who personally supervise and direct the work performed by the corporation. These people are required to be experienced engineers, and in Ontario their names are designated on the corporation's Certificate of Authorization. (In other provinces, the C of A may be called a "permit to practise.") Since Engineer X was designated on the C of A, and permitted his seal to be applied to the design drawings, he is responsible for the work. Clearly, Engineer X has been negligent in permitting deficient or unsafe work to be carried out and has failed to carry out his responsibility to supervise and direct the work. Engineer X will likely be subjected to a disciplinary action for negligence (see Chapter 14 of this text). Designation on a C of A is not a formality, nor is it a meaningless title.

   The other ABC engineers, who actually designed the deficient electrical and mechanical work, might also be subject to disciplinary action for incompetence. Although they work under the supervision of Engineer X, their design work is clearly inadequate.

Short-Answer Examination Questions

The Ontario PPE often includes a few short-answer questions in the Ethics part of the exam, such as those which follow, in the Ethics part:

9. Provide a definition of "ethics."

   Answer. Ethics is the study of right and wrong, good and evil, obligations and rights, justice, and social and political ideals. (see text, ch. 6)

10. In a few sentences, describe what a "profession" is.

    Answer. A profession is an occupation that requires specialized knowledge and skills, obtained by intensive learning and practice, and that is organized or regulated to ensure that its practitioners apply high standards of performance and conduct, commit themselves to continuing competence, and place the public good ahead of narrow personal interests. (see text, ch. 1)

11. Is your province's Code of Ethics for engineers enforceable under your professional engineering Act? Explain.

    Answer. Almost every province and territory (except Ontario) states or implies that the Code of Ethics is enforced under the Act, and violations of the code may result in disciplinary action. In Ontario the code is not enforceable, and a separate clause defines professional misconduct. (see text, ch. 6)

12. Explain what "conflict of interest" means.

    Answer. A conflict of interest occurs whenever a practitioner receives a benefit from more than one source when performing professional duties. For example, an engineer, paid by a client to supervise a contract, may be offered a commission by a supplier to use the supplier's materials on the contract. Concealing such a conflict of interest from the client is unethical. (see text, ch. 9)

13. Does your province's Professional Engineering Act explicitly restrict an engineer to practise in his or her branch of registration only? How does the Code of Ethics deal with the problem of practising outside of one's branch of registration?

    Answer. Although a few specialties are regulated in some provinces (especially the SER—the Structural Engineer of Record), professional practice is not generally limited to the branch of registration. However, every Code of Ethics forbids practitioners from accepting or performing work for which they are not qualified. Therefore, practitioners must obtain appropriate preparation—such as academic studies, on-the-job experience, and/or assisted practice—to develop adequate skill and knowledge in the new area. The question of competence is left to the judgment of the practitioner. In the event of a complaint, the practitioner would be expected to demonstrate evidence of adequate preparation in the new area. Failure to show adequate preparation would be a basis for disciplinary action as either negligence or incompetence. (See your Code of Ethics in Appendix CD–B.)

14. The Association of Professional Engineers is the self-regulating organization responsible for the practice of engineering in your province. What is the principal objective of this organization?

    Answer. The principal objective of each Association is to regulate the profession, so that the public interest may be served and protected. This is usually stated or implied near the start of the Act.

15. To become licensed to practise professional engineering in your province, you must meet certain requirements. Discuss briefly the five most significant of these.

    Answer. Since age is almost never a problem, the five most important are (1) citizenship (or permanent resident status); (2) adequate education—a university degree from an accredited engineering program (or equivalent); (3) examinations—typically the professional practice exam must be written; (4) adequate experience—typically four years; and (5) good character as determined from references. (see ch. 2)

Multiple-Choice Examination Questions

Some provinces administer the PPE in a multiple-choice format, as illustrated in the next five questions. A typical exam is two hours long, consists of one hundred multiple-choice questions, and is closed-book (that is, no aids are permitted).

16. According to the Code of Ethics, which of the following activities by a professional member would be considered unethical?
    a. not charging a fee for presenting a speech
    b. signing plans prepared by an unknown person
    c. reviewing the work of another member with that member's consent
    d. providing professional services as a consultant

    Answer. (b) It is unethical for professionals to sign plans not prepared by themselves or under their direct supervision or that they have not thoroughly checked. Since the plans were prepared by an unknown person, the analysis must be completely redone before they can be signed and sealed.

17. Which of the following is the most common job activity of top-level managers?
    a. writing and reading corporate financial reports
    b. developing and testing new products
    c. designing and implementing production systems
    d. directing and interacting with people

    Answer. (d) Most top managers spend most of their time interacting with other people.

18. The professional's standard of care and skill establishes the point at which a professional
    a. may or may not charge a fee for services.
    b. has the duty to apply "reasonable care."

c. may be judged negligent in the performance of services.

d. has met the minimum requirements for registration.

Answer. (c) The standard of care is used to judge whether a professional has been negligent in the performance of services.

19. Which of the following is a minimum requirement for registration as a professional engineer?
    a. Canadian citizenship
    b. experience in engineering work
    c. course work in engineering
    d. residence in the province

    Answer. (b) Of the items listed, engineering experience is the only requirement for registration that is absolutely essential. Each Act has provisions for people lacking one or more of the other three.

20. To effectively reduce liability exposure, the professional geologist should
    a. pursue continuing educational opportunities.
    b. work under the supervision of a senior geologist.
    c. maintain professional standards of practice.
    d. provide clients with frequent progress reports.

    Answer. (c) Maintaining professional standards of practice is the most effective way of reducing liability exposure.

True/False Examination Questions

The Associations no longer use the true–false format for the Professional Practice Exam, but the following five questions, taken from an older exam, are a good review, and readers are urged to answer them:

21. A person may assume the title "Professional Engineer" before being registered with the Association if working under the direct supervision of a registered Professional Engineer.

    False. Every engineering Act prohibits any unlicensed person from using the title (see text, Chapter 14).

22. If an employer knowingly engages a person for work that requires the services of a Professional Engineer and that person is not registered or licensed with the Association, both the employer and the employee are in violation of the professional engineering Act.

    True. This is stated explicitly in most Acts or Codes of Ethics.

23. A person convicted of a criminal offence under an act other than the Professional Engineering Act may be suspended from membership in the Association.

    True. However, most Acts contain the provision that the criminal offence must affect the member's suitability to practise. (see text, ch. 14)

24. Members of the armed forces stationed in your province are subject to the provisions of the professional engineering Act.

False. Members of the armed forces who are practising engineering on federal property are exempt from the provincial engineering Act. However, if they should work either full-time or part-time outside of federal property, then they would be subject to the full provisions of the provincial Act.

25. A Professional Engineer must be aware of all the related facts before publicly expressing an opinion on an engineering subject.

    True. Almost every Code of Ethics states this explicitly.

NOTES

1. Canadian Council of Professional Engineers (CCPE), *Guideline on the Professional Practice Examination*, Document G06-93, CCPE, Ottawa, Ontario, <http://www.ccpe.ca/e/guide_guidelines.cfm> (November 28, 2003), pp. 2–3. Reprinted with permission of CCPE.
2. Association of Professional Engineers, Geologists and Geophysicists of Alberta (APEGGA), *National Professional Practice Examination*, APEGGA, Edmonton, AB, <http://www.apegga.org/pdf/registration/PPExam.pdf> (November 30, 2003), p. 1. Reprinted with permission of APEGGA.
3. Association of Professional Engineers and Geoscientists of British Columbia (APEGBC), Professional Practice Examination <www.apeg.bc.ca/reg/ ProfessionalPracticeExaminationframe.html> (November 30, 2003).
4. J. Delaney, *How to Do Your Best on Law School Exams* (Bogota, NJ: J. Delaney Publications, 1990).
5. G.C. Andrews, *Study Guide for the PEO Professional Practice Exam*, 6th ed., University of Waterloo, Distance Education Department, 2003.
6. The author would like to express his appreciation to provincial Associations in Alberta, New Brunswick, and Ontario for their assistance in obtaining sample questions and their permission to reprint these questions from previous Professional Practice Exams.

APPENDIX A

PROVINCIAL AND TERRITORIAL ENGINEERING/ GEOSCIENCE ASSOCIATIONS

Canadian Council of Professional Engineers (CCPE)
180 Elgin St., Suite 1100
Ottawa, ON K2P 2K3

Tel: (613) 232-2474 / Fax: (613) 230-5759
info@ccpe.ca
www.ccpe.ca

Canadian Council of Professional Geoscientists (CCPG)
Suite 2200, Scotia Centre
700—2 Street SW
Calgary, AB T2P 2W1

Tel.: (403) 232-8511 / Fax: (403) 269-2787
contact@ccpg.ca
www.ccpg.ca

Association of Professional Engineers, Geologists and Geophysicists of Alberta (APEGGA)
1500 Scotia One
10060 Jasper Avenue NW
Edmonton, AB T5J 4A2

Tel.: (780) 426-3990 / Fax: (780) 426-1877
e-mail@apegga.org
www.apegga.org

Association of Professional Engineers and Geoscientists of British Columbia (APEGBC)
200—4010 Regent Street
Burnaby, BC V5C 6N2

Tel.: (604) 430-8035 / Fax: (604) 430-8085
apeginfo@apeg.bc.ca
www.apeg.bc.ca

Association of Professional Engineers and Geoscientists of the Province of Manitoba (APEGM)
850A Pembina Highway
Winnipeg, MB R3M 2M7

Tel.: (204) 474-2736 / Fax: (204) 474-5960
apegm@apegm.mb.ca
www.apegm.mb.ca

Association of Professional Engineers and Geoscientists of New Brunswick (APEGNB)

535 Beaverbrook Court, Suite 105
Fredericton, NB E3B 1X6

Tel.: (506) 458-8083 / Fax: (506) 451-9629
info@apegnb.com
www.apegnb.com

Professional Engineers and Geoscientists of Newfoundland and Labrador (PEG-NL)

P.O. Box 21207
Suite 203, Baine Johnston Centre
10 Fort William Place
St. John's, NF A1A 5B2

Tel.: (709) 753-7714 / Fax: (709) 753-6131
main@pegnl.ca
www.pegnl.ca

Association of Professional Engineers, Geologists and Geophysicists of the Northwest Territories and Nunavut (NAPEGG)

Bowling Green Building
201, 4817—49th Street
Yellowknife, NT X1A 3S7

Tel.: (867) 920-4055 / Fax: (867) 873-4058
napegg@tamarack.nt.ca
www.napegg.nt.ca

Association of Professional Engineers of Nova Scotia (APENS)

P.O. Box 129
1355 Barrington Street
Halifax, NS B3J 2M4

Tel.: (902) 429-2250 / Fax: (902) 423-9769
info@apens.ns.ca
www.apens.ns.ca

Association of Professional Geoscientists of Nova Scotia (APGNS)

P.O. Box 8541
Halifax, NS B3K 5M3

Tel.: (902) 420-9928
mooersjo@gov.ns.ca
www.apgns.ns.ca

Professional Engineers Ontario (PEO)
25 Sheppard Avenue West, Suite 1000
North York, ON M2N 6S9

Tel.: (416) 224-1100 / Fax: (416) 224-8168
webmaster@peo.on.ca
www.peo.on.ca

Association of Professional Geoscientists of Ontario (APGO)
67 Yonge Street, Suite 1500
Toronto, ON M5E 1J8

Tel.: (416) 203-2746 / Fax: (416) 203-6181
info@apgo.net
www.apgo.net

Association of Professional Engineers of Prince Edward Island (APEPEI)
549 North River Road
Charlottetown, PE C1E 1J6

Tel.: (902) 566-1268 / Fax: (902) 566-5551
apepei@apepei.com
www.apepei.com

Ordre des ingénieurs du Québec (OIQ)
2020, rue University, 18e étage
Montréal, QC H3A 2A5

Tel.: (514) 845-6141 / Fax: (514) 845-1833
dirgen@oiq.qc.ca
www.oiq.qc.ca

Ordre des Géologues du Québec (OGQ)
Bureau 912
1117, rue Ste-Catherine Ouest
Montréal QC H3B 1H9

Tel.: (514) 278-6220 / Fax: (514) 279-7139
nevillew@globetrotter.net
www.ogq.qc.ca

Association of Professional Engineers and Geoscientists of Saskatchewan (APEGS)
2255—13th Avenue, Suite 104
Regina, SK S4P 0V6

Tel.: (306) 525-9547 / Fax: (306) 525-0851
apegs@apegs.sk.ca
www.apegs.sk.ca

Association of Professional Engineers of Yukon (APEY)
3106 Third Avenue, Suite 404
Whitehorse, YT Y1A 5G1

Tel.: (867) 667-6727 / Fax: (867) 668-2142
staff@apey.yk.ca
www.apey.yk.ca

INDEX